GOD
THE SCIENCE
THE EVIDENCE

www.godthesciencetheevidence.com

Graphic design and layout: Caroline Hardouin

USA E.1
Copyright © 2025 Palomar éditions
Translation © 2025 by Rebecca M. West and Christine Elizabeth Jones

All rights reserved. No part of this book may be reproduced, stored in a retrieval system, or transmitted in any form or by any means, electronic, mechanical, photocopying, recording, or otherwise, without the prior written permission of the publisher.

Originally published in French as *Dieu, La science, Les preuves* by Tredaniel, 2021.
Translated into English by Rebecca M. West and Christine Elizabeth Jones.

Published by Palomar éditions
13, rue Beaumont, L-1219 Luxembourg, Grand-Duché de Luxembourg

Distributed by Abrams Books

ISBN 978-99987-824-0-2

Michel-Yves Bolloré
Olivier Bonnassies

GOD
THE SCIENCE
THE EVIDENCE

THE DAWN OF A REVOLUTION

Table of Contents

Table of Contents 7
The Book Everyone Is Talking About 9
Preface 19

INTRODUCTION 21
1. The Dawn of a Revolution 23
2. What is Evidence? 37
3. Implications Arising from the Two Theories: "A Creator God Exists" versus "Nothing Exists Beyond the Material Universe" 51

EVIDENCE WITHIN THE SCIENCES 61
4. The Thermal Death of the Universe: Story of an End, Proof of a Beginning 63
5. A Brief History of the Big Bang 83
6. Attempts at Alternatives to the Big Bang Model 111
7. Convergent Evidence for an Absolute Beginning to the Universe 119
8. The Big Bang, A Noir Thriller 127
9. The Fine-Tuning of the Universe 181
10. The Multiverse: Theory or Loophole? 215
11. Preliminary Conclusions: One Small Chapter for Our Book, One Giant Leap for Our Argument 221
12. Biology: The Incredible Leap from Inert to Living Matter 229
13. One Hundred Essential Citations From Leading Scientists 263
14. What Do Scientists Believe In? 295
15. What Did Einstein Believe In? 309
16. What Did Gödel Believe In? 317

EVIDENCE FROM OUTSIDE THE SCIENCES 335

Introduction 337
17. The Humanly Inaccessible Truths of the Bible 339
18. The Alleged Errors of the Bible, Which Are Not Errors 363
19. Jesus: Who Could He Be? 381
20. The Jewish People: A Destiny Beyond the Improbable 415
21. Fátima: Illusion, Deception, or Miracle? 447
22. Is Everything Permitted? 487
23. Philosophical Proofs Strike Back 495
24. Materialist Arguments Against the Existence of God 517

CONCLUSION 533

25. Materialism: An Irrational Belief 535

APPENDIX 539

Appendix 1: Timeline of the Universe 539
Appendix 2: Physical Measurements Great and Small 547
Appendix 3: Benchmarks of Orders of Magnitude in Biology 549

Glossary 551
Index of Personal Names 559
Acknowledgments 565
Detailed Table of Contents 569
Image Credits 581

THE BOOK EVERYONE IS TALKING ABOUT

15 INTERNATIONAL
PERSONALITIES FROM
ALL WALKS OF LIFE

JEAN STAUNE

Philosopher of science. Author of *Does Our Existence Have a Meaning?* and *Explorers of the Invisible*.

INTRODUCTION

You hold in your hands a true phenomenon of the French publishing world.

Very few experts could have imagined that 400,000 copies of a book of this size, addressing the age-old and widely-debated question of the existence of God, would sell within a few years!

This book has sparked numerous debates across a wide variety of circles.

First, a trial concerning its legitimacy: aren't science and religion two "separate magisteria," to use the famous expression of the American paleontologist Stephen Gould?

Additionally, critics of the book have taken advantage of the fact that it has been criticized by a few theologians. Unfortunately, people unfamiliar with this type of debate are unaware that, just as an honourable old English noblewoman confronted with Darwin's discoveries is said to have exclaimed, "We descended from monkeys, but for pity's sake, please don't let this get out," a few European theologians today adopt the motto, "God exists, but for pity's sake, please don't let this get out!"

This is why it is so important that, in the following pages, you will find a number of testimonies from top scientific personalities supporting the book, as well as representatives of Catholicism, Protestantism, Judaism, Islam, and Freemasonry.

These testimonies from all horizons demonstrate the legitimacy and credibility of the exceptional work of Michel-Yves Bolloré and Olivier Bonnassies. Their book restores the rich philosophical tradition of the West, from a time when questions of "how" and "why" were considered together.

■

"IF THE UNIVERSE HAD A BEGINNING, THEN WE CANNOT AVOID THE QUESTION OF CREATION"

This book is a very good presentation of the development of the Big Bang theory and its impact on our beliefs and our representation of the world. After reading the various chapters about cosmology, I found that this book offers a very interesting perspective on science, cosmology and their philosophical or religious implications. Although the general thesis of the authors, Michel Yves Bolloré & Olivier Bonnassies, both Engineers, that a higher mind could be at the origin of the Universe does not provide a satisfying explanation for me, I can accept its coherence. Even if my work as a cosmologist is limited to a strictly scientific interpretation, I can understand that the Big Bang theory may elicit a metaphysical explanation. In the Steady State case, as Professor Fred Hoyle, my cosmology teacher at Caltech assumed, the Universe is eternal and the question of its creation does not arise. But conversely, as the Big Bang theory suggests, if the Universe had a beginning, then we cannot avoid the question of creation.

"A REMARKABLE WORK"

A very welcome, interesting and easy to read book which argues convincingly that the existence of a creator God, far from being negated by the scientific understanding of the Universe, is rationally supported by it. This is a remarkable work with hundreds of references citing leading contemporary scientists.

"HIGHLY RECOMMENDED!"

> *The last century has offered an extraordinary amount of evidence in cosmology, physics and biology to support the reality of a supreme intelligence behind it all. This extraordinary synthesis begins with the emergence of Big Bang cosmology in the early 20th century, implying an actual creation event and refuting the long-held scientific assumption of an eternal and static universe. The immensely improbable fine-tuning of physical constants in the universe to such tight precision that allows for creation of subatomic particles, atoms, molecules, cells, living beings and planets and stars is revealed from the perspective of many scientists. The improbability of biological molecules like proteins and nucleic acids arising through simple chemical mechanisms alone further supports the intelligence required for our universe to exist and evolve as it does. Additional observations by Einstein and the most renowned logician of the 20th century, Kurt Gödel, combined with many statements made by working scientists about the existence of God, complete this compelling argument for the reality of an astonishing intelligence within the Universe itself - the reality of God. This book thus provides a strong, rational argument for the reality of a creator God that is very comforting to those with a scientific mindset. Highly recommended!*

"A WARRANTED BELIEF"

> *The amazing advances in science and cosmology over the past century raise fundamental questions about our existence. Is the Universe a closed or an open causal system? Are the remarkably precise values of so many parameters just brute facts or shall we one day understand them at a deeper level? Why are we here, and is there a purpose to life? Michel-Yves Bolloré and Olivier Bonnassies explore these questions in a highly readable way, and give a personal account of how their answers lead them to a warranted belief in God.*

"A VALUABLE CONTRIBUTION TO THE CONVERSATION BETWEEN SCIENCE AND RELIGION"

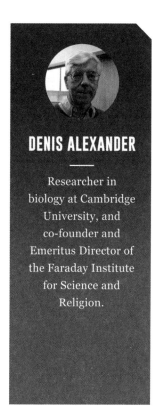

DENIS ALEXANDER

Researcher in biology at Cambridge University, and co-founder and Emeritus Director of the Faraday Institute for Science and Religion.

I'm delighted to note that this book takes both science and religion very seriously, exploring ways in which both fields of inquiry can interact positively together, but also not hesitating to address the points at which some readers might find tensions or even conflict. I agree with this book that it is perfectly rational to utilize evidence that lies beyond science in the justification of our beliefs. I hope by now that it's clear that personally, I find this book a valuable contribution to the science-religion conversation, and I hope very much that readers coming from many different backgrounds, whether religious or not, might find in these 600 pages much to stimulate thought and initiate fruitful conversation.

"SCIENCE PRACTICED IN A SINCERE QUEST FOR THE TRUTH BRINGS MAN CLOSER TO GOD"

LUC JAEGER

Professor of Chemistry and Biochemistry at the University of California Santa Barbara (UCSB). Specialist in RNA, the complexity of biological systems, and the origins of life.

As this book demonstrates remarkably well, science practiced in a sincere quest for truth brings humanity closer to God rather than creating distance. To perceive science as being opposed to the existence of God is a profoundly ideological notion that defies a healthy exercise of reason. While many scientific questions remain open—such as those concerning the emergence of life from matter and its increasing complexity—it is certain that answers to these questions cannot overlook the reality of the "miracles" around us, which point to transcendence.

"A UNIQUE RESOURCE!"

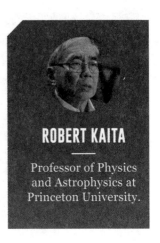

ROBERT KAITA

Professor of Physics and Astrophysics at Princeton University.

God, the Science, the Evidence clearly provides the scientific evidence for a Universe that began at the so-called "Big Bang," and with the "fine-tuning" of the physical constants needed for our universe to be. Accessible to the general reader, the book makes a persuasive case for how the science supports the existence of a creator God. The authors also recognize that science alone does not encompass all human understanding, and the second half of their work is devoted to the religious and philosophical reasons for belief in God. By presenting both realms, God, the Science, the Evidence is a unique resource for those with an interest in science and faith.

"THIS BOOK REVIVES THE MOST ESSENTIAL METAPHYSICAL QUESTION THERE IS"

Michel-Yves Bolloré and Olivier Bonnassies' book is written in the great metaphysical tradition that, from its origins, shows that it is essential to think of the world, life, and humanity from the starting point of an intelligent, living, and beautiful principle. It is God who proves the existence of God, not a proof. Even so, asking the question of what is the principle at the origin of everything is essential. The day we cease to ask this question, we will be less intelligent and less alive. Michel-Yves Bolloré and Olivier Bonnassies' book has the merit of reviving the most essential metaphysical question there is.

"A SOLID DEMONSTRATION OF THE EXISTENCE OF A CREATOR GOD!"

" *God, the Science, the Evidence builds a bridge between two perspectives that, for a long time, seemed opposed. This work demonstrates that the antinomic thinking of knowledge and belief, still widespread in the 21st century, is based on prejudice and cliché. The book highlights in a remarkable and rigorous manner how the latest scientific advances point towards a creator God. Wonderfully documented, it also introduces many renowned scientists who advocated for a dynamic harmony and osmosis between their discoveries and the idea of a creator at the origin of all things.* "

"THIS BOOK IS A PHENOMENON"

" *God, the Science, the Evidence is now a best-selling book and has generated a wide-ranging public debate on the question of the existence of God based on the scientific discoveries of the past 100 years. The authors are eminent figures: Olivier Bonnassies is a former student of the École Polytechnique; Michel-Yves Bolloré is a Master of Science and an Engineer. The book has sparked our interest—even if the God the authors believe in is not the God of Israel. Nevertheless, in a scientistic, disenchanted, and stupidly materialistic world, where schoolchildren have never opened the Bible, and at a time when the boat of Judeo-Christian Europe is sinking, we must challenge in all ways, whether we are believers or not, the modern desire to put an end to all forms of transcendence, and to look to a European culture nourished by Christianity and the Bible, as a treasure of human thought to be preserved at all costs.* "

DAVID REINHARC

Editor-in-Chief of *Israel Magazine*, which ran a feature story and put *God, the Science, the Evidence* on the cover of its January 2022 issue.

"An 'Unidentified Literary Object'"

STEVE ABD AL KARIM & PHILIPPE ROBICHON

Hosts of Beur FM, a French Muslim radio station that covers books and those who read them, welcomed Michel-Yves Bolloré and Olivier Bonnassies for a debate that had a far-reaching echo.

> We hosted a session of the 'book club' on Beur FM to share a 'ULO' with our listeners; that is, an 'unidentified literary object.' It's the publishing sensation of the year's end. This is a book that had gone unnoticed, flying under the radar, but it has become the number one bestselling non-fiction book in France. God is selling, after all! They approach the topic from a deist perspective. I really liked this book. But I also felt a lot of sadness while reading it. A lot of sadness for atheists. This notion of a grand reversal—could it be too harsh? And the image of fine-tuning is incredibly powerful as well.
>
> This book is very simple. It's 600 pages long but reads like a novel, like a thriller. It presents scientific evidence. And there are people who were killed for these beliefs. The Big Bang shattered many limiting beliefs held by humanity. Everything that has a beginning has a cause; now, the Universe has a beginning, so the Universe must have a cause. QED. This is one of the discoveries in their book, one of its proofs. It's a central idea.
>
> We encourage everyone to read this book because there is so much to gain from it, and it is absolutely fascinating.

"GOD, THE SCIENCE, THE EVIDENCE:
PUT TO THE TEST BY THE FREEMASONS"

> Michel-Yves Bolloré and Olivier Bonnassies' book has piqued the interest of even the Freemasons, at least those who do not slumber in the dogmatism of convenient skepticism. Freemasonry has always sought to understand this world.
>
> Of course, science can never prove the intervention of a "living" God who reveals Himself to humanity, but an honestly-conducted inquiry can gather converging evidence that creates a compelling effect: the world is not merely a powerful yet absurd mechanism. It is a deliberate, effective construct governed by laws that originate from a supremely ordered intelligence.
>
> The temple of the French National Grand Lodge was full—500 people gathered to listen to Michel-Yves Bolloré. As the national orator of the lodge, I engaged the speaker using the insights of the great Blaise Pascal.
>
> It was truly a moment of grace under our roof. Thanks to the authors who, amid one of the most formidable crises of earthly civilization, help us lift our eyes to the stars and find simple words to accompany the hymn of the Universe!

"An Invitation to Wonder"

This book represents an innovation. It demonstrates that it is no longer 'credible' to pit science against faith. God, the Science, the Evidence reaches people where they are, in a successful effort that provokes reflection. It argues that it is permissible to consider the question of God on the basis of science. The authors explore ideas that are so often stifled by the rationalism that infects our culture. Let's accept that science can ask questions about God! Everyone has the right to speak about God, whether they are a coal miner, an astrophysicist or... a philosopher. We can do so in the most 'secular' context that exists. The book God, the Science, the Evidence widens our field of reasoning and is an invitation to wonder.

"An Immense Work"

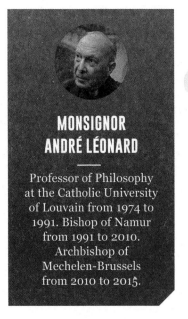

MONSIGNOR ANDRÉ LÉONARD

Professor of Philosophy at the Catholic University of Louvain from 1974 to 1991. Bishop of Namur from 1991 to 2010. Archbishop of Mechelen-Brussels from 2010 to 2015.

I was so grateful to read God, the Science, the Evidence, an immense work whose solid documentation, logical applicability, and pedagogical concerns I applaud. Its commercial success is fully deserved and makes me rejoice, because it will help undermine the atheism of convenience which often anesthetizes contemporary culture.

Preface

Dear Reader,

This book is the culmination of nearly four years of research conducted in collaboration with a team of around twenty high-level international specialists and scientists.

Its objective is unique: to shed light for you on the question of the existence or non-existence of a creator God, one of the most important questions of our lives which is being posed today in completely new terms.

Our hope is that by the end of this reading you will have in hand all the elements necessary to decide what seems most reasonable for you to believe.

We have conducted this work as a rigorous investigation. We have always used rationality as our only compass and have sought to avoid two major pitfalls of our time: on one side, that of creationism, which rejects modern discoveries and adheres to fanciful beliefs; on the other, that of an extreme materialism which refuses to consider the implications arising from the latest scientific advances.

Until recently, believing in God seemed incompatible with science. Now, unexpectedly, science appears to have become God's ally. Materialism, which has always been a belief just like any other, is seriously shaken as a result.

This book is a 180° panorama of current knowledge regarding the existence or non-existence of a creator God. It is not limited to science but also explores what philosophy, morality, history, and even some of humanity's great enigmas have to say on the matter, because the question is far too important to be confined solely to the realm of science.

To our knowledge, no other book like the one you are holding in your hands exists. We have aimed to make it easy and enjoyable to read for all, while ensuring accuracy. The chapters are independent, so readers can approach them in any order of their choice.

We have done our best; we leave the rest to your free judgment.

Michel-Yves Bolloré and Olivier Bonnassies

INTRODUCTION

1.

The Dawn of a Revolution

Never before have so many spectacular scientific discoveries emerged in such a short period of time. These discoveries have overturned our view of the cosmos and forcefully put the question of the existence of a creator God back on the table

Physics, much like an overflowing river, has surged beyond its banks into the realm of metaphysics. This powerful intersection reveals the necessity for an intelligent creator. For nearly a century, groundbreaking theories have sparked robust intellectual debate. Above all, this is the narrative we aim to explore in this book.

We are living through an amazing moment in the history of knowledge. Advances in mathematics and physics have turned questions once thought to be forever beyond the reach of human knowledge—questions about the nature of time and eternity, its beginning, its end, and the improbable unfolding of the fine-tuning of the Universe and of the appearance of life—into objects of scientific study.

These scientific advances, which began in the early twentieth century, brought about a complete reversal of the thinking of previous centuries, when science was considered incompatible with any discussion of the existence of God.[1]

---

[1]. In the French edition of this book, the authors deliberately used a lowercase "g" when referring to "god" to emphasize that they were discussing the concept in a generic sense.

However, in the English version, the authors have chosen to use a capital "G" when referring to God to align with the common practice in English.

In philosophical discourse, particularly when exploring concepts like existence, the nature of divinity, or a "prime cause for everything that is," the capital "G" is used to clarify that the discussion pertains to a specific (often monotheistic) concept of God, rather than to a general deity

Revolutionary discoveries and their impact

- **The thermal death of the Universe**: A result of the theory of thermodynamics, first developed in 1824 and confirmed in 1998 through observation of the accelerating expansion of the Universe, thermal death (also known as heat death) implies that the Universe had a beginning. But every beginning presupposes a creator.

- **The theory of relativity**: Developed by Einstein between 1905 and 1917 and since confirmed many times over, this theory affirms that time, space, matter, and energy are interrelated and that no single one of them can exist without the other. This finding implies that if a cause exists at the origin of our Universe, it is necessarily atemporal, non-spatial, and immaterial.

- **The Big Bang**: Theorized in the 1920s by Friedmann and Lemaître and confirmed in 1964, the Big Bang's account of the beginning of the Universe is so specific and spectacular that the theory sparked an explosive controversy in the world of ideas, so much so that in some countries, scientists risked their lives to study and defend it. We dedicate an entire chapter to exploring the overlooked or concealed persecution and execution of scientists. Tragically, these persecutions highlight the profound significance of the discoveries made by the scientists involved.

---

among other deities. In contrast, a lowercase "g" is typically reserved for general references to gods, such as those in polytheistic traditions, where multiple deities exist. The lowercase "g" is also used in more abstract or metaphorical contexts, where "god" might refer to anything worshipped or considered divine.

In English, the capital "G" is also customary in scientific discussions about God, especially in fields like cosmology, physics, or biology, where arguments concerning the existence of a specific kind of deity are often examined—one whose existence is described as independent and self-subsisting, i.e., as not being contingent upon anything else.

Furthermore, in interdisciplinary dialogue—particularly when empirical sciences engage with topics rooted in philosophy or theology, such as the existence of God—using a capital "G" ensures consistency with these other disciplines. This practice helps prevent confusion that might arise from using a lowercase "g", which could imply a discussion about gods in a more general, polytheistic, or pluralistic context.

Therefore, the reader should note that the use of "God" in this text is not intended to refer exclusively to the concept of God within the Abrahamic religions, but rather to an entity that, unlike created beings, which rely on external factors for their existence (as humans depend on parents for their existence), is the cause of its own existence, uncaused and unconditioned by anything external.

- **The fine-tuning of the Universe**: Generally accepted since the 1970s, the question of the fine-tuning of the Universe poses such a problem for materialist cosmologists that to evade it they are forced to invent purely speculative, unverifiable models of the Universe—such as multiple, successive, or parallel universes.

- **Biology**: Toward the end of the twentieth century, advances in biology showed that the Universe needed even more fine-tuning to create the conditions for inert matter to come to life. In fact, what we once thought of as a short step between the most complex inert matter and the simplest living matter now seems more like a leap across an immense chasm—a leap that certainly could not have been made

according to the laws of chance alone. And while we still do not know how that gap was bridged or, a fortiori, how to replicate such an event, we do know enough to appreciate its infinite improbability.

In earlier centuries, however, a series of scientific discoveries seemed to go against faith

Since the late sixteenth century, a series of scientific discoveries appeared to challenge the foundations of belief in God, unsettling the pillars of faith. Here's a brief historical overview:
- The discovery that the Earth revolves around the Sun, rather than the opposite (Copernicus 1543 – Galileo 1610).
- The mathematical description of a simple, clockwork Universe (Newton 1687).
- The Earth's extreme old age, much older than just a few thousand years (Buffon 1787 – Lyell 1830 – Kelvin 1862).
- The concept of a Universe entirely governed by physical laws eliminates the need for angels to maneuver the planets (Laplace 1805).
- The view that life emerged and developed over millions or even billions of years (Lamarck 1809, and others).
- The idea that this evolution was based on natural selection, not divine intervention (Darwin 1859).
- The rise of Marxist dialectical materialism, with its seductive promise of a world of equality and justice (beginning in 1870).
- The ideas of Freud (around 1890), who postulated that man is not even the master of his own thoughts and for whom this new science offered man a life "free from prejudices."

Somewhat smugly, the Viennese psychoanalyst Sigmund Freud spoke of the "three humiliations" that modern man had suffered thanks to Copernicus, Darwin, and himself. The wounds to our pride were indeed accumulating: humankind had lost its place at the geographic center of the Universe; it had lost its sense of superiority in discovering that, according to the proponents of evolution, it had descended from apes; and finally, it had even lost autonomy and responsibility for its deepest thoughts through the theory of the subconscious.

Thus, over the course of three centuries, from Galileo to Freud via Darwin and Marx, much of the knowledge that once seemed part of the bedrock of Western thought was shaken, causing confusion among religious believers. In the end, there was actually no reason to be disturbed by these new discoveries, as those which eventually proved to be true did not in fact contradict the faith in any way. Many believers, however, lacked the perspective and knowledge needed to see this, and as a result these scientific advances were greeted with disbelief and even hostility. After all, it takes a great deal of effort to abandon old certainties and redraw one's mental geography.

Materialists, on the other hand, embraced these discoveries with enthusiasm and used them to justify their theses. Their cause was greatly aided by the technological advances of the time, which simultaneously eradicated famine and epidemics in the West, cured most diseases, extended life spans, reduced infant mortality, and provided an unprecedented abundance of material goods. Science forced religion to retreat, while material abundance made turning to a god to solve humanity's problems seem unnecessary.

It was in this favorable climate that materialism came to reign supreme in the intellectual world of the first half of the twentieth century.

Under these circumstances, many believers in the West found it even easier to abandon their faith, as it now seemed superficial and implausible. Those who remained often developed an inferiority complex in the face of the ascendant rationalism. Avoiding scientific and philosophical debate, they retreated into their own private sphere, where they were expected to stay—lest they face mockery, disdain, or hostility from the intellectually dominant materialist class.

The second half of the twentieth century saw the twilight of this seemingly irresistible materialist trend

Until the middle of the twentieth century, human reason remained boxed in by three analytical viewpoints that cut it off from any spiritual aspiration: Marxism, Freudianism, and Scientism. However, cracks

- During the first half of the twentieth century, belief in a simple, mechanical, and deterministic world was shattered by a more precise confirmation of the principles of quantum mechanics and its postulates of indeterminacy.

- In 1990, the failure and collapse of the Soviet Union and the parallel abandonment of Marxist economic doctrine by the communist bloc in Asia proved to the world the falsity of Marxist dialectical materialism. Eventually this collapse would expose the economic, political, and human horrors to which it had given birth, such as the gulags, where death tolls ranged in the millions.

- This disillusionment coincided more or less with a reevaluation of the theories of Freud. For example, French author Catherine Meyer's influential *Le Livre noir de la psychanalyse [The Black Book of Psychoanalysis]*,[2] published in 2005, took a critical look at the rise and fall of this intellectual idol of the mid-twentieth century. Though Freud has been knocked from his pedestal, his legacy lives on in the concepts of permissive education and sexual liberty. All this has significantly shaped the modern West.[3]

The simultaneous destruction of these three intellectual pillars of materialism was not accompanied by a return to faith, but it did considerably weaken the materialist system of thought. The system was further undermined by the cosmological discoveries referenced earlier, which provide compelling scientific evidence supporting the existence of a creator God. As a result, these theories initially faced significant resistance from atheist scientists, who sought to deny this evidence for as long as possible, beginning in the 1930s.

---

2. Catherine Meyer, *Le livre noir de la psychanalyse* (Paris: Édition des Arènes, 2005). For an English language work that treats similar topics, see Samuel Bendeck Sotillos, *Psychology Without Spirit: The Freudian Quandary* (Chicago: The Institute of Traditional Psychology, 2018).

3. A column published in the French magazine *L'Obs* in the autumn of 2019 and signed by sixty psychiatrists and psychologists, called for the exclusion of psychoanalysts from France's universities, public hospitals, and judicial institutions.

We will devote a long chapter to this resistance on the part of certain materialists, a resistance that has taken various forms, from support for alternative speculative theories to counter the Big Bang (such as the Big Crunch or multiple universes), to the deportation and even execution of a number of scientific experts in the USSR and Germany. Such reactions say much about people's limited ability to accept scientific theories that challenge their beliefs.

We need this review of intellectual history to situate our discussion within its historical and ideological context. If it was challenging for believers to accept Galileo and Darwin—whose discoveries were not inherently at odds with their faith—how much more arduous must it be for materialists to accommodate concepts like the thermal death or fine-tuning of the Universe into their belief systems? These revelations present formidable challenges to their worldview. Accepting these facts necessitates not merely an update in thinking, but a complete reevaluation of their internal framework.

Emotions can get in the way of accepting the truth

Our ability to accept a claim, scientific or otherwise, depends on more than rational evidence. Our emotions also come into play when we evaluate a claim and its conclusions.

Some questions science asks today are emotionally neutral—why the dinosaurs went extinct, the origin of the moon, how water on Earth came to be, or why the Neanderthals suddenly disappeared. Of course, these questions spark lively debate among scientists, who argue for differing, even opposing, solutions. But in the end, everyone accepts the results of their intellectual efforts since these subjects do not tend to bring our emotions into play.

All that changes as soon as we step onto more delicate ground. Issues like global warming, the environment, nuclear energy, and Marxist economics certainly demand scientific answers, but these topics are so politicized that our intellects are not free to think about them in a normal way. Our passions, political opinions, and self interest get in the way of our reason.

The phenomenon is particularly acute when one broaches the subject of the existence of God, because what is at stake is not just some point of scientific data but the very meaning of our life. The prospect of having to recognize, at the end of a long search, that they are creatures created by and dependent upon a creator is seen by many as a fundamental threat to their autonomy and freedom.

For many people, the desire to be free and autonomous, to be able to determine their course of action entirely on their own with "neither God nor master," takes precedence over everything else. Their inmost self recoils from this idea of God: to defend itself, it mobilizes all its intellectual resources to oppose the search for truth and to protect its own perceived independence and freedom.

SETI radio telescopes in New Mexico

It is not surprising, then, that instead of stimulating thoughtful discussion, the subject often provokes reactions ranging from annoyed indifference to ridicule, contempt, and even violence.

It is telling that we invest vast amounts of time and resources in searching for extraterrestrial life, as the Search for Extra-Terrestrial Intelligence Institute (SETI) program does, rather than dedicating even a fraction of that effort to exploring the hypothesis of a creator God. If God exists, is he not, in a sense, a supreme form of extraterrestrial? And yet, God's existence is more likely and more readily admissible, and the traces of his action in the universe are far more tangible, than those of potential extraterrestrials. This imbalance points to a kind of fear. For a materialist, collecting traces of far-off extraterrestrial life is definitely thrilling, but it does not demand a reevaluation of the meaning of existence. Becoming aware that God exists, on the contrary, carries the risk of serious interior disruption.

So it is that ideology and emotion can prevent us from accepting facts and keep us from a clear-headed examination of evidence that might require us to revolutionize our understanding of the world.

To make our intentions perfectly clear from the outset: the purpose of this book is not to militate for a particular religion, much less to engage in an analysis of the nature of God or his attributes. The goal of this book is to gather into one volume the most up-to-date rational arguments for the possible existence of a creator God.

Defining what scientific evidence is

In order to firmly establish the value of the evidence we present, we must begin by looking at the nature of scientific evidence. To that end we will examine the nature of evidence in general and in the field of science in particular.

Next, we will consider the consequences that flow from two opposing theses or viewpoints: belief in the existence of a creator God versus belief in a purely material world. Materialism is, after all, a belief just like any other. We will see that these two beliefs have a variety of implications, each of which can be easily proved or disproved by measuring them against our observations of the real world.

Part One: An inventory of the most up-to-date scientific evidence

This portion of the book will consider the revolutionary discoveries mentioned earlier: the heat death of the Universe, the Big Bang, the fine-tuning of the Universe and the anthropic principle, and finally the question of the leap from inert to living matter. We will treat each of these topics in depth.

Part Two: Evidence from reason outside the realm of science

Part two will study fields of knowledge other than science which nonetheless pertain to reason. In science, as in history or philosophy, it is always worth paying attention to anomalies or contradictions, i.e., to facts that have no reasonable explanation in a purely material world. We will ask questions like: "What is the basis for the Bible's often puzzling claims?" "Jesus: who could he be?" "Can the destiny of the Jewish people be explained in purely natural terms?" "What really happened at Fatima in 1917?" or "Does mankind have complete freedom to define good and evil?"

We will also take a look at the current status and value of philosophical proofs for the existence of God and the renewed interest taken in them by mathematicians like Kurt Gödel.

These considerations will furnish the reader with a varied survey of probative arguments that, taken as a whole, constitute a converging and convincing body of evidence.

Part Three: Putting an end to the usual objections

We will conclude by addressing a range of arguments, both historical and contemporary, that assert the impossibility or unknowability of the existence of a creator God. We will consider claims like, "If there were proof for God's existence, we would already know it;" "We don't need God to explain the Universe;" "The Bible is nothing more than a collection of old legends riddled with historical errors;" "Religion causes wars;" and "If God exists, how can we explain evil?"

We will address these questions sincerely, even if they are clichéd, and provide thoughtful responses that rectify these common intellectual missteps.

A sign of the times

You may notice that the greater part of the knowledge upon which we base our arguments dates from after the beginning of the twentieth century. This was not a deliberate choice on our part but rather serves as a confirmation that times are changing and that we are on the cusp of an intellectual revolution.

A project based primarily on reason

The makeup of this book might seem unusual. Some readers may be surprised to find modern scientific knowledge alongside reflections on the Bible or the story of a miracle in Portugal.

However, each of these concepts has its place in this book. The theory that "nothing exists beyond the material Universe" implies that miracles do not occur and that even the most astonishing events in human history must be explained without relying on supernatural explanations. If we observe evidence to the contrary in the real world, then this would serve as strong evidence that the materialist theory is flawed.

In the end, either God exists or he doesn't. The answer to these stark alternatives exists independently of us and it is binary: either yes or no. Until now, our lack of knowledge has been a serious obstacle to answering the question with confidence. But the emergence of a vast array of convergent evidence from various independent fields of knowledge has shed new and perhaps decisive light on the question.

THE GREAT

Five centuries of scientific discovery

Copernicus (1543) Heliocentrism

Galileo (1610) Strengthening heliocentrism

Newton (1687) Universal gravitation

Buffon (1787) Calculating the Earth's age

Laplace (1805) Determinism

Lamarck (1809) Evolution

Darwin (1859) Natural selection

Marx (1870) ...tical materialism

Freud (1896) ...oanalysis

1500

REVERSAL

behind the rise and fall of materialist thought

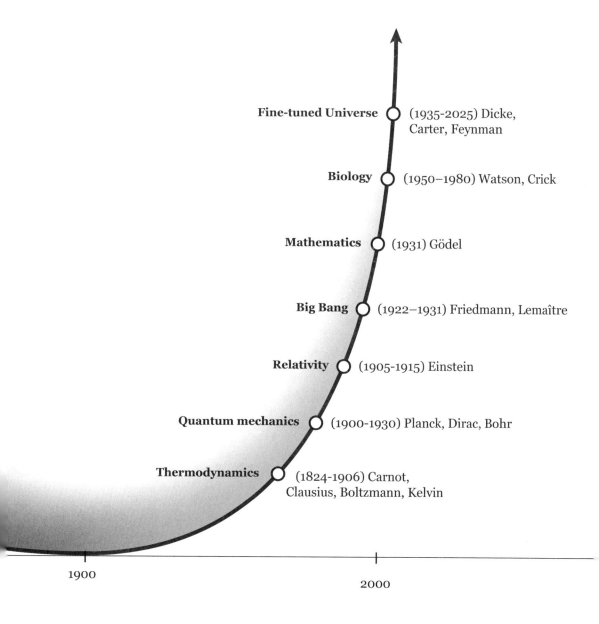

2.

What is Evidence?

The purpose of our book is to provide an overview of the current evidence for the existence of God. It seems natural to start with evidence emerging from recent scientific discoveries, since science is universally regarded today as the most reliable source of knowledge.[4]

However, we will not limit ourselves to scientific evidence alone, as the importance of this question requires that we explore other areas of reason to see if they also support the arguments provided by science.

While the meanings of the words *God* and *Science* are generally clear,[5] the terms *evidence* and *proof* are more complex.

According to the *Oxford English Dictionary,* "evidence" is defined as "the available body of facts or information indicating whether a belief or proposition is true or valid." However, further clarification is needed, since evidence can be categorized into two types:

Absolute proofs (often referred to simply as "proof") exist in theoretical or formal domains such as mathematics, games, and logic.[6] These proofs are definitive and incontestable because they result from a finite set of clearly defined principles, axioms, and rules.

---

4. Also because, for a long time, science was considered the primary rational discourse opposing the idea of God's existence, making the shift over the past 100 years particularly remarkable.
5. Definitions for the terms "God" and "science" can be found in the glossary of this book.
6. It should be noted that the well-known philosophical proofs for the existence of God, discussed in Chapter 23, are considered by many to be absolute proofs, even if this view is not shared by everyone. Additionally, the renowned mathematician Kurt Gödel developed a mathematical proof for the existence of God, intending for it to be an absolute proof. While this proof remains debated, its reasoning has, to date, not been found faulty (see the chapter on Gödel).

Relative proofs (often simply referred to as "evidence") are those that exist in the real world. They are never absolute, since they rely on our inherently imperfect and incomplete understanding of reality.

The purpose of this chapter is to clarify the nature and logical mechanisms of evidence and to assess the level of conviction or certainty that evidence can provide, especially in relation to our subject of investigation.

This chapter is written to be as accessible as possible; however, readers who find it too technical may skip it, as it is not essential for understanding the subsequent chapters.

Absolute proofs

Absolute proofs are familiar to us from our school mathematics, where we encountered examples like the Pythagorean theorem. This theorem demonstrates that, in Euclidean geometry, the square of the hypotenuse of a right triangle equals the sum of the squares of the other two sides. Demonstrated over 2,500 years ago, this proof has remained unchanged ever since. This example helps us immediately understand the characteristics of absolute proofs:

- They are accepted by everyone.
- One absolute proof is enough to demonstrate a proposition; additional proofs are not necessary.
- They are universal.
- They are unchanging over time.
- They exist only in theoretical domains, such as mathematics, games, and pure logic—fields of abstraction where there is complete knowledge of rules, axioms, and data.

The principle of logic asserts that correct reasoning, when applied to data that are fully-known and exact, leads to a valid conclusion.

For example, in Euclidean geometry, it can be demonstrated that a triangle with three equal sides has three equal angles, or that, in a particular chess scenario, a checkmate in three moves is inevitable.

An absolute proof remains absolute, regardless of the number of reasoning steps required to reach it.[7]

However, absolute proofs do not exist in the real world, as we will soon explore.

Relative proofs

Relative proofs, or "evidence," cannot be absolute because, in the real world, correct reasoning applied to accurate data does not necessarily yield a correct conclusion. Ignorance of this counterintuitive fact has, unfortunately, frequently led to terrible mistakes.

Reaching a correct conclusion requires that accurate reasoning be applied not only to some correct data, but to all data relevant to the problem. Unfortunately, in the real world, we never have complete knowledge of reality; as a result, we never possess all the necessary data. Even if we did have all the data, the huge volume would be too vast to fully account for.

A true and tragic anecdote illustrates this surprising reality

In the 1950s, wheat harvests in China were very poor. Agricultural officials explained to Mao Zedong, the founding leader of the People's Republic of China and chairman of the Chinese Communist Party, that sparrows were consuming a significant portion of the sown seeds—and this was true. Mao then reasoned that if the sparrows were eliminated, these seeds would no longer be eaten, which was also true. He concluded that the harvest would increase by the equivalent quan-

---

7. The one hundred and twenty-five pages required by English mathematician Andrew Wiles to prove Fermat's famous theorem are a remarkable example of mathematical dedication. To demonstrate this theorem—stating that $x^n+y^n=z^n$ is impossible if n is greater than or equal to 3—Wiles worked for years, weaving together arguments from various branches of mathematics, before publishing his proof in 1995. Each argument was rigorously examined by experts in the relevant fields, and the proof's accuracy was unanimously accepted by the scientific community. Yet, an experimental verification of this theorem remains impossible, making this achievement all the more extraordinary.

tity of wheat that would have come from the spared seeds. But this proved to be completely wrong. Mao's reasoning overlooked just one piece of the data: while sparrows do eat some seeds, they also consume worms and insects that are far more destructive to crops. The decision to eradicate sparrows was implemented nationwide in 1958 during the "Great Leap Forward," without prior experimentation. The result was a catastrophic famine that led to millions of deaths.

This story illustrates how a single ignored piece of data can lead to outcomes that are the exact opposite of expectations.

Consequently, in the real or empirical domain, absolute proofs do not exist; or, if they do, they are generally inaccessible to us. What we have instead is evidence of varying degrees of strength, the accumulation of which can nevertheless lead us to a conviction beyond all reasonable doubt.[8]

Since evidence in the empirical world is not absolute, it is essential to address this limitation by increasing the quantity of evidence and diversifying the sources. This approach helps to more firmly establish the thesis that the evidence is meant to support.

The Four Pests campaign

An example of empirical evidence: the realm of justice

The realm of justice, which depends on the presentation of evidence, serves as an excellent example due to its universal familiarity.

---

8. Unlike absolute proofs, relative proofs ultimately require a judgment, an intellectual leap.

In a criminal court trial, the prosecutor must provide evidence of the accused's guilt. This evidence may be physical or circumstantial, and its strength can vary. It might include DNA traces, blood group analysis, fingerprints matching those of the accused, or footprints at the scene. There may also be evidence stemming from a motive, along with testimonies placing the accused at the crime scene. The credibility of a testimony depends on factors such as the witness's personality and reputation. Testimonies from multiple independent witnesses—particularly those who are unrelated and unfamiliar with one another—are generally considered more reliable than those from witnesses within the same family group.

No single piece of evidence can be considered absolute; even strong material evidence, such as a trace of DNA, could theoretically be part of a well-executed conspiracy. However, when the evidence is abundant, compelling, convergent, and independent, jurors may reach an inner conviction beyond any reasonable doubt, allowing them to decide on the accused's guilt.

This example demonstrates that in the empirical domain, it is essential to gather as much convergent and independent evidence as possible to achieve a conviction beyond any reasonable doubt.

Evidence in science

We have explored two types of proof that are familiar and relatively easy to understand: absolute proofs in mathematics and relative proofs—or evidence—in the realm of justice.

Now, we will examine how proofs are approached in science, a field of particular importance for us, as these are the proofs we will use in this book.

The natural sciences belong to the empirical domain, where, as we have seen, absolute proofs do not exist.

According to Karl Popper, *"in the empirical sciences, which alone can provide information about the world in which we live, proofs do not exist, if by "proof" we mean a fact that establishes the truth of a theory once and for all."*[9]

Thomas Kuhn suggests that the scientific community operates within "paradigms"—widely accepted theories that are not absolutely proven. These paradigms persist until a crisis arises, prompting a "revolution" that leads to the adoption of new paradigms that provide a better explanation of reality.[10]

Therefore, in seeking to explain a phenomenon, the scientific approach proceeds as follows:

- The first step, when observing a phenomenon that we want to explain, is to create a theory. This theory is a representation of the universe, and it contains its own logic, which usually generates "implications." These implications are essential because they allow for testing the theory's validity or invalidity.[11]

- The second step is to compare these implications with observable data from the real universe. If observed reality contradicts the predicted implications, it can be concluded that the theory is false. If the implications align with reality, it can be concluded that the theory is possibly true.

- The more confirmed implications a theory has, the more firmly it can be considered established.

  These first two steps form the minimal foundation of any scientific theory. Fortunately, in many cases, it is possible to go further.

---

9. Karl Popper, *The Open Society and Its Enemies* (Abingdon, UK: Routledge, 2011), 229–30.

10. Thomas Kuhn, *The Structure of Scientific Revolutions* (Chicago: University of Chicago Press, 1962). Using Thomas Kuhn's schema, it could be said that since the time of Copernicus, the prevailing paradigm has held that science can explain the world without invoking the hypothesis of God. Discoveries in the 20th century have begun to challenge this view, placing the paradigm under scrutiny.

11. For example, the theory that the Earth is round has many implications, one of which is that by traveling in a straight line, one would eventually return to the starting point. This implication is verifiable and was confirmed by Magellan's voyage, providing a final proof to complement all the relative evidence already known since antiquity. This implication also allowed Christopher Columbus to set out from the Canary Islands to reach the Indies.

- The third step (which is not always possible) is to construct a mathematical model of the theory to study its functioning and examine its predictions. These predictions are then compared to real-world observations. If the model aligns with reality, the level of proof is greatly strengthened—particularly if the model predicts unexpected outcomes that are subsequently verified.

- The fourth step is experimentation. This final step offers a very high level of proof if the theory can be repeatedly verified.

An example: Newton's theory of gravity

Newton's theory of gravitation provides a clear example of this four-step process. According to the story, Newton observed an apple falling and wondered why it fell straight to the ground.

The Theory of Gravity, conceived by Newton, proved to conform to reality

Step 1: The Theory — From this initial observation, Newton formulated a theory suggesting that bodies attract each other with a force dependent solely on their mass and the distance between them.

Step 2: The implications — As the first verifiable consequence of his theory, Newton observed that it was the apple that fell to the ground, not the other way around, since the Earth's larger mass attracted the smaller apple. Additionally, an apple in the Southern Hemisphere also fell toward the Earth, even though it appeared "upside down" from the perspective of an observer in the Northern Hemisphere. Thus, the initial implications of his theory aligned with observable reality.

Step 3: The Mathematical Model — Newton developed a mathematical model for his theory, proposing that the force of attraction between two bodies was proportional to their masses and inversely proportional to the square of the distance between them, expressed as $F = Gmm/d^2$.

Using this model, Newton calculated planetary orbits, predicting elliptical shapes—an insight that neither Copernicus nor Galileo had envisioned, though Kepler had observed them in Mars's orbit. By refining his model, Newton ultimately created a predictive calendar of lunar and planetary eclipses.

Step 4: Experimentation — Newton's calculations and predictions, which were verifiable at the time, proved to be remarkably accurate. The theory's alignment with reality was successful, and, even more impressively, it yielded unexpected predictions that were later confirmed. As a result, the scientific community regarded Newton's theory as proven to the highest possible extent within its domain.

Later, Newton's theory of gravitation was refined by Einstein's theory of relativity. While Newton's model remains an excellent approximation in most cases, Einstein's framework provided a more precise understanding. This progression illustrates how scientific models evolve, with each new theory building upon and improving the accuracy of its predecessor.

Scientific theories can be classified into groups based on the level of evidence supporting them

The strength of a theory is determined by the number of validation steps it has successfully passed. Depending on whether it is validated by two, three, or all four of the steps listed previously, a theory's strength can be categorized into different groups, from Group 2 (the strongest) to Group 6 (the weakest), with Group 1 reserved for absolute proof.

- **Group 1: Absolute proof**
  This type is found only in theoretical or formal domains.

- **Group 2: Theories that can be tested against reality, that are mathematically modelable, and that are subject to experimentation**
  This group encompasses fields like physics, mechanics, electricity, electromagnetism, and chemistry. The evidence in this category is so compelling that it approaches absolute proof and is rarely contested, though it may be refined by future convergent models.

- **Group 3: Theories that can be tested against reality, that are modelable, but that are not subject to experimentation**
  This group includes fields like cosmology, climatology (especially climate change research), and econometrics. While these theories may not be testable in the traditional sense, they can be modeled, and the predictions generated by these models can be verified. The level of evidence supporting theories in this group is strong.

- **Group 4: Theories that can be tested against reality, that can be subject to experimentation, but that are not mathematically modelable**
  This group includes fields like physiology, pharmacology, and biology. Although these theories may not be modelable, their reliability is reinforced through repeated experimentation, which provides strong verification. The level of evidence in this group is high, as with Group 3, but achieved through different means.

- **Group 5: Theories that can be tested against reality, but that are neither modelable nor subject to experimentation**
  While lacking the probative strength of previous groups, this category encompasses many widely-accepted scientific fields. Examples include Darwinian evolution (which, for a long time, was neither modelable nor subject to experimentation), paleontology (e.g., theories about dinosaur extinction and Neanderthal disappearance), and origin-related questions (e.g., the origin of life on Earth, the

Moon, and water on our planet). These theories are validated solely by comparing their implications with observable aspects of the real world. This group also includes the thesis "there is a creator God" and the thesis "nothing exists beyond the material universe."

Although neither of these theories is modelable or testable, they do carry logical implications that can be compared to reality. This topic will be explored further in the next chapter.

- **Group 6: Theories without implications, and that are neither modelable nor subject to experimentation**
  These are purely speculative hypotheses. They generate no observable implications that can be tested against reality and, therefore, hold no level of validation.

  One example in this group is the multiverse theory, also known as "parallel universes." This theory has no observable implications. Another example is the theory of the existence of extra-terrestrials, which likewise has no observable implications and therefore remains pure science fiction.

Table summarizing the six groups of evidence

The table below summarizes the two domains and the six groups of evidence, ranging from absolute proof in the formal domain to the absence of proof in the final group, based on a feasible classification.

This approach aligns with the views of Austrian-born philosopher of science Karl Popper (1902–1994). According to Popper, for a thesis to be considered scientific, it must stem from a theory grounded in observation and must be potentially refutable; in other words, it should have enough observable implications that could prove it false. For Popper, refutability is the essential criterion for determining whether a theory or thesis qualifies as scientific or not.[12]

---

12. Imre Lakatos, in *Proofs and Refutations: The Logic of Mathematical Discovery* (Cambridge: Cambridge University Press, 1976), proposed a synthesis of Popper's and Kuhn's approaches. While Popper emphasized falsifiability, Kuhn focused on comprehensive paradigms. Lakatos introduced the concept of a "research program," which suggests a gradual evolution of scientific theories

Type of Domain	Type of Reasoning				Force of Proof
Theoretical or formal domain					
Group 1: mathematics, games, logic, algorithms	Demonstration				Absolute Proof
Real or empirical domain	Theorizable	Can be tested against reality	Can be modeled	Subject to experimentation	
Group 2: physics (mechanics, quantum mechanics, electricity, electromagnetism), chemistry	Yes	Yes	Yes	Yes	Very strong proof, nearly absolute
Group 3: cosmology (the Big Bang, thermal death of the Universe, anthropic principle), climatology, etc.	Yes	Yes	Yes	No	Strong proof
Group 4: physiology, pharmacology, medicine	Yes	Yes	No	Yes	Strong proof
Group 5: theory of evolution, paleontology, origin of life on Earth, origin of the Moon, origin of water, existence of a creator God	Yes	Yes	No	No	The strength of the proof depends on the quality and number of correspondences between the implications of the theory and observable reality
Group 6: parallel universes, multiverses, theories about what happened before the Big Bang	Yes	No	No	No	Not proof at all, pure speculation

According to Popper's criteria, multiverse theories would not be considered scientific, since they lack observable implications. In contrast, the thesis of God's non-existence would be, as it has many implications.

It is worth noting that many scientists and philosophers also consider the existence or non-existence of God to be a scientific question. Richard Dawkins, a leading figure in contemporary atheism, argues this view in his bestselling book *The God Delusion*.[13] He writes, *"The God hypothesis is a scientific hypothesis about the universe, which should be analysed as sceptically as any other"* (p. 18). He further asserts, *"Either [God] exists, or he doesn't. It's a scientific question; one day we may know the answer, and meanwhile we can say something pretty strong about its probability"* (p. 54). Dawkins adds, *"Contrary to Huxley, I shall suggest that the existence of God is a scientific hypothesis like any other. Even if hard to test in practice, it belongs in the same TAB or temporary agnosticism box as the controversies over Permian and Cretaceous extinctions"* (p. 56).

Following this in-depth exploration of the nature of evidence, and recognizing that real-world evidence is never absolute,[14] we find that

---

centered around a "hard core" of foundational ideas. These core ideas are supported by auxiliary hypotheses that scientists continuously adjust and assess, allowing for some tolerance of anomalies. This approach leads us to an important principle in science: inference to the best explanation (IBE). IBE, widely used to establish and validate scientific theories, suggests that when faced with data, the most justified hypothesis is the one offering the most satisfactory explanation among alternatives.

Rather than solely verifying or falsifying a hypothesis (as in Popper's approach) or adhering rigidly to a paradigm (as in Kuhn's), scientists use IBE to determine which hypothesis best explains the available facts. This approach is crucial for selecting theories when absolute proof is unattainable. It's the very method used in this book to explore the question of God's existence. Our methodology also aligns with Kuhn's perspective, which holds that paradigms are accepted within a given era until accumulating challenges eventually push the prevailing theory into crisis.

13. Richard Dawkins, *The God Delusion* (Boston-New York: Mariner Book, 2008).

14. Within a single theory, there can simultaneously exist valid evidence supporting opposing views, creating an apparent paradox. A well-known example of this is Newton's theory, which was confirmed by precise evidence, such as the flattening of the Earth's poles, the elliptical orbit of the Moon around the Earth, and the return of Halley's Comet in 1758. However, in 1919, Eddington's experiment produced contrary evidence, demonstrating that Newton's theory was not exactly correct. Thus, evidence existed both for the accuracy of Newton's theory and for its inaccuracy, a finding that is not entirely surprising, given that Newton's model was mathematically accurate to 99%.

ultimate conviction arises from the presence of multiple, independent, and converging pieces of evidence.

This is why our research cannot rely solely on scientific evidence; it must also incorporate insights from philosophy, morality, history, and even historical enigmas. The chapters in the second part of this book are therefore not digressions from the scientific field but rather reflect a deliberate intent to gather rational evidence from a variety of independent perspectives.

At this point, the reader should understand that the purpose of this book is not to offer absolute proof that compels universal agreement on the existence of a creator God. Rather, it seeks to present numerous, converging, and independent pieces of evidence that can inspire conviction beyond all reasonable doubt.

Ultimately, at the end of this investigation, everyone must reach their own conclusions. If they find the evidence compelling enough, they will act as both judge and jury and make the leap to conviction.

With the nature of what evidence is now clarified, the next step is to explore whether the thesis of a creator God has observable implications in our world.

It is not abnormal for there to be evidence supporting both sides. If we understand this well, we understand what evidence truly is. Evidence will never be an absolute demonstration, but a rational mind will be convinced by a body of independent and converging pieces of evidence.

Sometimes, contradictory evidence arises due to a lack of knowledge. Two historical examples illustrate this point. During the debates about the Earth's rotation and its orbit around the Sun, some people provided seemingly credible evidence that the Earth did not rotate. They argued, for instance, that if the Earth rotated, an arrow shot vertically by an archer should fall slightly west of his feet, not directly at them. This experiment was conducted, and instead, the arrow landed at the archer's feet. Likewise, two cannons fired in opposite directions, east and west, should have sent cannonballs different distances if the Earth rotated—yet they landed equally. In the Renaissance, people were unaware of the principle of inertia, so this evidence for Earth's non-rotation, though incorrect, was convincing enough to mislead minds, at least temporarily.

3.

Implications Arising from the Two Theories: "A Creator God Exists" versus "Nothing Exists Beyond the Material Universe"[15]

When we reflect on the existence of the Universe, two possible theories are on the table: one, the materialist view, asserts that nothing exists beyond space, time, matter, and energy, while the other postulates the existence of a creator God.[16] As discussed in the previous chapter, neither theory can be subject to experimentation or mathematically modeled; therefore, the only way to decide between them is by comparing their implications with what can be observed in the real world.

But do such implications exist?

---

15. The materialist thesis holds that the material universe consists exclusively of time, space, matter, and energy.

16. These two theories do not encompass all possibilities. In addition to them, there is a widespread belief in "spirits" or "spiritual forces" that do not stem from a creator God. Notable examples of this belief can be found in primitive religious forms like animism and shamanism, as well as in Asian religious and philosophical traditions such as Buddhism, Hinduism, and Brahmanism, along with New Age perspectives. However, these beliefs are rarely rationally articulated and offer little insight into the nature and origin of their supposed spirits or spiritual forces. As a result, they remain elusive, and even their adherents struggle to present them coherently. Therefore, it is difficult to categorize these beliefs as logically reasoned approaches.

For instance, the Buddha advises his followers to refrain from conjecturing about certain concepts, stating: *"There are these four Inconceivables that are not to be conjectured about, and anyone who does so will be driven to madness and vexation... Conjecture about [the origin, etc., of] the world is an inconceivable that is not to be conjectured about..."* (Anguttara Nikaya Catukkanipata 4:77). He further explains, *"What have I left undeclared? 'The world is eternal,' 'The world is not eternal,' 'The world is finite,' 'The world is infinite'... I have left this undeclared because it is unbeneficial; it does not pertain to the fundamentals of holy life and does not lead to disenchantment, dispassion, cessation, peace, direct knowledge, enlightenment, or Nibbana"* (*Majjhima Nikaya 63: Culamalunkya Sutta 63: 7-10*). For further exploration of this topic, see Peter Harvey's *An Introduction to Buddhism: Teachings, History, and Practices* (Cambridge: Cambridge University Press, 2013), 32–38, which discusses "Rebirth and Cosmology."

The table below presents the implications associated with each of the two theories:

If nothing exists beyond the material Universe, which constitutes the materialist thesis, then:(*)	If the Universe originates from a creator God, then:
1. The Universe cannot have had an **absolute beginning**.	1. The Universe can be expected to have a **purpose**.
2. The Universe cannot end in **thermal death**.(**)	2. It is likely to be **ordered, beautiful, and intelligible**.
3. The laws of nature, arising solely from chance and necessity, **have no particular reason to favour the emergence of life**.	3. It can be expected to have had a **beginning**.
4. **Miracles** cannot exist.	4. **Miracles** are possible.
5. **Prophecies and revelations** cannot exist.	5. **Prophecies** and **revelations** are possible.
6. **Good** and **evil**, having no transcendent source, are subject to limitless democratic decision-making.	
7. "**Spirits**"--the devil, angels, demons, etc. do not exist.	

(*) In this table, we have limited ourselves to seven implications, though other points of interest exist. Thus: if nothing exists outside the material Universe, the materialist must accept several challenging propositions, such as:

- The law of **conservation** of energy and matter was not upheld at the moment of the Big Bang.
- The law of continually increasing **entropy** was also not respected in the infinite past of the Universe.
- The realm of **Mathematics** is a human invention and cannot exist independently of our Universe.
- **Consciousness** is solely a product of matter and cannot exist independently of the brain's material activity; therefore, "near-death experiences" (NDEs) are necessarily illusions.
- **Beauty** is subjective; if it were objective and derived from rational harmonies, it would be highly improbable, making the common perception of nature's universal beauty inexplicable.

(**) Thermal death is incompatible with an infinite time in the past.

IMPLICATIONS RESULTING FROM TWO THEORIES

There is a common, though mistaken, belief that the existence or nonexistence of a creator God would have no observable implications in the real world.

This chapter will argue that such implications do, in fact, exist, that they are both numerous and significant, and that some of them, once thought to lie beyond human understanding, have now entered the realm of scientific discussion.

Upon examining the two corresponding columns (see previous page), three key observations emerge:

First, the reader will notice that the number of observable implications is significantly high, which increases our chances of deciding between the two opposing theories. The more implications we can identify, the greater the likelihood of confirming or refuting one of the two theories.

Second, the implications derived from these two theories are not equally attractive. The materialist theory—that 'nothing exists beyond the material universe'—is particularly interesting since it generates a larger number of implications that are both clear and precise. Therefore, the most promising way to affirm the existence of a creator God is to try to disprove the materialist theory. Since the two theories are mutually exclusive, disproving one would validate the other.

Third, while neither theory can be directly mathematically modeled or tested, both have implications that can be compared with observable reality, placing them in Group 5 of our evidence framework. In this group, theories are evaluated based only on how well their implications align with observable facts. Thus, the theories—"there is nothing beyond the material Universe," and "the Universe originates from a creator God"—can legitimately be included in Group 5.

Each of these implications will now be analyzed in depth to assess their origin and value.

I. Study of the implications of the thesis "nothing exists beyond the material universe"

If there is nothing other than the material Universe, then:

1. The Universe cannot have had an absolute beginning

This is true for two reasons: one philosophical and the other scientific.

a. Philosophical: As early as 450 B.C., the Greek philosopher Parmenides argued that *"from absolute nothingness, nothing can emerge,"*[17] (*ex nihilo nihil fit*) and no other philosopher has ever seriously challenged this principle.

b. Scientific: One of the most established laws of the Universe states that *"nothing is lost, nothing is created,"*[18] that matter and energy are linked, and that their total remains constant. Consequently, any variation in total mass-energy is deemed impossible.[19] Therefore, the emergence of mass-energy from nothing at the beginning of the Universe would directly violate this law.

This implication is critical to our study because it presents a binary choice. If we accept the proposition that *"if there is nothing beyond*

---

17. *Parmenides, Le Poème: Fragments*, ed. and trans. Marcel Conche (Paris: Épiméthée, PUF, 1996), See also B. Brunor's article in *Thinkable Clues*: "It was the great Parmenides, around the year 500 BCE, who had this simply brilliant idea. He said to himself: in fact, when you think about it, there has never been absolute nothingness. Why do you say that, Parmenides? his interlocutors ask. — Because if there had been absolute nothingness, it would still be there, and nothing would exist." (in *Chance Doesn't Write Messages*, Volume III: [http://www.brunor.fr/PAGES/Pages_Chroniques/25-Chronique.html][http://www.brunor.fr/PAGES/Pages_Chroniques/25-Chronique.html]).

18. "Nothing is lost, nothing is created, everything is transformed" is an apocryphal quote attributed to Antoine Lavoisier on the conservation of mass during changes in the state of matter. See *Elementary Treatise on Chemistry* (Paris: Cuchet, 1789). This statement is very close to that of the philosopher Anaxagoras, who wrote in his *Fragments* (5th century BCE): "Nothing is born nor perishes, but things already existing combine, then separate again." Einstein later confirmed the accuracy of this principle with the law of conservation of total mass and energy.

19. This is, however, not entirely accurate in the case of an expanding Universe like ours because the resulting creation of space leads to the generation of vacuum energy and causes photons traveling over long distances to lose energy (redshift).

the material universe, it cannot have had an absolute beginning" as true, the corollary follows: *"If the universe had an absolute beginning, then a creator exists."* Although this idea has been known since time immemorial, it was long dismissed as having little merit, since it was regarded as undecidable and beyond the reach of human knowledge.

What is remarkable today is that the question of the Universe's beginning has become a scientific matter, supported by substantial evidence, which will be discussed in Chapter 7. This development has transformed it into a question that can be answered.

It is unsurprising that materialist scientists contest these statements, as they are driven by necessity to do so. An absolute beginning to the universe implies the existence of a creator, an unacceptable conclusion for many. Consequently, they often favor alternative hypotheses, even those lacking a solid scientific foundation, such as the multiverse theory.

2. The Universe cannot end in thermal death, since such an end implies an absolute beginning

The second law of thermodynamics, as established by Carnot and Clausius, states that in any closed[20] system, entropy increases over time unless there is an external input of energy or information.[21] This law applies to the Universe in the same way it applies to a candle that gradually burns down and will eventually be entirely consumed if left to burn. Looking forward, entropy increases, leading to a future in which the Universe is completely 'worn out.' Conversely, when we trace entropy backward in time, we see that order increases—but only up to a finite point. This suggests that it is inconceivable for an isolated system to have been degrading indefinitely.

---

20. https://fredericthomasusa.com/about-detail-mary-mcaveney.php.

21. See "entropy" and all other technical terms in the glossary at the end of the book.

As materialists like Ernst Haeckel[22] noted in response to the findings of Boltzmann and Kelvin, if the second law of thermodynamics holds, it implies a beginning for the Universe. If the Universe had existed eternally, it would already be 'used up.'

Convinced by this line of reasoning, the Marxist philosopher Friedrich Engels wrote to Karl Marx on March 21, 1869: *"The original state of great heat from which everything cools is absolutely inexplicable; it is even a contradiction and thus presupposes the existence of a God."*[23] Engels further argued that the second law of thermodynamics must be false because accepting it would imply a beginning to the Universe—and, by extension, a creator—an idea incompatible with his dialectical materialism.[24]

3. Deterministic laws apply universally, and things are distributed randomly

If there is no creator God and the Universe is purely material, it must be governed by immutable laws, excluding any purposeful design. All processes within the Universe would therefore operate solely through chance and necessity, and, as a result, the laws of the Universe could not be particularly favorable to human existence.

---

22. Ernst Haeckel (1834-1919) claimed, "*I consider as supreme, the most general of the laws of nature, the true and unique fundamental cosmological law, the law of substance [. . .]. The chemical law of the conservation of matter and the physical law of the conservation of force form an inseparable whole. For all eternity, the infinite Universe is and will remain subject to the law of substance. Space is infinitely vast and unlimited. It never ends but is filled everywhere with substance. Time is likewise infinite and unlimited, without beginning or end; it is eternity. Substance is found everywhere and at all times in a state of motion and unceasing transformation. The eternal movement of substance in space is this eternal cycle, these phases of evolution that repeat indefinitely. If this theory of eternal existence were accurate, then this end of the world predicted by entropy could never happen, as a minimum of entropy would exist at the start of the Universe's differences in temperature. These two ideas, from our monistic and rigorously logical conception of an eternal cosmogenetic process, are as inadmissible as each other. All doubts contradict the law of substance. The world has neither started nor will it end. Even if the Universe is infinite, it will remain eternally in motion. The second proposition of mechanical heat theory contradicts the first and must be sacrificed.*" Ernst Haeckel, *The Enigmas of the Universe* (New York: Harper & Brothers, 1899).

23. Frederick Engels, *Collected Works*, vol. 43, Marx and Engels: 1868–1870 (New York: International Publishers, 1988), 285.

24. *Dialectic of Nature*, 1st manuscript (1875).

Consequently, the existence of fine-tuning and the anthropic principle can only be explained by adopting the highly speculative multiverse theory.

4. Miracles cannot exist
If the laws of the material Universe apply universally, miracles are impossible, and reported miraculous events must be attributed to errors in judgment or deception.

5. Prophecies and revelations cannot exist
For similar reasons, prophecies or revelations—foretelling improbable and unpredictable future events—are impossible. Any that do arise can only be attributed to luck, credulity, or scheming.

6. Good and evil are not absolute and are therefore open to unlimited democratic decision-making

7. The spirit world—including devils, angels, evil spirits, possessions, exorcisms—does not exist

II. Study of the implications of the thesis "a creator God exists"

Conversely, if the Universe originates from a creator God, then:

1. The Universe can be expected to have a purpose or end goal
If the Universe's creation is the result of intelligent intent, it would be logical for its evolution to be guided by an underlying order, progressing toward a predetermined direction.

2. The Universe can be expected to be ordered, beautiful, and intelligible
If a perfect and intelligent God created the Universe with the intention of fostering complexity and the emergence of humanity, it is logical to find order, harmony, and intelligibility within it.

3. The Universe can be expected to have had a beginning
If the Universe originates from a creator, it is reasonable to assume it had a beginning.

4. Miracles are possible

Miracles could occur either through primary causes (overriding the usual laws of the Universe) or through secondary causes (providential occurrences).

5. Prophecies and revelations are possible

A creator God, being omniscient, would have knowledge of the future; therefore, prophecies and revelations would be possible.

III. What a coherent materialist must accept as true

Based on the implications we have reviewed, a coherent materialist must first believe in the exclusively material nature of the Universe and reject the existence of any spiritual beings, such as God, devils, and souls—beliefs that are relatively simple and easy to uphold. However, to remain consistent, a materialist must also accept and uphold a series of observable and demanding statements, such as:

- The Universe had no absolute beginning.
- The Universe is not heading toward thermal death, despite widely accepted views to the contrary.
- Billions of parallel or successive universes exist, even though this hypothesis lacks serious scientific support, as it is the only way to account for the fine-tuning of the Universe and its exceptionally favorable conditions for human life.
- Some of the greatest laws of physics, considered universal and immutable, have been violated (e.g., the principle of conservation of mass-energy at the Universe's beginning).
- Philosophically, in the absence of a transcendent basis, the concepts of good and evil can be defined indefinitely and solely through democratic consensus.
- Miracles, prophecies, and revelations are merely illusions or forms of deception.

Coherent materialists, reflecting on these points, will realize the numerous extraordinary assertions and propositions they are required to believe, making the term "unbeliever" totally inappropriate in their case.

We will now begin by examining the two main beliefs underlying materialism: that the Universe did not have an absolute beginning, and that it is not heading toward thermal death—two statements that, as we will see, are quite likely to be false.

EVIDENCE WITHIN THE SCIENCES

4.

The Thermal Death of the Universe: Story of an End, Proof of a Beginning

From the fireplace to the stars: conceptualizing the thermal death of the Universe

The Universe is like a fire crackling away in a fireplace. Subject to the laws of thermodynamics, both are destined to burn themselves out over time.

As we watch the fire, we notice that the logs burn up and are gradually consumed by the flames. Eventually, only a handful of embers are left smoldering, until they too begin to go out, one by one. We can deduce that in a few hours' time only a heap of cold ashes will remain in the fireplace.

But there is another important conclusion we can draw from our observations: since the fire is being consumed at a measurable rate, it cannot have been burning forever. If it had always existed, then it would also have burned down and gone out an infinitely long time ago.

We can deduce, therefore, that the fire was kindled by someone at some particular point in the past.

The same goes for the Universe, which is consuming itself at an observable rate: if it had always existed, it would already have arrived at its exhaustion point, and therefore its end. This is why the heat death of the Universe implies that there was an absolute beginning.

Introduction

Paradoxically, this fascinating—even explosive—subject of the heat death of the Universe has caused less of a stir among intellectuals than even the Big Bang, itself a source of much debate and controversy. Is it perhaps more anxiety-inducing, even subconsciously so, to study the death of the Universe than its causes? Yet the discovery of heat death has several crucial implications and constitutes one of the strongest pieces of evidence for a beginning to the Universe. It brings to the fore the complex notion of entropy, which is linked to the irreversibility of time, the common thread that runs through this chapter.

What if, to better understand its origin, we leafed through the script of the Universe, starting with its denouement?

What does the future hold for our Universe?

After two centuries of scientific advances bearing on this question, the near-unanimous scientific consensus is that the Universe will completely end in an unavoidable heat death. Our Sun, which has existed for 4.5 billion years, will continue burning for about that long again before transforming into a red giant (engulfing the Earth and Mars) and then a white dwarf before finally going out. It's the same story for all stars: all over the Universe they will run out of fuel and burn out, like logs slowly burning down in a fireplace.[25]

---

25. All stars burn their hydrogen, irreversibly transforming it into helium. The Sun burns 620 million tons of hydrogen every second, transforming it into 615 million tons of helium. The other 5 million lost tons are the energy the Sun radiates into space.

Though this important discovery, made in the second half of the nineteenth century, took several decades to become established, it has since been confirmed by all subsequent theories and observations. It logically leads to a profound change in our vision of the world.

I. History of the discovery of the thermal death of the Universe

Sadi Carnot founds the new field of thermodynamics (1824)

It all started in 1824, in Paris. In his first work, *Reflections on the Motive Power of Fire and on Machines Fitted to Develop that Power*, the twenty-eight-year-old Sadi Carnot laid the foundations of an entirely new discipline. Though the term thermodynamics was not coined until more than a century later by William Thomson (later known as Lord Kelvin), Carnot is the true founder of this field of science, which ultimately proved as fundamental from a theoretical point of view as it is productive in its practical applications. To take just one example, we can thank Carnot for the first scientific account of heat engines, the theoretical basis for automotive and jet engines. He died very young in 1832, at the age of thirty-six, leaving behind very promising and fundamental, but unfinished, work.

Sadi Carnot (1796-1832)

Rudolf Clausius takes up the torch, defining the second law of thermodynamics (1865)

Utilizing the theoretical "toolbox" left behind by Carnot and building upon his work on thermal machines, Rudolf Clausius integrated the contributions of Lord Rumford, who demonstrated that friction produces heat, and Hermann von Helmholtz, who established the principle of conservation of energy, which includes heat as a form of energy. In 1865, Clausius postulated a new universal law that asserts that, without the input of new information or external energy, every isolated system will experience an irreversible increase in what he termed "entropy" (derived from the Greek word for "transformation") as it evolves from an initial state toward a state of final equilibrium. As a consequence, any return to the original state is impossible.[26] Clausius' daring theory became known as the second law of thermodynamics,[27] and scientists were eager to test the validity and universality of this law.

Rudolf Clausius (1882-1888)

Ilya Prigogine, the 1977 Nobel Prize winner in Chemistry, is undoubtedly the modern thinker who has thought more than anyone else about the second principle and its profound consequences: *"Therefore, the dynamic eternally opposes the 'second principle of thermodynamics,' the law of irreversible growth of entropy formulated by Rudolf Clausius in 1865; concerning the determinism of dynamic trajectories, a completely inexorable determinism of the processes that level out the differences of pressure, temperature, and chemical concentration, and which irreversibly lead the isolated thermodynamic system to its state of equilibrium, of maximal entropy."*[28]

In other words, the notion of entropy is valuable for giving a new momentum, a new perspective on how to envisage the end of the Universe.

---

26. That is to say, so infinitely improbable that it is a practical impossibility.

27. The first principle of thermodynamics is that of the conservation of energy during any transformation, defined in 1847 by Hermann Helmoltz, based on the work of Robert Mayer and Joule.

28. Ilya Prigogine and Isabelle Stengers. *Entre le temps et l'éternité (Between Time and Eternity)*, (Paris: Fayard, 1988, rééd. coll. Champs, Flammarion, 2009), 34.

Ludwig Boltzmann creates a model of entropy and arrives at strong conclusions (1878)

Building on the work of Clausius and his own research into the kinetic theory of gases, Boltzmann demonstrated that every isolated system can be modeled as an increasing function (S) as the system moves toward equilibrium, according to the formula S = k. log W.[29]

S stands for entropy, k is "Boltzmann's constant," and W is the myriad possible states of all atomic or microscopic elements.

Ludwig Boltzmann (1844–1906)

Einstein called this equation, which is engraved on Boltzmann's tombstone, "*the most important formula in physics.*"[30] It is in fact a revolutionary idea which proves that the disorder characterizing this or that object can only statistically increase. Never otherwise. Never the opposite.

Helmholtz (1854) and Lord Kelvin develop the idea of a "thermal death" of the Universe

In an 1854 article, Prussian scientist Hermann von Helmholtz went straight to the point.[31] The Universe has only one possible end: "heat death,"[32] when the temperature of the whole Universe will approach absolute zero. Von Helmholtz explained that the stars will progres-

---

29. Clausius designated entropy with the symbol "S" in honor of Sadi Carnot.

30. Einstein called thermodynamics "*the only physical theory of universal content concerning which I am convinced that...it will never be overthrown.*" Albert Einstein, "Autobiographical Notes," in *Albert Einstein: Philosopher-Scientist*, ed. Paul Arthur Schilpp (London: Harper and Row, 1959), 1:33. On this subject, see A. Einstein, "Eine Theorie der Grundlagen der Thermodynamik," *Annalen der Physik*, ser. 4, XI (1903): 170–187 (CP 2:77–94).

31. Von Helmholtz first spelled out the principle of the conservation of energy in 1847, in H. Helmholtz, *Ueber die Wechselwirkung der Naturkäfte* (Königsberg: von Gräfe & Unzer, 1854).

32. As thermodynamic conditions tend gradually to equilibrium, with the hottest places losing their energy, the Universe as a whole will approach absolute zero.

sively go out, one by one, and the temperature of the entire cosmos will decrease until *"the universe would be condemned to a state of eternal rest."*[33] A few years later Lord Kelvin went on to flesh out this idea of the Universe's "thermal death."[34]

Lord Kelvin (1824–1907)

Arthur Eddington summarizes the concept as "the arrow of time" (1928)

Sir Arthur Eddington demonstrated that the second principle of thermodynamics implies the existence of an "arrow of time," indicating that time flows in one direction. In other words, there is a forward directionality to time, which can be likened to the trajectory of an arrow. We have seen how entropy must increase in an isolated system. Therefore, if we measure the entropy of an isolated system at two different moments, we can tell which moment preceded the other and therefore which direction the thermodynamic "arrow" of time is pointing. If we apply this concept on a universal scale,[35] we can talk about a "cosmological arrow of time." This was a profound and radical discovery that carried with it important metaphysical and philosophical implications. In one stroke it invalidated the many cyclical views of the universe (such as the myth of the eternal return in ancient and Hindu metaphysics).

The novelty of the concept and its striking consequences go some way towards explaining why this great law of the Universe was received

---

33. H. Helmholtz, *Ueber die Wechselwirkung der Naturkäfte*, 24.

34. In "On the Age of the Sun's Heat," by Sir William Thomson (Lord Kelvin), *Macmillan's Magazine*, vol. 5 (March 5, 1862): 388-393 (online: https://zapatopi.net/kelvin/papers/on_the_age_of_the_suns_heat.html). Helmholtz was not the first to write on this topic. Lord Kelvin was actually the first to mention the possibility of a "Big Freeze." Kelvin's ideas were further developed over the following decade by Helmholtz and Rankine. We therefore refer to the article by Lord Kelvin.

35. Assuming, of course, that our Universe is a single, closed system.

with great reticence and even met with stiff opposition. In spite of its important claims, it took more than fifty years for it to be recognized and accepted by the whole scientific community. These implications certainly challenge what we thought we knew.

The first explosive consequence for metaphysics: the Universe must have had a beginning

From a materialist perspective, the Universe is a giant, closed system that contains everything there is. Nothing exists outside of it. That means the Universe is just like our fire blazing in the fireplace or the candle that slowly burns down. If we look toward the future, all energy will sooner or later be completely used up. By this logic, conversely, if we look toward the past, the Universe must have had a beginning because it is impossible to imagine a closed system that has been endlessly consuming itself over an infinitely long span of time. If that were the case, it would have been used up an infinity of time ago, and there would be no one left to tell of it. To put it in mathematical terms, "$\infty - T = \infty$." Whatever finite time (T) we subtract from infinity, we are always left with infinity. Therefore, if the Universe had been expending its usable energy for an infinitely long time, it should already have exhausted it. However, this is not the case. The Universe therefore must have had a beginning.

According to the Kalam cosmological argument,[36] this absolute beginning to the Universe must have a transcendent cause:

- Whatever begins to exist has a cause.
- The Universe had a beginning.
- Therefore the Universe has a cause.

---

36. The ideas of Christian philosopher and Aristotelian commentator John Philoponus (6th c. Egypt) on the impossibility of an infinite time stretching into the past—expressed in *De aeternitate mundi contra Proclum* (Leipzig: Teubner, 1899)—were later developed by the Muslim philosophers Al-Kindi (9th c.) and Al-Ghazali (12th c.), as well as by Saint Albert the Great and Saint Bonaventure (13th c.). These arguments are now known by the Arabic term *Kalam* (meaning 'discussion' or 'debate').

A second, even more revolutionary consequence: from its origin, the Universe was highly ordered

In 1878, Boltzmann recognized that the entropy of the Universe must have been extremely low at its inception. His insights were later confirmed by the work of Roger Penrose. Remarkably, at the moment the Universe reached its lowest entropy, everything in the primordial cosmos was likely organized with an almost unfathomable precision. This raises the question: how could such order arise? Despite Boltzmann's extensive research, he could not provide a formal proof for this phenomenon, and it took more than a century for the concept of "fine-tuning" to develop. Tragically, even though Boltzmann was undeniably ahead of his time, the intense criticism he faced from some colleagues took a toll on his already fragile psyche, leading to his death by suicide in 1906.

A chorus of opposition against Boltzmann's discoveries

- Henri Poincaré
  Boltzmann's calculations implied that time was irreversible. However, for Poincaré and the majority of scientists at the time, there was nothing at all about mechanics that implied the irreversibility of time.
- Ernst Mach
  The great Austrian scholar Ernst Mach, whom Boltzmann was once expected to follow as head of the departments of philosophy and history of science at the University of Vienna, swore to *"silence this petty researcher whose ideas are so dangerous for Physics."*[37] A committed materialist and supporter of the Paris Commune, Mach would not accept Boltzmann's work and contradicted all his public remarks.

---

37. On this subject, see V. Kartsev, "The Mach-Boltzmann Controversy and Maxwell's Views on Physical Reality," in *Probabilistic Thinking, Thermodynamics, and the Interaction of the History and Philosophy of Science: Proceedings of the 1978 Pisa Conference on the History and Philosophy of Science*, vol. 2, ed. Jaakko Hintikka, David Gruender, and Evandro Agazzi (Dordrecht: Springer, 1981).

- **Ernst Haeckel**

  Significantly, other experts began to realize that this new concept of the irreversibility of time implied the Universe had a beginning, challenging their core beliefs. This realization motivated the esteemed biologist Ernst Haeckel to devote a considerable portion of his academic career to disputing the validity of the second principle of thermodynamics.

- **Friedrich Engels**

  Friedrich Engels, the co-founder of Marxism, maintained that time was cyclical. He fought with all his might against the idea that there was an origin to the Universe, since this idea threatened his arguments for dialectical materialism.

- **Svante Arrhenius**

  In his *Evolution of Worlds*, the great Swedish chemist and 1903 Nobel Prize winner betrayed his philosophical prejudices when he wrote: "*If Clausius were right, however, this heat-death, we may object, should have already occurred in the infinitely long space of time that the universe has been in existence. Or we might argue that the world has not yet been in existence sufficiently long, but that, anyhow, it had a beginning. [...] in that case all the energy would have originated in the moment of creation. That is quite inconceivable.*"[38]

  This position is surprising coming from a major scientist, and a Nobel laureate to boot. His prejudice against the existence of God leads him to claim that since nothing can have been created, the Universe could not have had a beginning, and therefore, that the discoveries of thermodynamics are incomprehensible. The hypothesis that a god both exists and created the Universe clashes with his principles.

- **Marcellin Berthelot**

  The famous French chemist Marcellin Berthelot joins our long list of famous detractors. He developed an account of chemical phenomena in terms of the general laws of mechanics and was violently opposed

---

38. Svante Arrhenius, *Worlds in the Making: The Evolution of the Universe* (New York, London: Harper, 1908), 193.

to Pierre Duhem, who proposed, in collaboration with Willard Gibbs, a theory of chemical reactions based on the first and second laws of thermodynamics, an approach that has come to be called chemical thermodynamics. Since Duhem's conclusions were in conflict with his own, Berthelot, who was then minister of public education, did not allow Duhem to defend his doctoral thesis, and he was forced to choose another topic. Berthelot later made amends for holding Duhem back.

- **The early Albert Einstein**
  Albert Einstein was also long opposed to the idea of the irreversibility of time and universal expansion, but once the evidence became undeniable, he changed his mind on both subjects. At the beginning of the twentieth century, Planck used the second law of thermodynamics to explain black-body radiation. Einstein, who when studying the photoelectric effect in 1905 had interpreted these packets of energy as quanta, ended up declaring the second law of thermodynamics *"the first law of all science."*[39]

More consequences of the second law of thermodynamics

Conscious of the philosophical implications of the discoveries of Carnot and Clausius, by the end of the nineteenth century, many intellectuals had taken a radically pessimistic view regarding the destiny of the cosmos, which seemed doomed to slow and inevitable decay. But others saw this in a more positive light, thinking that the arrow of time and the impossibility of eternal cycles made our history much more interesting : *"Nevertheless, it would be an error to think of the second law of thermodynamics as only a source of pessimism and anguish. For certain physicists, such as Max Planck and above all Ludwig Boltzmann, it is also the symbol of a decisive turn. Physics*

---

39. A. Einstein, "Eine Theorie der Grundlagen der Thermodynamik," *Annalen der Physik*, ser. 4, XI, 1903, 170-187 (CP 2, 77-94). English translation available as "A Theory of the Foundations of Thermodynamics" in *The Collected Papers of Albert Einstein: English Translation. Volume 2: The Swiss Years: Writings, 1900-1909.* Translated by Anna Beck (Princeton, N.J.: Princeton University Press, 1987), 48-67.

could finally describe nature in terms of becoming; like the other sciences, physics would be able to describe a world open to history."[40]
History, it seems, is not doomed to repeat itself, after all.

Universal recognition of the Second Law of Thermodynamics

The principle of the physical world's increasing entropy, which evolves globally in a single direction, has never been seriously challenged. In fact, today this is one of the most solidly established laws of physics. And with the help of statistics it is possible to demonstrate that the probability of a return to an earlier state is all but infinitely low, growing less and less probable the larger the system is. The eminent astrophysicist Arthur Eddington is famous for his categorical commentary on the subject: *"I believe that the law that posits that entropy is perpetually increasing occupies a superior position among the laws of nature. If one of you demonstrates that your theory disagrees with the equations of Maxwell... too bad for the equations of Maxwell. But if your theory contradicts the second principle of thermodynamics, I can offer you no hope."*[41] Eddington's absolute conviction is evident beneath his irony.

Curiously, at the time when entropy was discovered, most scientists were convinced that the Universe was eternal

Surprisingly, none of the logical consequences were drawn from the discovery and acceptance of the second law of thermodynamics. The general belief remained that the Universe is eternal, and there was no large scale discussion of the theoretical hypothesis of a beginning. This definitely had something to do with philosophical *a priori* assumptions, but it was also due to the simple fact that, in the years 1915-1925, cosmology was not yet considered a real science. The following anecdote

---

40. Ilya Prigogine and Isabelle Stengers, *Entre le temps et l'éternité* (Between Time and Eternity) (Paris: Fayard, 1988; repr., Champs, Flammarion, 2009), 34.

41. Arthur Eddington, *The Nature of the Physical World* (New York: Macmillan, 1928), 74.

is revealing: Ernest Rutherford, then one of the most brilliant physicists in Europe, forbade all discussion of cosmology in his laboratory on the grounds that it was pseudo-science.[42] Hubble's discovery in 1924 that other galaxies exist, along with Lemaître's initial works on universal expansion published in 1927, rooted in Einstein's theory of General Relativity, were pivotal for establishing cosmology as a recognized scientific field. However, it took many more years before a cosmologist was nominated for the Nobel Prize. Friedmann, Lemaître, Hoyle, and Gamow deserved one, but the Nobel committee did not yet recognize cosmology as a science. This did not begin to change until 1953, the year Hubble died unexpectedly of a stroke, while a candidate for the Nobel Prize. Furthermore, between 1931 and 1965, lively debates on the hypothesis of the Big Bang overshadowed all other approaches, so much so that during these years, the consequences of the second law of thermodynamics were not the subject of debate and the idea of the beginning of the Universe remained unimaginable to most scientists.

Edwin Hubble (1889–1953)

Specialists in entropy are not the only ones with something to say about the beginning of the Universe

Theoreticians of entropy were not the only ones to imagine that the Universe had a beginning. Many others had foreseen it, including astronomer Heinrich Olbers in 1823;[43] writer Edgar Allan Poe in his fascinating essay

---

42. Professor Steven Weinberg of Harvard University, Nobel prize-winning scientist, explains that "*in the 1950s, the study of the early universe was widely regarded as not the sort of thing to which a respectable scientist would devote his time. Nor was this judgement unreasonable. Throughout most of the history of modern physics and astronomy, there simply has not existed an adequate observational and theoretical foundation on which to build a history of the early universe.*" Steven Weinberg, *The First Three Minutes* (London: Deutsch & Fontana, 1977), 13-14.

43. Author of the famous eponymous paradox that asks why night is dark, for if the Universe were stable and infinite, as was then believed, then every single way we look should point towards the

Eureka, published in 1848; astronomer François Arago; mathematician Bernhard Riemann in 1854; and astronomers Vesto Slipher (who taught Georges Lemaître) and Willem de Sitter. Their hypotheses were dismissed as fantasies and baseless daydreams that could not withstand the weight of contemporary certitudes and *a priori* judgments.

Upon the confirmation of the Big Bang in 1964, partisans of the eternal Universe propose the Big Crunch

The Big Bang hypothesis was confirmed in 1964 with the discovery of cosmic microwave background radiation (CMBR), which closely matched the predictions made by Gamow and his colleagues.[44] This finding appeared to decisively refute the arguments for an eternal universe, suggesting that the Big Bang represented an absolute beginning. Consequently, this led to the development of the "Big Crunch" theory, which posits a final collapse that is the opposite of the Big Bang. According to Einstein's equations of Relativity, the speed and acceleration of the expansion of the Universe depend on its density, its pressure, its spatial curvature, and the value of the cosmological constant. If the density of the Universe is large enough, Big Crunch scientists suggested, the Universe will at first inflate and expand but then enter a period of contraction and retraction. For decades, this idea of repeated cycles seemed the most promising hypothesis for preserving the idea of an eternal Universe. Everyone wanted to predict how the expansion of the Universe would slow, and they tried to calculate quite precisely the critical figures beyond which the Big Crunch was inevitable.

A dramatic turn in 1998: the accelerating expansion of the Universe (1998) and the collapse of the Big Crunch hypothesis

Contrary to all the hopes and efforts of materialist scientists, it was proved in 1998 by Saul Perlmutter, Brian Schmidt, and Adam Riess

---

surface of a star, and the night sky should therefore be bright.

44. See Chapters 5 and 6

that the expansion of the universe was not slowing down as everyone imagined, but rather accelerating.

This groundbreaking discovery earned the authors the Nobel Prize in Physics in 2011, and their findings have been further validated by the WMAP (2001) and Planck (2009) missions, which showed consistent measurements of an extremely small spatial curvature of the universe. As a result, it is now difficult to support the Big Crunch hypothesis. If we accept the notion of a permanent cosmological constant, the equations of General Relativity align perfectly with the phenomenon of accelerating expansion. This is consistent with all our scientific observations, collectively suggesting that the expansion of the universe will continue indefinitely.[45]

The necessary hypotheses of Dark Matter and Dark Energy, a repulsive force

Our understanding of the evolution of the Universe as a physical system is based on the theory of General Relativity. According to the equations involved, the expansion rate of the Universe is a function of its average energy density as well as of one of its geometric properties—its spatial curvature—and of the famous cosmological constant already mentioned. How can these be determined? Astrophysicists have observed that the structure and motion of galaxies do not align with expectations based on their measured mass. Much of the mass required to explain galactic motion appears to be missing. This paradox has led to the idea that most of the mass in galaxies exists as so-called "dark matter." To explain the observed acceleration of the Universe's expansion, the cosmological constant has to exist. Thus, it has been calculated that known matter (observable atoms) make up only 4% of the Universe, while a little less than 26% is this "dark

---

[45]. Even if the Universe contracted and then rebounded into a new period of expansion, it could not be eternal. The second law of thermodynamics in effect stipulates that each new cycle in time increases entropy, which results in longer and longer cycles. Consequently, by going back in time, the cycles would have been increasingly shortened until they got to the point where a cycle couldn't be any shorter, thus marking the beginning of the Universe. Cf. Alan H. Guth and Marc Sher, "The Impossibility of a Bouncing Universe," *Nature,* 302, (April 1983): 505-506.

matter," a mysterious form of matter that would explain the "missing" gravitational attraction. The remaining 70% is then a no less mysterious "dark energy" (or "vacuum energy"), which corresponds to the famous cosmological constant and acts as a repulsive force opposed to gravity, explaining the acceleration of the Universe's expansion. The existence of this dark matter and dark energy appears certain, from the measurements and calculations that allow us to define them, but their nature remains totally unknown. Given that the Universe is most definitely and observably expanding, the mystery behind the missing matter and energy just goes to illustrate the fact that a considerable part of the Universe remains unknown.

An end that seems inevitable: the heat death of the Universe

While we still lack complete certainty about the nature of dark matter and dark energy, the best currently available data agrees that unless the laws of nature change over time, the Universe will end in heat death. All stars will eventually be extinguished, every source of energy will be consumed, and the universe will grow colder and colder until it tends towards a temperature of absolute zero and a state of maximum entropy, in which no further thermodynamic reactions are possible. It is estimated that this "Dark Era" will occur around 10^{100} years from now. However, even before that, approximately 10^{30} years from now, the remaining energy will be insufficient to support life. There are several alternative versions of this final act, such as the "Big Rip," a speculative hypothesis proposed in 2003 by three American researchers. According to this model, the final curtain goes down a mere 22 billion years in the future. But whatever the precise details of this vast time scale, nearly everyone agrees that the eternally accelerating expansion of the Universe will lead to a state of total darkness, cold, and eventual death.

Conclusion

At the start of the twentieth century, our understanding of the Universe underwent a significant transformation. This change was prompted by two groundbreaking insights from thermodynamics: first, that the Universe had a definitive beginning, and second, that this beginning was characterized by a state of minimum entropy, or maximum order—essentially, a highly fine-tuned condition. These revolutionary ideas that Boltzmann deduced from the laws of thermodynamics have been verified, and no one has ever seriously called them into question. And yet, even after crossing this crucial line, many great scientific minds hesitated to take one more step toward an ultimate conclusion that, while not scientific in the strictest sense, is nonetheless demanded by reason: if the Universe has a beginning in time, there must have been a cause that preceded it.

II. The scenario with the greatest consensus today

The accelerating expansion of the Universe is confirmed by observation and widely accepted by cosmologists

The observable diameter of the Universe is estimated at around 93 billion light years and its age about 13.8 billion years. Observations carried out since 1998 by Saul Perlmutter and Brian Schmidt, winners of the 2011 Nobel prize in physics, have proven that it is expanding at an ever-accelerating rate.

Most cosmologists believe that the Universe is undergoing this process of expansion and that it has done so since its origin in accordance with the natural course of the development of the Universe as envisioned by astrophysicists.

From all this, there is a nearly universal consensus regarding the Thermal Death of the Universe

The heat death of the Universe is a logical outcome of applying the second law of thermodynamics to an ever-expanding space. Currently, no widely accepted scientific theory offers a plausible alternative to this scenario, although the modeling of the process and the timeframe for its conclusion still require refinement.

10^{30} years from now: the end of stars and of all life

Currently, we estimate that four to five stars are formed each year in our galaxy, corresponding to roughly 300,000 new stars each second in the 2,000 billion galaxies of the observable Universe.

In 4.5 billion years (10^9), the Earth will be destroyed when our Sun transforms into a red giant with a diameter reaching all the way to Mars. The Sun will then consume all its hydrogen and go out.[46]

In 1,000 billion years (10^{12}), all the galaxies outside our local cluster, which will by then consist of a single galaxy called Milkomeda, a contraction of the Milky Way and Andromeda, will have passed to the other side of the horizon due to the acceleration of the Universe's expansion. Any human civilization remaining in the galaxy will believe itself alone in the Universe.[47]

In 1,000 billion to 100,000 billion years (10^{12} to 10^{14}), stars will stop forming and then begin to go out as they run out of the gas needed to keep burning.

In 100,000 billion years (10^{14}), all stars will be extinguished: all white dwarfs and neutron stars will cool, leading to the extinction of all life.

In 100,000 trillion years (10^{23}), the dead stars will be absorbed by the black hole at the center of the galaxy.

In 100 to 10,000,000,000 billion billion billion (10^{29} to 10^{37}), the Universe will be made up of 90% dead stars, 9% supermassive black holes formed by collapsing galaxies, and 1% atomic matter, mostly hydrogen.

---

46. Eric Betz, "Here's What Happens to the Solar System When the Sun Dies," *Discover* (Feb 6, 2020).

47. See Paul Gilster, "What will astronomers see a trillion years from now?" *Gizmodo* (April 16, 2011). We use the term "horizon" to refer to the limit of all possible observation because light coming from beyond this limit cannot reach us, as it is going slower than the expansion of the Universe.

10^{30} to 10^{38} years from now: the possible disintegration of protons would cause neutrons to disappear

In 10,000 billion billion billion years, particle physics suggests that protons may disintegrate,[48] leaving only neutrons that will then quickly disintegrate into protons, their autonomous existence only lasting fifteen minutes. Their disappearance will liberate protons, electrons, and antineutrinos, filling space with a gas so thin that the distance between two particles would be approximately the diameter of our current galaxy.

10^{100} years from now: the likely end of black holes

Some scientists think that in 10^{68} to 10^{102} years, black holes will dissipate. This astonishing hypothesis on the end of black holes was advanced by Stephen Hawking, based on his research in quantum mechanics. John Wheeler was one of the first to examine more closely the notion of entropy in cosmology, shifting the discussion into the realm of the physics of black holes. Inspired by his reflections, Jacob Bekenstein and Stephen Hawking eventually concluded that a black hole has an entropy proportional to the square of its mass and that it emits a radiation (Hawking radiation) through quantum tunnelling that will eventually lead to it breaking apart.[49]

Beyond 10^{100} years: the probable advent of a "Dark Era" of complete thermal death

Beyond 10^{100} years from now we arrive at the complete heat death of the Universe. Having reached a state of extreme dilation through its process of expansion, the Universe will reach a state of maximum entropy

---

48. Stephen F. King et al., "Confronting SO(10) GUTs with Proton Decay and Gravitational Waves," *Journal of High Energy Physics*, no. 10 (October 28, 2021): 1–38 and https://www.kavlifoundation.org/news/the-enduring-quest-for-proton-decay.

49. Fred Adams and Greg Laughlin, *The Five Ages of the Universe: Inside the Physics of Eternity* (New York: Free Press, 1999), 107-152.

that implies the end of all thermodynamic activity. Then what is quite justly called the "Dark Era" will begin, when mostly photons will be left, floating in a gigantic, ever-colder space tending towards absolute zero.

5.

A Brief History of the Big Bang

The Big Bang theory didn't emerge with a bang. There was no definitive "eureka" moment or overwhelming evidence that shocked the scientific community. Instead, its development was a long and painstaking journey, marked from the outset by skepticism. This was followed by a gradual shift in established opinions, accompanied by hesitation and a frantic search for alternative hypotheses, as some scientists grappled with the metaphysical implications of this initial singularity.

However, the first thing that the Big Bang shattered was a host of certainties and a priori assumptions surrounding the representation of our Universe.

I. The Big Bang and the birth of cosmology in the twentieth century

As we have seen, before Einstein, cosmology was not a science, and until 1912–1925 it had little claim to scientific legitimacy. Most researchers believed that the Universe was fixed, unchanging, and immense, without limit in time or space. At the beginning of the twentieth century, the discussion was considered closed. No one dared to imagine that the Universe had undergone a series of major changes that could be known through reason and observation. This dogma would soon be shattered by the discoveries of a certain young scientist.

Einstein's theory of relativity: a giant leap for our understanding of the Universe

For Einstein, then a nobody at the Bern patent office, 1905 was the *annus mirabilis* (miracle year). It was in this year that he published his innovative theories as a series of four articles in the journal *Annalen der Physik*.

The third of these articles, entitled, "On the Electrodynamics of Moving Bodies," holds that the speed of light is a constant and an absolute that cannot be surpassed in our universe. Time and space, however, are relative, contracting or expanding based on the frame of reference of the observer. This was a major conceptual revolution. Einstein set the capstone on his work in 1915 when he presented his theory of gravitation, commonly known as General Relativity, which encompasses, corrects, and surpasses Isaac Newton's theory of universal gravitation. He theorized that space, time, and matter are linked and that the presence of matter or energy distorts space-time. To put it concretely, in the language of relativity, the planets do not "revolve" around the Sun but move "straight ahead" in a space locally curved by the Sun's gravitational field.

Experimental testing of Relativity

These daring theories—and the experimental tests that later confirmed them—rocked the scientific world. The great astronomer Sir Arthur Eddington was the first to measure the curvature of space by observing the alteration of the apparent position of the stars visually close to the Sun during a 1919 solar eclipse. Given the mass of the Sun, he was able to verify with great precision that this alteration takes place according to the angle predicted by Einstein's theory.

For Einstein, the idea of an expanding Universe was inconceivable, so he proposed the cosmological constant to sustain the static Universe model

The distortion of space-time was tested in 1954, a year before Einstein's death, by atomic clocks onboard a jet plane designed to fly at great speed at very high altitudes, where the gravitational field is weaker. By the end of the flight, the clocks had advanced a few millionths of a second compared to clocks located at ground level, demonstrating the accuracy of Einstein's concept and calculations concerning the dilation or slowing down of space-time in a gravitational field.

Experiments carried out in the early 1960s made it possible to verify that, from the point of view of an external observer, the local time of an object moving at very high speed expands, stretches and therefore lasts longer, or slows down, as predicted by Einstein's theory of special relativity. Thus, particles from secondary cosmic rays in the upper atmosphere, which have such a short lifespan that they are unobservable if they do not move, can in fact be observed, because they move at the speed close to that of the light relative to the observer. Einstein is stubborn, and the facts are too, and they all support the theory of Relativity. Cosmology can then take off on completely new and well-established foundations.

An unjustified cosmological constant

In 1921, Einstein's fame grew yet again when he won the Nobel Prize in Physics for his work on the photoelectric effect, in addition to the already partially confirmed Theory of Relativity. But if we follow the implications of his theory, we arrive at an a priori non-stable Universe, which was, for Einstein, inconceivable. To solve the problem, he set the cosmological constant in his 1917 equations to a value that yielded a static Universe. Einstein did so because he found the alternative unimaginable. He had already dared to postulate an enormous conceptual leap with his new theories, but he was not at all ready to imagine the next leap, according to which the Universe could be expanding. It took the audacity of a young Russian researcher to help him take the plunge.

Alexander Friedmann versus Einstein: a duel over the expansion of the Universe

As early as 1922, the need for this cosmological constant was being called into question by a thirty-three-year-old Russian mathematician named Alexander Friedmann, who was the first ever to publish on the theory of an expanding Universe. He mailed his article to Einstein, whose work he had used as the basis for his own arguments,[50] but the father of the theory of General Relativity reacted very poorly, declaring: *"The circumstance [of an expanding universe] irritates me."* In another letter, he wrote: *"To admit such possibilities seems senseless."*[51] An exasperated Einstein drafted a terse letter denouncing Friedmann's "miscalculations," which was published in the premier journal of theoretical physics, the *Zeitschrift für Physik*: *"The results obtained in [Friedmann's] paper regarding a nonstationary universe seemed suspect to me. In fact, it turns out that the solution given does not agree with the field equations."*[52]

Friedmann was deeply hurt by Einstein's reaction, which he did not understand. When he again took up his pen to ask his illustrious correspondent where his error lay, he got no response. Fortunately, his friend Yuri Krutkov stepped in. With the help of Paul Ehrenfest, Krutkov's former physics professor and a good friend of Einstein, the three collaborated on re-formulating the problem and submitted it to him again the following year, 1923. Einstein then apologized: Friedmann had not been mistaken. Without entirely conceding the idea of an expanding Universe, he did publish an honest retraction of his earlier article, in which he acknowledged that Friedmann's calculations were correct and opened up *"new avenues of research."* Unfortunately, Friedmann was not able to explore these avenues, as he died prematurely two years later.

Alexander Friedmann (1888–1925)

---

50. Alexander Friedmann, "Über die Krümmung des Raumes," *Zeitschrift für Physik* 10, issue 1 (June 29, 1922), 377–386.

51. Norman L. Geisler and Peter Bocchino, *Unshakable Foundations* (Minneapolis, MN: Bethany House Publishers, 2001), 97–98.

52. Albert Einstein, "Comment on A. Friedmann's Paper: 'On the Curvature of Space,'" *Zeitschrift für Physik* 11 (September 18, 1922), in *The Collected Papers of Albert Einstein, Volume 13: The Berlin Years: Writings & Correspondence January 1922–March 1923* (English translation supplement), ed. Diana Kormos Buchwald, József Illy, Ze'ev Rosenkranz, & Tilman Sauer; trans. Ann M. Hentschel & Osik Mose (Princeton, NJ: Princeton University Press, 2012), 271–272.

Georges Lemaître (1927): priest, cosmologist, and visionary

A few years later, in 1927, a little-known young scientist began to traverse these "new avenues." A Catholic priest with a PhD from MIT, Georges Lemaître had also studied Albert Einstein's work and felt compelled to follow all its consequences to their logical conclusions. His thesis, published in the *Annals of the Scientific Society of Brussels* and entitled "A homogeneous universe of constant mass and increasing radius, accounting for the radial velocity of extragalactic nebulae,"[53] presented a theory of universal expansion. Specifically, Lemaître derived the precise value for the law of proportionality between a galaxy's distance from the Earth and the speed at which it is moving away from it, a law Edwin Hubble would verify in 1929 with the new telescope at the Mount Wilson Observatory equipped with a 100-inch lens—by far the largest in the world at the time. [54]

Georges Lemaître (1894–1966), priest and cosmologist, suffered the taunts of Einstein and other colleagues. Later, they had to concede that his theories about the expansion of the Universe were correct

Georges Lemaître's article made a splash. Albert Einstein read it with wonder, but remained trapped in his prejudices against the idea of the expansion of the Universe. *"Your calculations are correct but your physics is atrocious,"*[55] he told Lemaître that year at Solvay, a famous congress in Brussels that gathered together the greatest contemporary minds in physics.[56] In private, he disparaged Lemaître's theory as

---

53. In *Monthly Notices of the Royal Astronomical Society* 91, no. 5 (March 1931): 483–490, https://doi.org/10.1093/mnras/91.5.483.

54. In conformity with a 2018 request by the International Astronomical Union, the law is now called the Hubble-Lemaître law.

55. A. Deprit, "Monsignor Georges Lemaître," *The Big Bang and Georges Lemaître*, ed. A. Barger (Dordrecht, Netherlands: D. Reidel, 1984), 370.

56. See a summary of this episode by J. P. Luminet in his introduction to a book by Alexander Friedmann and Georges Lemaître, *Essais de Cosmologie* (Paris: Seuil, "Sources du Savoir," 1997): *"From October 24 to 29, 1927, the Fifth Solvay Physics Congress was held in Brussels [. . .]*

Installed in 1917, the Hooker telescope on Mt. Wilson, with its 100-inch lens, remained the world's largest telescope until 1949

the *"physics of a priest."*[57] Likewise, most scientists at the time did not accept Lemaître's hypothesis. Sir Arthur Eddington, Lemaître's former teacher, believed that the expansion of the Universe was *"so absurd and so incredible"* that he felt almost *"outraged that anyone could believe it."*[58]

Vesto Slipher (1912) and Edwin Hubble (1929): A decisive observation

In 1929, American astronomer Edwin Hubble, building on the work of Vesto Slipher, made a game-changing discovery: he noticed that the

---

devoted to the new discipline of quantum mechanics, whose problems disturbed many physicists. Among them, Einstein. For Lemaître, it was an opportunity to talk to the father of relativity. He himself later recounted this meeting: 'While walking through the alleys of Leopold Park, [Einstein] told me about an article, little noticed, that I had written the previous year on the expansion of the universe and that a friend had made him read. After some favorable technical remarks, he concluded by saying that from the physical point of view it seemed completely atrocious to him.'

57. For Hubble's confirmation of Lemaître's work, see https://www.physicsoftheuniverse.com/scientists_lemaitre.html.

58. See J. Stachel, "Eddington and Einstein," *The Prism of Science*, ed. E. Ullmann-Margalit (Dordrecht: D. Reidel, 1986). The February 2014 edition of *Nature* magazine has this to say on the subject: *"Other leading researchers, such as the eminent Cambridge astronomer Arthur Eddington, were also suspicious of the Big Bang theory, because it suggested a mystical moment of creation."* https://www.nature.com/articles/506418a.

light coming from distant galaxies always shifted towards the red part of the electromagnetic spectrum. He concluded that this "red shift" must be a Doppler effect resulting from the fact that the light sources are receding relative to us. Since the velocity at which a galaxy recedes from the Earth depends only on its current distance from us, Hubble postulated that galaxies are moving farther apart from one another, picking up speed as they go. With this extraordinary observation, Lemaître confirmed that the Universe was expanding just as Friedmann had predicted in 1922 (and he himself forecast five years later), based on Einstein's theory of General Relativity.

The lens of the telescope at Mt. Wilson

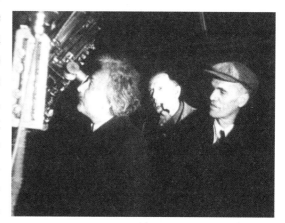

Einstein with Hubble at Mt. Wilson

"*Of all the great predictions that science has ever made over the centuries,*" John Wheeler exclaims, "*was there ever one greater than this, to predict, and predict correctly, and predict against all expectation a phenomenon so fantastic as the expansion of the universe?*"[59]

---

59. J. Wheeler, "Beyond the Black Hole," *Some Strangeness in the Proportion: A Centennial Symposium to Celebrate the Achievements of Albert Einstein*, ed. H. Woolf (Reading, Berkshire: Addison-Wesley, 1980), 354. See on this topic Tarek Osman's article, "Our Participatory Universe": https://medium.com/@tarek_osman/our-participatory-universe-ce640fed6585.

Faced with the evidence, leading scientists rally around Lemaître's ideas

Hubble's observation so strongly confirmed the theory of universal expansion that within just a few years it led to a complete reversal of the scientific consensus. In order to convince himself, Einstein traveled to Mount Wilson in 1931, where he spoke with Edwin Hubble. He emerged defeated and confessed that introducing the cosmological constant into his calculations because of his philosophical preconceptions was *"the biggest blunder he had made in his entire life."*[60] Sir Arthur Eddington also radically altered his opinion, declaring Lemaître a genius and proclaiming to the whole world that the first serious cosmological argument for the expansion of the Universe had arrived at long last. Georges Lemaître finally got the recognition he deserved. In 1933, American newspapers showered him with praise, and in 1934 he received the Prix Francqui, Belgium's most prestigious scientific award. But this general acceptance of the idea that the Universe is expanding did not lead to an immediate consensus as to how it began, even though "rewinding" the history of the Universe is all it takes to show that the Big Bang is a logical necessity.

The theory of the primeval atom calls this fresh consensus into question: the great scientific minds rebel again

Scientific consensus on the expansion of the Universe had implications for the study of its origin. These consequences were not lost on Georges Lemaître, who proposed a second theory. In 1931,[61] he put forward a hypothesis that would prove even more impossible for scientists of his day to accept. According to his findings, the Universe had originated in a "primeval atom." This dense kernel, containing what would become all the matter and energy of the entire Universe, suddenly entered a period

---

60. Words attributed to Einstein by Gamow in "The Evolutionary Universe," *Scientific American* 195, issue 3 (September 1956), 140. See also Cormac O'Raifeartaigh, "Investigating the legend of Einstein's 'biggest blunder,'" *Physics Today* (October 30, 2018), https://physicstoday.scitation.org/do/10.1063/PT.6.3.20181030a/full/.

61. Georges Lemaître, "The Beginning of the World from the Point of View of Quantum Theory," Nature 127, no. 706 (1931): 706, https://doi.org/10.1038/127706b0.

of expansion, and at once space and time were created. Lemaitre summarized his conclusion thus: "*We can reason that space began with the primeval atom and that the beginning of space marked the beginning of time.*"[62] What an outrageous idea!

Outrage and disbelief

They called him a genius for the expansion of the Universe; they called him crazy for the primeval atom.

Georges Lemaître's new theory was so scandalous that it threatened his hard-won reputation. No one accepted the primeval atom. Lemaître's revolutionary idea pushed scientists up against a conceptual barrier they absolutely refused to cross. "*Philosophically, the notion of a beginning to the present order of Nature is repugnant to me,*"[63] explained Arthur Eddington, who described Lemaître's hypothesis as "*forbidding.*" Anytime someone mentioned the primeval atom, Albert Einstein exclaimed: "*No, not that, it smacks too much of creation!*"[64]

Georges Lemaître's priestly vocation brought him under suspicion: some took it as proof of his partiality. He was accused of being a "concordist," of inventing a theory of the Universe's origins that corresponded at all costs to the biblical account of creation ex nihilo, "*from nothing,*" as described in Genesis.[65] His critics all too quickly forgot that the father of the heliocentric model, Nicholas Copernicus, was a cathedral canon, and that Mendel, founder of the study of genetics, was a Catholic monk. The habit might not make the monk, but neither does the ecclesiastical state make a bad scientist. Be that as it may, atheists soon made the Big Bang enemy number one.

---

62. Georges Lemaître, *L'Hypothèse de l'atome primitif: essai de cosmogonie* (Neuchâtel: Griffon, 1946).

63. Arthur S. Eddington, "The End of the World: From the Standpoint of Mathematical Physics," *Nature 127* (1931): 447–453.

64. Georges Lemaître, "Rencontres avec A. Einstein," *Revue des Questions Scientifiques* 129 (1958).

65. The expression comes from 2 Mac 7:28.

From 1947 onwards, world-famous English astrophysicist Fred Hoyle headed the resistance against the theory of the primeval atom and opposed every theory that envisioned anything like a creative act. Instead, he argued for the Steady-State model, which imagines the Universe as infinite, eternal, and slowly expanding, with a density that always remains the same as new galaxies are created, atom by atom, essentially *ex nihilo*. Hoyle launched a media campaign to discredit Lemaître, opposing and ridiculing his theory in the *Times* and over the airwaves of the BBC.

Cosmic microwave background radiation (CMBR): long-ignored evidence in support of Lemaître's theory

George Gamow was one of the many students of Alexander Friedmann who managed to leave the USSR for the United States—in his case, on the pretext of attending a 1933 scientific conference. In 1948, with his pupil Ralph Alpher, he published a foundational article that argued that hydrogen, helium, and deuterium atoms must have been created within the first several minutes of the Universe's existence.[66] As a small aside: Gamow included the name of Hans Bethe, future Nobel laureate in Physics, as a coauthor of the article, despite the fact that Bethe had not really contributed. The inclusion of Bethe meant that the article was authored by the first three letters of the Greek alphabet: alpha, beta, gamma (Alpher, Bethe, Gamow). Definitely a scientist's idea of a joke!

Following his research with Gamow and with the help of fellow student Robert Herman, Alpher concluded that the first light that the emerging Universe put out—corresponding to the electromagnetic radiation emitted by a blackbody in thermal equilibrium at 3,000 kelvin—should still be detectable at every point in the Universe but at a temperature one thousand times lower, since the Universe is

---

66. R. A. Alpher, H. Bethe, and G. Gamow, "The Origin of Chemical Elements," *Physical Review* 73, no. 7 (April 1, 1948): 803-4.

now one thousand times larger.[67] He ultimately calculated this at a temperature of 5 kelvin (in the microwave frequency range). Gamow presented his findings in a book called *The Creation of the Universe*, but only a few noticed his sensational prediction, and most cosmologists undervalued his contribution.

Hoyle coins the term "Big Bang" (1949) to disparage the concept

Fred Hoyle, then president of the Royal Astronomical Society, remained one of the staunchest opponents of the theories of Alexander Friedmann and Georges Lemaître (and later those of George Gamow). He used the phrase "Big Bang" for the first time in 1949, live on BBC radio, to poke fun at Friedmann and Lemaître, and subsequently fell back on it often, to devastating effect. Once, as Fr. Georges Lemaître arrived at a conference in Pasadena in 1960, Hoyle introduced him as, "*the Big Bang man.*"[68] The term was a smashing success and helped sow contempt for the idea from the start.

The Big Bang sinks into oblivion (1953)

For a long time, the idea of the Big Bang sat shelved. The ironclad prejudices of the scientific authorities had outlasted the tenacity of the few researchers who defended the concept. By 1953, the battle seemed lost. With a heavy heart, the three discredited pioneers gave up their research on the Big Bang. Gamow gradually moved away from physics. Alpher left the university, wounded by the contemptuous mockery of his colleagues. His student Herman changed careers, entering the automotive industry. Expansion? Who cares! The Big Bang? A preposterous hypothesis, without the slightest proof. Galaxies are moving in space, sure. But the cosmos itself? Come on!

---

67. "Remarks on the evolution of the Expanding Universe" (1949): https://doi.org/10.1103/PhysRev.75.1089

68. See M. Lachièze-Rey and J- Luminet, *Celestial Treasury: From the Music of the Spheres to the Conquest of Space* (Cambridge: Cambridge University Press, 2001), 154.

An unexpected rebirth, thanks to the chance discovery of Cosmic Microwave Background Radiation (CMBR) (1964)

The Big Bang had been discredited, but a few tenacious researchers still thought it worth their while to examine Gamow's research. Among these were Robert Dicke and Jim Peebles, researchers at Princeton, who resolved to try to spot the famous cosmic microwave background radiation predicted by Gamow. While they were at work one day in 1964, they got a phone call from two engineers at Bell Laboratories, only thirty miles away, where Arno Penzias and Robert Wilson (both former students of Fred Hoyle) were putting the finishing touches on what was then to be the world's largest directional antenna. While trying to improve the reception of signals sent from the earliest satellites, they had identified a strange signal, a "parasite" coming from all directions of the Universe at 2.725 kelvins. They had already carefully checked off, one by one, all the possible causes of this parasite, going so far as to climb into their huge horn-shaped antenna to expel a couple of resident pigeons they suspected were the culprits. Without the slightest intention of doing so, Penzias and Wilson had in fact managed to pick up the echoes of the electromag-

In 1964, Arno Penzias and Robert W. Wilson unintentionally discovered the Cosmic Microwave Background Radiation predicted by George Gamow in 1948

netic signal emitted after the Big Bang. This discovery earned them the Nobel Prize in Physics in 1978. George Gamow was pleased to learn of this result three years before his death. As for Georges Lemaître, only a few weeks before his death in 1966, his friend Odon Godart told him about the discovery of this extraordinary fossil radiation which Lemaître had called "*the vanishing brilliance of the origin of the worlds.*"[69] From his hospital bed, where he had already lain for two weeks dying of leukemia, he simply said: "*I am glad now, we have the proof.*"[70]

Except for pockets of residual resistance, the theory is unanimously accepted

Faced with this experimental observation (soon widely confirmed) and an accumulating body of proof,[71] most scientists finally surrendered to the evidence. But the opponents of the Big Bang would not completely lay down their arms for the three decades between 1950 and 1980. Scientists of the opposition camp seemed most alarmed by the existential questions raised by the Big Bang. In 1963, the Chair of Cosmology at the Collège de France in Paris, Alexandre Dauvillier, said about the Big Bang: "*This is rubbish. The universe has no beginning; to think that the universe has a beginning is no longer physics, it is metaphysics.*"[72] As late as 1976, the Swedish astronomer Hannes Alfvén, winner of the Nobel Prize in Physics, compared Big Bang cosmology to Ptolemy's mythical geocentric model and criticized those who defended it, claiming that "the prevailing

---

69. In J.-P. Luminet, "The Rise of Big Bang Models, from Myth to Theory and Observations," Invited talk at conference "Antropogenesi: Dall'energia al fenomeno umano," 19-21 October 2007, Portogruaro (Italy), 9: "*The evolution of the world can be compared to a display of fireworks that has just ended: some few red wisps, ashes and smoke. Standing on a cooled cinder, we see the slow fading of the suns, and we try to recall the vanishing brilliance of the origin of the worlds.*" https://www.astro.uvic.ca/~jwillis/teaching/astr405/reading/luminet.pdf

70. In J-P Luminet, "The Rise of Big Bang Models, from Myth to Theory and Observations," 12.

71. The composition of the oldest gas clouds in the Universe was analyzed in the 1960s, and the "abundance measurement" corresponded exactly to the predictions (75% hydrogen, 25% helium, with traces of deuterium, lithium, beryllium, and boron). This was a second brilliant confirmation of the Big Bang theory.

72. Our translation of Dauvillier's comment as reported by Claude Tresmontant, *Comment se pose aujourd'hui le problème de l'existence de Dieu* (Paris: le Seuil, 1966), 20.

attitude is to ignore all objections to the Big Bang theory."[73] But today, the Big Bang is unanimously accepted. Ironically, even its fiercest detractor, Fred Hoyle, ended up moderating his criticism and, after years of atheism, died a deist. Alexander Friedmann, Georges Lemaître, and George Gamow were right: the Big Bang had happened, and it looked like an absolute beginning.

"The face of God" (1992)

In the decades since, a series of confirmations have verified what is today called the "*Standard Model*" of the Big Bang. The COBE, WMAP, and Planck satellites made it possible to draw up an increasingly precise picture of the Universe at the time it emitted its first light, as measured today by the CMBR (Cosmic Microwave Background Radiation). The image reveals a Universe in almost perfect thermal equilibrium, with minute variations that are at the origin of all future developments of the Universe. George

Image of "Cosmic Microwave Background Radiation (CMBR)" the first light released by the Universe, 380,000 years after the Big Bang. This map, examined from every possible angle, is the main source of everything known about the Big Bang (see Chapter 9 on the anthropic principle for more details)

---

73. Hannes Alfvén and Gustaf Arrhenius, *Evolution of the Solar System* (Washington, D.C.: Scientific and Technical Information Office, National Aeronautics and Space Administration, 1976). See also Hannes Alfvén, "Cosmology: Myth or Science?" *Journal of Astrophysics and Astronomy* 5 (1984): 91. The Big Bang is indeed a cosmology of the same character as the Ptolemaean: absolutely sterile. Will it have the same life expectancy?

Smoot was the first to publish this image in 1992, and it earned him the 2006 Nobel Prize. In his reception speech, directed at his colleagues in the American Physical Society, he projected photos of the first cosmic light on the screen and used this phrase: "*It's like seeing God.*"[74]

The failure of all alternative theories only strengthens the standard Big Bang model

Half a century after the standard account of the Big Bang was confirmed by the discovery of CMBR, there is still no alternative theory supported by experimental observation (see Chapter 6 on the subject). We wait in vain.

The failure of alternative theories led American philosopher William Lane Craig to conclude the following in his book *Reasonable Faith* (2008): "*The history of twentieth-century cosmogony has, in one sense, been a series of failed attempts to craft acceptable non-standard models of the expanding universe in such a way as to avert the absolute beginning predicted by the Standard Model. This parade of failures can be confusing to the layman, leading him mistakenly to infer that the field of cosmology is in constant flux, as new theories of the universe's origin continually come and go, with no assured results. In fact, the Standard Model's prediction of an absolute beginning has persisted through a century of astonishing progress in theoretical and observational cosmology and survived an onslaught of alternative theories. With each successive failure of alternative cosmogonic theories to avoid the absolute beginning of the universe predicted by the Standard Model, that prediction has been corroborated. It can be confidently said that no cosmogonic model has been as repeatedly verified in its predictions and as corroborated by attempts at its falsification, or as concordant with empirical discoveries and as philosophically coherent, as the Standard Big Bang Model.*"[75]

---

74. George Smoot and Keay Davidson, *Wrinkles in Time* (New York: William Morrow, 1993), 289.

75. William Lane Craig, *Reasonable Faith: Christian Truth and Apologetics* (Wheaton: Crossway, 2008), 139–140.

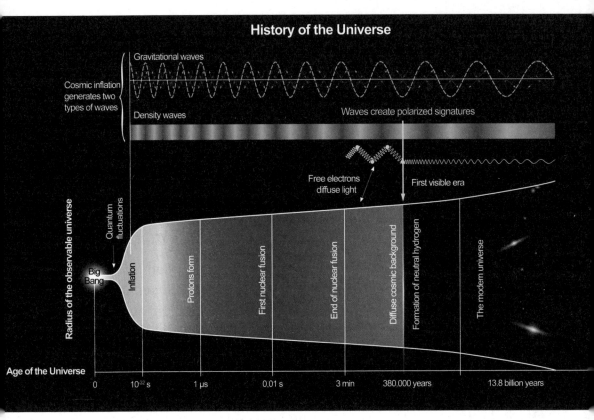

The Big Bang was actually an extremely organized development over several phases. It was no random, messy, or dangerous explosion

II. In the beginning was the Big Bang

From polemical to popular

Fred Hoyle didn't quite get what he expected when he coined the term "Big Bang" in 1949. His crude alliteration was intended to mock Lemaître's concept of the primeval atom. He thought the term "Big Bang" sounded childish and unscientific. Ironically, this derogatory label, meant to belittle the idea, played a significant role in its eventual success. "Big Bang" was easy to remember! Once Lemaître and Gamow's theories were confirmed, the term became widely recognized by both scientists and the general public: "Big Bang."

But this familiarity confuses our idea of how the Universe came to be almost as badly as Hoyle's sarcasm did. The Big Bang was nothing like the random, messy, or dangerous explosion its nickname suggests. It was in fact an extremely organized, multi-phased development. All the elements that constitute our Universe came to be through an amazingly regulated and gradual process.

A moment impossible to imagine

According to the Standard Model of the Big Bang, there was neither time, nor space, nor matter "before" the event. All physical space and all the elements of the Universe, or more precisely, their precursors, originate from a "primeval atom" that expands, stretches, and spreads out. Imagine the Universe as the surface of an inflating balloon. Then think of the matter and energy particles of the Universe as stickers applied to the surface of the balloon. The expansion of the Universe is like the expansion of the balloon as it inflates. As the balloon expands, the stickers move away from each other. In the same way, particles in an early expanding Universe, and galaxies today, all move apart at rates roughly proportional to the distances between them. And those rates are constantly increasing—the bigger the balloon gets, the faster the stickers move apart.

According to the standard model, this is the beginning of space, time and matter

From the point of view of physics and the standard conception of time, there was no time 'before' the Big Bang because physical time—the kind we measure on a watch—was created instantly at that moment of the event, as Georges Lemaître realized in 1931.[76] Neither was there any place outside

---

76. George Lemaître described this first instant thus: *"This origin appears to us, in space-time, as a background that defies our imagination and our reason, opposing them with a barrier that they cannot break through. Space-time appears to us like a conical cup. We progress towards the future by following the generating lines of the cone towards the outer edge of the cup. We move through space by traveling along a circle perpendicular to the generating lines. When we think our way back through time, we approach the base of the cup, that unique instant that wasn't*

it. No one could have observed the Big Bang, since the only physical space that exists is our own, which emerged concurrently with time and matter (then in the form of energy), starting out extremely dense. Indeed, according to Einstein's theory of General Relativity, space, time, and matter are so intimately linked that we speak of space-time, and it is this space-time that arose all at once alongside the energy that would give rise to matter.

This point is crucial because if science confirms that time, space, and matter had an absolute beginning, it becomes clear that the Universe proceeds from a cause that is neither temporal, spatial, nor material. In other words, it proceeds from a transcendent, non-natural cause at the origin of all that exists and, as we shall see, at the origin of the extreme fine-tuning of the Universe's initial parameters and the laws of physics and biology, which are indispensable for the existence and evolution of atoms, stars, and complex life.

A fantastic process as finely tuned as a professional orchestra

This extraordinary event began 13.8 billion years ago and took place according to several different very ordered chronological phases, the main outlines of which are given below:

- **The First Instant** - The instant the Big Bang happened is impossible to describe in terms of the laws of physics ($t = 0$). During the so-called Planck era which is the first unimaginably small fraction of a second after the Big Bang, 10^{-43} s, or ten million billion billion billion billion times less than one second, much shorter than a blink of an eye in the 13.8 billion years of life of the Universe—the four fundamental interactions (electromagnetism, weak interaction, strong interaction and gravitation) were possibly unified. Our models are incapable of

---

there yesterday, because yesterday space did not exist. The natural beginning of the world, an origin for which thought cannot conceive of a pre-existence, since it is space itself that begins, and we cannot conceive of anything without space. Time seems able to stretch at will towards the past and the future. But space can begin, and time cannot exist without space; we could therefore say that space strangles time and prevents it from extending beyond the bottom of 'space-time'" (Translated from Georges Lemaître, "L'hypothèse de l'atome primitif," *Revue des Questions Scientifiques* 61 (1948): 321-339, 338-339.

describing these interactions, since General Relativity and quantum physics are incomplete theories which can only be mobilized when gravitation and quantum effects are studied separately.

In addition, "Planck's instant" is the smallest unit of time that is physically meaningful according to our current understanding of physics. And so it is that the ordinary notions and laws of physics in our Universe do not allow us to describe time zero. Indeed, whatever came "before" the Planck epoch is inaccessable to physics, and some think that it will remain forever so.

So is there anything we can say about it? Yes and no...

No, because as we have seen, while it is possible for science to affirm that the cause of the Big Bang is non-spatial, non-temporal and non-material (which is already a lot), the approach science takes is "apophatic;" that is to say, we are talking about realities that we can only deduce indirectly and qualify only in a negative way, without having the slightest knowledge of the phenomenon at stake.

By relying on the principle of causality, though, which is an integral part of science, we can, in a very scientific way, conclude that our Universe is incomplete.[77]

One cannot therefore have *a priori* knowledge of "before the Big Bang," and this state before space-time no doubt will always remain outside the realm of experimental science.[78] Nevertheless, this does

---

77. If we were to examine footprints in the sand, we might affirm that according to the rules of physical science there must be a cause for these traces, though the marks themselves do not prove the natural interactions of physical forces. Similarly, when Alain Aspect concluded that there was a (quantum) entanglement between two particles that were fourteen meters apart but dialoged instantaneously, it became clear that within the realm of physics there is another layer of reality, one that eludes our own space-time. This was again the case when Gödel, working through logic and logical mathematics, concluded that non-demonstrable truths necessarily exist, truths that fall outside the purview of mathematics. The same "apophatic" reasoning applies also to the Big Bang, which falls inside the bounds of cosmological science.

78. In the current state of science and method, it is also the case that what happened 10^{-11} seconds after the Big Bang will remain beyond experimental science, because the energies in play are too elevated and impossible to reproduce, even with the extraordinary CERN particle accelerators in Geneva.

not mean that there was "nothing" before Planck's Wall. One such theory was developed around 2010 by Sir Roger Penrose of Oxford University, 2020 Nobel laureate in Physics. This longtime intellectual companion of celebrated Cambridge physicist Stephen Hawking wrote a series of scientific articles on the subject, as well as a fascinating book entitled *Cycles of Time*, with "What Came Before the Big Bang?" blazoned across the cover of the first English edition.[79] Also braving the obstinate skepticism of his colleagues, astrophysicist George Efstathiou, director of the prestigious Kavli Institute for Cosmology at Cambridge University, claimed in 2013: *"It's perfectly possible that there was some phase of the universe before the Big Bang actually happened where you can track the history of the universe to a pre-Big Bang period.*[80] As early as 1993, thirteen years before he won the Nobel Prize—George Smoot posed a risky but prophetic question in his *Wrinkles in Time*: *"What was there before the big bang? What was there before time began?"*[81]

An excellent question! But what to make of it?

Here is where our investigation really gets interesting. Since physicists rightly believe that space, time, and matter, as we know them in our Universe, came into existence together, this means that before the Planck era, time, space, and matter did not yet exist. This logical conclusion is shared by all scientists versed in the standard model of physics. If we push the reasoning a little further, we deduce that instead of time only something atemporal could exist before the Big Bang. Likewise, since matter did not yet exist, anything we found to exist "before the Big Bang" would be immaterial. But how can we understand this timeless and immaterial reality? What type of entity has attributes like these? Is not the existence of an

---

79. Roger Penrose, *Cycles of Time: An Extraordinary New View of the Universe* (New York: Alfred A. Knopf, 2011).

80. In "Space: Planck Maps the Dawn of Time," *Euronews* (March 21, 2013). https://www.euronews.com/2013/03/21/planck-maps-the-dawn-of-time.

81. George Smoot, *Wrinkles in Time* (New York: W. Morrow, 1993), 291.

immaterial creator God located outside space and time the most natural explanation?

Some scientists have believed it possible to sketch out some answers…

Let's begin with time. What might it have looked like before the Big Bang? The response given in particular by Stephen Hawking in the 1980s is surprising: for him, before the Big Bang, time was not real, but perhaps imaginary. Everyday time, the time we measure on our watches, is measured in numbers that mathematicians call "real," that is, numbers whose square is always positive. For example, 2 or -2 squared is always 4. On the other hand, mathematics has postulated numbers whose square is always negative. As early as the seventeenth century, the philosopher Descartes called these unique numbers "imaginary."[82] In the twenty-first century, some scientists, including Stephen Hawking, have employed this concept of imaginary time to rule out a cosmic beginning or temporal singularity. However, Hawking's use of imaginary time does not achieve this. Imaginary time is only a neat way to solve one of Einstein's equations, temporarily obscuring the temporal singularity. But in an intermediate step where the imaginary numbers are converted back into real numbers, as they must be for the equation to pertain to reality, the singularity reappears. Regardless, it is not unreasonable to consider that time was fundamentally different before the Big Bang, possibly of a nature that could be measured using imaginary numbers. In this scenario, matter, as we understand it, would not have existed. What might we conceive instead? Perhaps something intangible, like information. Before the Big Bang—more precisely at "instant zero"—time would have been purely imaginary, and reality at that stage could have existed only as pure information, resembling an essential mathematical code.

This primordial information would have "*programmed*" the birth of the Universe at the time of the Big Bang and then its evolution

---

82. Imaginary time has widespread application in theoretical physics, in particular in quantum physics. Stephen Hawking was the first to propose that the time close to the Big Bang was imaginary.

throughout billions of years with dizzying precision. This reasoning leads us to a clearly legitimate question: if there was mathematical information before the Big Bang, who is the incredible "programmer" behind such a code? This question comes up regularly, and we will return to it frequently later.[83]

- **After 10^{-43} seconds (Planck time)**, we can begin to describe the evolution of the Universe with some precision, using inflation theory, which, though it still lacks formal confirmation, is held to by a majority of cosmologists), the theory of General Relativity to explain the action of gravity, and the standard model of particle physics to account for the other three forces. With the Universe crammed into a tiny space of 10^{-35} meters, at the inconceivable temperature of 10^{32} kelvin, and an energy of 10^{19} GeV, there is no stable matter. This extremely dense space contains only a form of pure energy. Since this moment, the primordial quantity of mass-energy in the Universe is fixed and varies only under the effect of the expansion of the Universe (the energy of the vacuum of space created by expansion brings a slight increase, balanced by the loss of energy linked to the red shift of photon emissions).

- **After 10^{-35} seconds (Inflationary epoch)**, the tremendous energy contained in the hypothetical "inflaton field" causes space to enter into accelerating expansion. The Universe expands by a factor of at least 10^{26} between 10^{-35} and 10^{-32} seconds.[84] During this period, the strong nuclear force separates from the electroweak force (the combination of the electric and weak nuclear forces). This separation ushers in the electroweak epoch, when that force dominated.

- **After 10^{-12} seconds (Quark epoch)**, the electroweak force splits into electromagnetic interaction and weak interaction. This weak

---

83. Especially in Chapter 9.

84. Different inflationary models vary in their timelines. For the timeline described here, see Louis Lessenger, "How Can We Determine that inflation Occurred from 10^{-35} to 10^{-32} second after the Big Bang, and that the universe grew 10^{26} in that time?" *Astronomy* (February 2019), https://www.astronomy.com/magazine/ask-astro/2019/01/inflation.

interaction is probably behind the nano-asymmetry between matter and antimatter which lasts to the present day. The four fundamental forces are then definitively separated. The Universe is filled with a hot, dense quark–gluon plasma, containing quarks, electrons and other leptons, and their antiparticles.

- **Between 10^{-6} and 10^{-4} seconds (Hadron epoch)**, the quarks combine to form the total amount of hadrons—that is, protons and neutrons—that will constitute all matter. Neutrons, which are unstable and live for only fifteen minutes on their own, must rapidly associate with protons in stable nuclei. After this short hadron phase, the creation of new protons and neutrons will never again be possible in the Universe.[85]

- **At one second (disappearance of antimatter)**, the nano-asymmetry between matter and antimatter leads to the near-total destruction of antimatter.

- **Between one second and fifteen minutes after the Big Bang (first nucleosynthesis)**, the nucleosynthesis of all light elements takes place once and for all: hydrogen (75% of the Universe's current matter), helium (25%), and lithium, beryllium, and boron (in small quantities) come into existence. These elements could only be created in the extreme conditions of the Big Bang's first minutes.

- **At 15 minutes (matter setting),** the composition of the material Universe is almost determined. It is essentially made up of hydrogen nuclei (92%), a bit of helium (8%), traces of deuterium (0.002%), and lithium in infinitesimal quantities ($1:10^{12}$). Once the temperature drops below one billion degrees, the number of these light nuclei is set for all time.

- **Between 15 minutes and 380,000 years (Photon epoch)**, the rapid expansion of the Universe is dominated by the agitation of high-energy photons that slow the creation of atoms by preventing the stable association of electrons with existing nuclei.

---

85. Edward W. Kolb and Michael S. Turner, *The Early Universe* (New York: Addison-Wesley, 1994).

- **By 380,000 years (formation of the first atoms and light)**, the temperature has dropped to 3,000 kelvins, and photons no longer have enough energy to dissociate atoms formed when existing nuclei capture electrons. At this point, the Universe becomes transparent to light and therefore observable, as photons can now travel freely in a straight line in space. With visible light comes the first radiation, which is emitted from all points of the Universe in all directions. This primordial radiation is the famous Cosmic Microwave Background Radiation (CMBR) discovered by Penzias and Wilson in 1964. We may not realize it, but we are surrounded by these particles from the Big Bang: 411 photons in every cubic centimeter of space. Now that we know that a small percentage of the interference on a cathode ray tube television set goes back to the Cosmic Microwave Background, we'll remember those "bug races" in a new light! These photons from the CMBR are now at a temperature of 2.725 K because the Universe has increased in volume by a factor of about one thousand since the release of the first visible light, and so the temperature has decreased by the same proportion.

- **Between 380,000 years and 1 billion years (first cosmic dark age and first stars)**, the Universe enters a first dark age, but slight variations in the density of matter (an anisotropy factor of only 1:100,000) lead gradually to the production of clumps of dust that eventually, after 150 to 200 million years, will make it possible to ignite the first stars. These early stars group together in the first galaxies (observable today as quasars).[86]

- **After 3 to 5 billion years (formation of heavy elements)**, the first generations of stars end their lives as supernovas, creating the conditions for the formation of the heavy elements in Mendeleev's periodic table.[87] Besides the hydrogen, helium, deuterium, and

---

86. David H. Lyth, *The History of the Universe* (Boca Raton, FL: CRC Press, 2015).

87. The nucleus of the iron atom is the most stable. Lighter chemical elements (helium, carbon, oxygen, chlorine, potassium, calcium, titanium, silicon, etc.) are formed in nuclear fusion re-

lithium that formed in earlier ages, all the atoms that make up our planet, our body, and the things we use every day come from this "stardust."

- **After 9 billion years (Sun)**, our Sun begins to form, as a third-generation (Population I) star.

And this is how the solar system, our Earth, life, and each and every one of us came into being, after a history of almost 13.8 billion years.

The "Standard Model of the Big Bang": a very solid theory, tested and confirmed by observation

The Standard Model of the Big Bang is thus supported by such a large number of probative observations that the vast majority of cosmologists now accept its central claims. It very precisely describes a Universe whose structure came to be at the dawn of time. This Universe has not existed for all eternity: it came into existence at a finite point in the past. The origin of the Universe suggested by this model is an absolute beginning. This is a radical claim whose implications deserve to be fully investigated: matter and energy, even time and space, came into existence in this instant of initial cosmological "singularity." Is it really such a leap, then, to entertain the possibility of a creative act behind this singularity?

Where is God in all this?

The Big Bang forces us into a corner. To put it bluntly, it brings us face-to-face with the idea of God. The absolute beginning of the Universe is where physics meets its maker, a creative cause exterior to the Universe. Now that science has largely resolved the mystery of the Universe's origin from the Planck moment, it is necessary to ask the question concerning the before and the why.

---

actions inside stars, but heavier elements (nickel, copper, zinc, silver, tin, platinum, gold, lead, uranium, etc.) can only form under the extreme conditions of the explosion of supernovae that mark the end of these massive stars.

For those who believe that the Universe was created, there are, in theory, two possibilities:

- Either we live in a static Universe that emerged in a single instant, just as it is today; or:
- The Universe was created to be dynamic and has evolved from an absolute beginning of space, time, and matter. In this case, we might expect that everything began from a single point.

But now we know that the Universe is not static and that it unfolds in an extremely precise and organized manner, much like a plant, animal, or human develops from a single cell. Further, the initial laws and data of the Universe give structure to and inform all future development.

We dare to say that the picture painted by the Big Bang model is remarkably consistent with the idea of the creation of the Universe by God. The fact that we can't really think about the time before the Big Bang, as the categories of time, space, and matter simply do not apply, lends credence to the idea of a creative act. Finally, let's pause to admire the researchers who made these discoveries: to be able to imagine our Universe at 10^{-43} seconds of life and to describe it precisely from one ten-billionth of a second, or ten to the power of minus ten seconds onwards.[88] They have brilliantly solved the question of how our Universe was formed, and this description corresponds perfectly to what believers have always claimed: a creative will made the Universe arise from nothing.

A few quotations on the subject

Arno A. Penzias, 1978 Nobel Prize in Physics: *"To be in accord with our observations, we must understand that not only is there a creation of matter but also the creation of space and time. My argument is that the best data we have are exactly what I would have predicted,*

---

88. At CERN in Geneva, it is now possible to reconstruct the energies that existed one second after the Big Bang, which makes it possible to verify experimentally the accuracy of theories on this period. Before 10^{-10} second, the theories are not verified, and, before 10^{-43} seconds, we do not have and may never have a theory.

had I had nothing to go on but the five books of Moses, the Psalms, the Bible as a whole."[89] "Astronomy leads us to an unique event, a universe which was created out of nothing."[90]

Alexander Vilenkin, physicist at Tufts University, co-author with Arvind Borde and Alan Guth of the great cosmological theorem confirming that the Universe had a beginning: *"It is said that an argument is what convinces reasonable men and a proof is what it takes to convince even an unreasonable man. With the proof now in place, cosmologists can no longer hide behind the possibility of a past-eternal universe. There is no escape: they have to face the problem of a cosmic beginning."*[91]

Max Planck: *"All matter originates and exists only by virtue of a force which brings the particle of an atom to vibration and holds this most minute solar system of the atom together. We must assume behind this force the existence of a conscious and intelligent mind. This mind is the matrix of all matter."*[92]

Igor and Grichka Bogdanov: *"But then, if the singularity at the origin of the Universe is now well established scientifically, why does it arouse so much passion, not to mention rejection? Surely it is because it forces us to make an impossible choice: between a Universe without cause, or the face of God."*[93]

---

89. In Malcolm W. Browne, "Clues to Universe Origin Expected," *New York Times* (March 12, 1978), https://www.nytimes.com/1978/03/12/archives/clues-to-universe-origin-expected-the-making-of-the-universe.html. On this subject, see Steven Weinberg, *The First Three Minutes: A Modern View of the Origin of the Universe* (New York: Basic Books, 1977).

90. Arno Penzias, quoted by Walter Bradley in "The Designed 'Just-so' Universe," 1999, http://www.leaderu.com/offices/bradley/docs/universe.html. See Arno Penzias and Robert Wilson, "A Measurement of Excess Antenna Temperature at 4080 Mc/s.," *The Astrophysical Journal* 142 (July 1965): 419–421.

91. Alexander Vilenkin, *Many Worlds in One* (New York: Hill and Wang, 2006), 176. See also Vilenkin's 2013 interview "Many Worlds in One" for the Copernicus Center for Interdisciplinary Studies: https://www.youtube.com/watch?v=ZHEp855NS6c.

92. Max Planck, "Das Wesen der Materie" [The Nature of Matter]," speech at Florence, Italy (1944), in the Archiv zur Geschichte der Max-Planck-Gesellschaft (Boltzmannstraße 14, D-14195 Berlin-Dahlem). On this subject, see: "Max Planck Biographical," *Nobel Lectures, Physics 1901–1921*, (Amsterdam: Elsevier Publishing Company, 1967).

93. Translated from Igor and Grichka Bogdanov, *Le Visage de Dieu* (Paris: Grasset, 2010), 117.

As you can imagine from reading these quotes, the kind of beginning proposed by the Big Bang theory simply did not sit well with materialists. One way to get around the problem was to create alternative theories. And that is indeed what happened, as we will see in the pages that follow.

6.

Attempts at Alternatives to the Big Bang Model

As well established and supported as it is, the standard model of the Big Bang continues to annoy some scientists, much like a metaphysical pebble in their materialist shoe. How do they get rid of it? By developing alternative theories! None of these has withstood the weight of evidence and verification. Each has either been found invalid the outset or has advanced so little beyond the first stages of conceptual formulation that even beginning to seek confirmation for it is impossible.

A quick look at the most popular alternative theories

1. The **tired light hypothesis**, proposed by Fritz Zwicky in 1929, sought to challenge the notion of an expanding Universe, which was inferred by Hubble and Lemaître from the redshift of distant galaxies. This redshift is a result of the Doppler effect, indicating that the sources of light are moving away from us. In contrast to this theory, Swiss-American astronomer Fritz Zwicky suggested that the redshift occurred due to energy loss in photons as they traveled vast distances. This led to the term "tired light," which was coined by Richard Tolman the same year. Frenchman Jean-Claude Pecker, member of the Academy of Sciences and the Rationalist Union, defended this theory until 1978. He stubbornly clung to his opinion in the face of mounting evidence to the contrary, and as late as 2004 even signed a letter against the standard model of the Big Bang; however, observations carried out in the 1990s by the COBE satellite have definitively disproved the theory of tired light.

2. Around 1940, the **steady-state model** assumed the existence of a phenomenon of continuous creation of matter, making it possible to imagine a Universe in eternal and immutable expansion. Proposed by Fred Hoyle, Thomas Gold, and Hermann Bondi, the idea was debated in the 1960s but has fallen into obscurity today.

3. In 1957, the **many-worlds interpretation** attempts to explain the phenomenon of "wave function collapse," according to which, once measured, a physical system sees its state reduced to what was measured. Before it is measured, what we call "reality" has an infinite number of possible states. But once we measure it, it is reduced to whatever state it had when measured. Where do the other possible states go? To resolve this difficulty, Hugh Everett devised the many-worlds interpretation. He postulated that parallel universes are created at every moment and with each measurement, evolving independently from that point onward. The hypothesis has not been taken seriously, though the idea of parallel universes has flourished in science fiction.

4. In 1960, **plasma cosmology** was formulated as an alternative to the Big Bang. It suggests that electromagnetic phenomena play a more important role than the force of gravitation in explaining the structure of the Universe. It suggests that expansion takes place in only a small part of the Universe. Despite the fame of its principal supporter, Swedish Nobel Laureate Hannes Alfvén, the model, which shares certain similarities with the tired light theory, was quickly abandoned.

5. In the 1970s, after the discovery of fossil radiation (which confirmed the Big Bang), the **Big Crunch hypothesis** came into vogue. It predicted that the expansion of the Universe would slow down and eventually enter a phase of contraction, ultimately returning to the size of the primitive atom. Friedmann had the idea in 1922, and Lemaître echoed it by introducing the term "phoenix Universe" to describe a world that oscillated between expansion and contraction. Following them, some scientists investigated the possibility of cyclic "Big Bounces," but this hypothesis was abandoned. In fact, it was observed that the spatial curvature of the Universe was insufficient; more importantly, in 1998 we discovered that the expansion was even accelerating. The Borde-

Guth-Vilenkin theorem, published in 2012, challenged the theory of the Big Crunch by confirming again, and in another way, the impossibility of perpetual inflation in an infinite past.[94]

6. In 1980, **cosmic inflation theory** proposed a scenario of very rapid expansion of the Universe at its beginning in order to explain why it is surprisingly flat, homogenous, and isotropic. This interesting explanation is now accepted by a majority of cosmologists. Developed by Alan Guth, the first of these theories also postulated, however, that the size of the Universe is at least 10^{23} times greater than we can observe; in that as-yet-unaccounted-for space, other universes are developing like bubbles, separate from our own Universe and governed by different physical laws. Following in Guth's footsteps, more than fifty variants of inflation theory have been developed, with more popping up all the time, but none of them have received experimental confirmation.

7. In 1983, the **boundless Universe theory** was formulated by James Hartle and by Stephen Hawking. According to this model, the Universe, while not infinite in space and time, nevertheless has no temporal boundaries, and thus, perhaps, no beginning. It depicts the Universe as something like the surface of a sphere, which has no clear starting point. To defend this model, also sometimes called "quantum cosmology," Hawking and Hartle solved the Wheeler-Dewitt equation, which describes possible universes that could exist or emerge from the cosmological singularity.[95] Though Hawking's popular work suggests that his use of imaginary time in his mathematical calculations might eliminate the need to invoke such a singularity, the technical papers he co-authored with Hartle presupposed the reality of such a singularity and never eliminated it.[96]

---

94. Cf. *MIT Technology Review*, April 24, 2012.

95. J. B. Hartle and S. W. Hawking, "Wave Function of the Universe," *Physical Review* D 28, no. 12 (December 15, 1983): 2960–75. In addition, cosmologist Alexander Vilenkin has shown that the entire approach of using imaginary time to solve relevant cosmological equations is flawed because it does not lead to an unambiguous solution.

96. Ibid. See also Alexander Vilenkin and Masaki Yamada, "Tunneling Wave Function of the Universe," *Physical Review* D 98, no. 6 (August 6, 2018).

8. In 1984, early **string theory** picked up 1970s research, hoping to resolve the fundamental problem of "quantum gravity." The researchers hoped to unite the part of Einstein's General Relativity that describes the force of gravitation with the insights of quantum mechanics, which deals with the three other fundamental interactions on the microscopic level. They wanted to understand the physical laws that govern "singularities"—like the Big Bang or black holes—on both the infinitely large and infinitesimally small scales.

9. In the next twenty years, four more versions of **string theory** or **superstring theory** were elaborated, each demanding extremely complicated mathematics. String theory requires us to imagine the existence of "supersymmetric particles" and a Universe in ten dimensions: time, the three known dimensions of space, and six other hypothetical invisible dimensions that are "folded back" onto themselves. String theories have gained popularity in recent decades, but they are extremely complex and highly speculative. Some of their specific predictions have been disproved by experiments at the Large Hadron Collider (LHC) in Geneva, and others are unverifiable on a human timescale. Currently, the number of researchers who believe in them and work on them is diminishing.

10. Around 1990, **quasi-steady-state theory** again postulated that the universe creates matter over time but introduced the innovation that this expansion occurs cyclically. This theory, put forward by Jayant Narlikar, Fred Hoyle, and Geoffrey Burbidge, was swept aside by the analysis of cosmic microwave background radiation in 1992, and its lack of simplicity and predictive power led to a general rejection of it.[97] In addition, the discovery in 1998 that the expansion of the Universe is accelerating put a final nail in its coffin.[98]

---

97. Staunch atheist Fred Hoyle, who opposed George Lemaître and coined the term Big Bang to mock him, eventually became a deist himself after the discovery of the fine-tuning of the Universe (see Chapter 9). He was, however, one of the few who continued to believe in the Steady State theory he had championed, against all evidence, right up to his death in 2001. Proof, as if any were needed, that people do not change their minds easily.

98. Helge Kragh, "Quasi-Steady-State and Related Cosmological Models: A Historical Review," January 17, 2012.

11. Around the same time, a new model called **chaotic inflation** put forward the hypothesis that new universes could spontaneously appear at different spots in an existing universe, and these, in turn, could naturally generate other universes, doing this in an eternal way, without limit to the process. In this model, which was defended by Andrei Linde, a Russian scientist working at Stanford University after the fall of the Soviet Union, this chaotic inflation was initially thought to have occurred forever in the past. Consequently, the Universe might not have had a beginning. But this possibility was refuted by the Borde-Guth-Vilenkin theorem, which demonstrated that inflationary cosmologies, as well as any other realistic cosmology, might be eternal in the future, but not into the past, since the process must have had an absolute beginning.[99]

12. In 1995, **M-theory** was developed by Edward Witten as an attempt to unite the five different versions of superstring theory. The *M* signifies *"magic, mystery, mother, monster, or membrane...according to taste."*[100] It speculates about the existence of bidimensional membranes that vibrate in a space of eleven dimensions. The theory has the advantage for its supporters of conceptual simplicity and coherence from the mathematical point of view, but its calculations are highly complex and incomplete, and it has never been confirmed. In addition, attempts to apply string theory to cosmology cannot eliminate the beginning due to the Borde-Guth-Vilenkin theorem.[101]

13. Beginning in the late 1980s, **loop quantum gravity** (LQG) was developed as another attempt to formulate a theory of quantum gravity that unifies general relativity and quantum mechanics. Introduced by Abhay Ashtekar, a researcher in theoretical physics from India, the theory modeled space as a network or fabric of miniscule granules of finite size. This quantization of space limits how small the Universe can

---

99. Arvind Borde, Alan Guth, and Alexander Vilenkin, "Inflationary Spacetimes Are Incomplete in Past Directions," *Physical Review* Letters 90, no. 15 (April 15, 2003).

100. Edward Witten, "Magic, Mystery, and Matrix," *Notices of the AMS* 45, no. 9: 1129. Cf. https://en.wikipedia.org/wiki/Introduction_to_M-theory.

101. John Horgan, "Why String Theory is Still Not Even Wrong," *Scientific American* (27 April 2017).

become, eliminating a spatial singularity and leading to a "rebound" when, in a Big Crunch-type contraction of the Universe, a particular density (5×10^{96} kg/m³) impossible to surpass is reached. Nevertheless, today the theory remains only an unconfirmed attempt, and though it has advanced over the decades of its development, it has not managed to eliminate a temporal singularity. It thus leaves the question of the ultimate origin of the Universe unanswered.

14. In 2005, **conformal cyclic cosmology** (CCC) was proposed by renowned cosmologist Roger Penrose. The theory postulates that after approximately 10^{100} years, once all black holes have evaporated, the universe will be reborn as a new universe in a kind of Big Bang event. This process is eternally cyclical. At the end of each cycle only photons and massless particles remain, floating in the same eternal space in which the cycle began, and in which the next will begin. Penrose attempted to verify CCC by analyzing the cosmic microwave background radiation, but most cosmologists have dismissed CCC due to a lack of convincing evidence and other problems. [102]

15. In 2010, in his book *The Grand Design*, **Stephen Hawking** argued that "*because there is a law like gravity, the universe can and will create itself from nothing.*"[103] Despite the media interest it garnered, the theory did not convince anyone in the scientific community, because gravitation needs a universe to exist.[104] But Hawking's statement reveals a confusion about what the laws of nature are and what they can do. As philosopher of science Stephen Meyer explains in his book *The Return of the God Hypothesis*, "*Hawking's statements about the laws of nature explaining how the universe originated betray a confusion of categories—a philosophical misunderstanding about*

---

102. Ethan Siegel, "No, Roger Penrose, We See No Evidence Of A 'Universe Before the Big Bang'," *Forbes* (October 8, 2020).

103. Stephen Hawking and Leonard Mlodinow, *The Grand Design* (New York: Bantam Books, 2010), 180.

104. Stephen Hawking's strange position earned him the following caustic remark from his Oxford colleague John Lennox: "*What all this goes to show is that nonsense remains nonsense, even when talked by world-famous scientists.*" John Lennox, *God and Stephen Hawking: Whose Design Is It Anyway?* (Oxford: Lion Hudson, 2011), 32. cf. https://www.youtube.com/watch?v=gS72Pf2MK7c.

what the laws of nature do and don't do. The laws of nature are our descriptions, typically framed in mathematical terms, of what nature ordinarily does. [. . .] they have no objective existence in the universe independent of our minds [. . .] still less would they cause the origin of the universe itself. Saying they do is like saying that the longitude and latitude lines on the map explain how the Hawaiian Islands popped up in the middle of the Pacific Ocean." [105]

16. In 2012, **Lawrence Krauss** published his book *A Universe from Nothing*, in which he proposed that the Universe could have come from a "deep nothing." The book was endorsed by militant atheist Richard Dawkins for purportedly eliminating the need for a creator God to explain the Universe. But Krauss's model was really a popularization of a cosmological model developed by Alexander Vilenkin, in which Vilenkin attempted to show that the Universe is the result of some preexisting physical law.[106] Vilenkin's model was similar to that of Hawking and Hartle in that it explained not how the Universe came from nothing but how it emerged from a preexisting mathematical law or, more precisely, a preexisting mathematical description of possible universes that could exist. Yet neither Vilenkin nor Hawking ever explained how this realm of mathematical abstraction could cause an actual Universe to come into existence.[107]

17. In 2023, **Thomas Hertog**, the last scientist to work closely with Stephen Hawking, laid out Hawking's final theory on the "origin of time." Hertog explains that the big question that obsessed Stephen Hawking was *"the mysterious biophilia of the Universe,"* that astonishing fact of the fine-tuning of the laws of the Universe, which needs an explanation. However, according to Hawking, as quoted by Hertog, *"it's obvious that*

---

105. Stephen Meyer, *Return of the God Hypothesis* (New York: HarperOne, 2021), 372–75. Hawking memorably expressed the problem thus: *"What is it that breathes fire into the equations and makes a universe for them to describe?"* S. Hawking, *A Brief History of Time*, Tenth Anniversary edition (New York: Bantam, 1998), 190.

106. Alexandre Vilenkin, *Many Worlds in One* (New York: Hill and Wang, 2006).

107. Hawking, *A Brief History of Time* (New York: Bantam, 1988), 175.

the Multiverse doesn't explain anything."[108] and he even asserts that for Hawking *"scientific explanations like the idea of the Multiverse or a theory of everything"* are *"dead."* Hence the need for materialists to find something else. That's why Stephen Hawking came up with a new theory, in which time does not exist, a theory based on the hypothesis that our human observations modify the origin of the Universe and thus enable "Darwinian selection" of its laws. But nobody has followed up on this surprising thesis…

Though the theories we've reviewed above are certainly creative, we've seen that some are completely off base, and others are just pure speculations without any confirmation.[109]

Though this research may be of intellectual and scientific interest, it is important to notice the incredible amount of brainpower and time that has been put forth in the effort to topple or circumvent a simple, verified, and well-documented theory through complex alternatives that completely lack convincing arguments. This is another indirect argument that allows us to affirm that the Standard Model of the Big Bang can be counted as evidence for the existence of a creator God. The need to work through so many alternative theories has slowed the start of metaphysical inquiry. But in the domain of understanding the Universe, the Big Bang remains the Big Boss.

---

108. See also *Science & avenir 920* (October 2023): 30–41.

109 It must be emphasized that these theories have three qualities in common that radically distinguish them from the Big Bang standard model: (1) They are pure speculations originating in the scientist's imaginations, (2) they do not enjoy the slightest scientific confirmation, and (3) they have not garnered any scientific consensus.

7.

Convergent Evidence for an Absolute Beginning to the Universe

The growing conviction that the Universe had an absolute beginning stems not only from the confirmation of the Big Bang theory, which is only a single piece of evidence, but also from the convergence of many discoveries made in various fields of knowledge. This chapter aims to present these discoveries in a comprehensive way and give the history behind them to demonstrate their strength and coherence, without diving again into those we have already discussed in detail.

Five different fields of thought and science today lead us to the inescapable fact that the Universe had an absolute beginning.

1. Thermodynamics

If our Universe is heading towards its thermal death and this death has still not happened yet, this means that time in the past is not infinite (this point has been discussed in detail in Chapter 4).

2. Cosmology

The expansion of the Universe and the Big Bang naturally point towards a beginning to the Universe. Certainly, many scientists still contest this idea, but their position has become especially difficult to maintain since the work of Borde, Guth, and Vilenkin. After Penrose and Hawking's work on singularities, these scientists demonstrated in 2003 that any universe that has been, on average, in expansion throughout its entire history cannot be infinite in the past and therefore must have a limit in its past space-time. Consequently, even if other *"singularities"* happened in the past (i.e., other Big Bangs), the Universe, which is expanding, cannot be infinite in the past and must necessarily have had a beginning (cf. Chapters 5 and 6).[110]

3. Philosophy

In sixth-century Egypt, a Christian commentator on Aristotle named John Philoponus was the first to confirm that an infinite time in the past is impossible.[111]

His reasoning is simple yet completely valid. It states that we can imagine infinity into the future with today as our starting point, but we will always remain in a *"potential infinity."* That is, if we count 0, 1, 2, 3… without ever stopping, we will stay on track towards infinity, but we will never reach it. The number will increase without ever hitting a maximum. Likewise, just as it is impossible to set out right now to reach infinity in the future, it is impossible to depart from infinity in the past to reach our time. Traversing an infinite greatness is no more possible than coming from a place to which one can't return. Progress that cannot be made in one direction cannot be made in another. An infinite time in the past is therefore a logical impossibility (cf. Chapter 23, section III).

This line of reasoning was developed further by Muslim thinkers Al-Kindi (9th century) and Al-Ghazali (12th century). Their methodology is known today by the Arabic word *kalam* (*"discussion"*). This method of reasoning reached maturity in the West in the thirteenth century with Saint Bonaventure and Saint Albert the Great. However, after a long debate,

---

110. The solutions proposed by Hawking, Penrose, and Ellis in the 1960s also require the Universe to have a beginning. As Meyer explains, *"They end up providing more rigorous proofs of the singular at the beginning of the universe. A temporal singularity and a spatial singularity. Showing there was a beginning to time and space if the theory of general relativity is true."* Sean McDowell and Scott Rae, "The Return of the God Hypothesis, with Stephen Meyer," *Think Biblically: Conversations on Faith and Culture*, Biola University, June 17, 2021, https://www.biola.edu/blogs/think-biblically/2021/return-of-god-hypothesis.

111. *De aeternitate mundi contra Proclum* (Hildesheim: Olms, 1963; originally published by Leipzig: Teubner, 1899).

Albert's student Saint Thomas Aquinas chose not to support it. Due to this, kalam did not enjoy significant recognition in the West.

Saint Thomas Aquinas holds that real infinity cannot exist,[112] but he also maintains that an infinite time in the past does not imply a real infinity, an argument which would later be disputed.[113] Indeed, if there were an infinite time in the past, the number of hours elapsed would add up to a truly infinite number, which is not possible. The assertion that the infinite past does not imply the existence of a real infinity, but only of a potential infinity, is therefore impossible.[114] In sum, real infinity does not exist; only potential infinities exist. However, a potential infinity is an infinity we can only move towards bit by bit, never reaching it. But the past is frozen. It's impossible for it to grow further and further away from the current present. There can therefore be neither real infinity nor potential infinity in the past.

4. Mathematics

In the field of mathematics, the famous mathematician George Cantor (1845-1918) developed a transfinite mathematics manipulating infinite

---

112. *Summa Theologica*, Ia, q. 7, a. 2: "*Things other than God can be relatively infinite, but not absolutely infinite.*" Ia, q. 7, a. 3: "*Now it is manifest that a natural body cannot be actually infinite.*" Ia, q. 7, a. 4: "*It is impossible for there to be an actually infinite multitude.*"

113. One can read a defense of Bonaventure against Thomas Aquinas by Fernand van Steenberghen, "Le mythe d'un monde éternel," *Revue Philosophique de Louvain* 30 (1978): 157–179.

114. In the question of the *Summa theologica* (S. Th., I, q. 46, a. 2) where he takes up this position, it seems that Saint Thomas defends his position with questionable arguments. Notably, in the response to objection #6, the "On the contrary" rests on a poorly-based principle. This is the responding argument : "*Hence it cannot be demonstrated that man, or heaven, or a stone were not always.*" Yet this is false. As solution #4 affirms: "*Those who hold the eternity of the world hold that some region was changed an infinite number of times, from being uninhabitable to being inhabitable and "vice versa,*" and this is equally untenable and false. Finally, in the response to objection #6, he writes : "*Passage is always understood as being from term to term. Whatever bygone day we choose, from it to the present day there is a finite number of days which can be passed through. The objection is founded on the idea that, given two extremes, there is an infinite number of mean terms.*" For this argument, the fact that every point in the past is a finite distance from the present implies that the entire past can be crossed over. This is obviously false because, even if each portion of the past [-n; 0] could be crossed whatever the value of n might be, it absolutely does not follow that the whole of the infinite past could be crossed. It is a sophism of composition that William Lane Craig denounced in 1979 in his thesis on *kalam*, see *The Kalam Cosmological Argument* (London: MacMillan, 1979).

sets, demonstrating that some infinite sets are infinitely larger than others. But this idea of an abstract infinity that works in mathematics cannot be translated into the real world, as David Hilbert and the mathematicians at the conference of the Mathematical Society of Westphalia confirmed in June of 1925, stating: *"Our principal result is that the infinite is nowhere to be found in reality. It neither exists in nature nor provides a legitimate basis for rational thought...The role that remains for the infinite to play is solely that of an idea."*[115] Otherwise, one arrives at absurdities, like Hilbert's infinite hotel, which has an infinite number of rooms: even when all of the rooms are occupied, it is always possible to find a solution that allows us to have an infinite number of empty rooms.[116]

5. Physics

Physicists tend to chase down infinities: every time they encounter equations that diverge and head off towards infinity, they say to themselves, *"No, this is not physics. The models must be revised."* In the twentieth century, these intuitions would be confirmed again and again:

- In the realm of the infinitely large, the Theory of Relativity leads to the conclusion that infinite speed does not exist: the speed of light cannot be exceeded.
- Conversely, in the realm of the infinitely small, it is not possible to divide quantities such as distance an infinite number of times: the process is limited by quanta.
- Finally, even if our data is not yet precise enough to assert this with certainty, it is likely that the Universe is a hypersphere in shape (the 4-dimensional equivalent of the surface of an ordinary, 3-dimensional sphere). In fact, the most recent measurements from the Planck satellite, taken in 2013, seem to confirm "with a confidence level of

---

115. P. Benacerraf and H. Putnam, *Philosophy of Mathematics: Selected Readings* (Cambridge: Cambridge University Press, 1984), 201.
116. Just ask each occupant to move to the room whose number is twice his own. In this way, an infinite number of rooms are freed up: all those with an odd number.

99%" the notion of a very small but slightly positive spacial curvature to the Universe, which does indeed mean a spherical Universe.[117]
- By analogy, we might note that the same problem could have been raised in the past about the Earth,[118] which at a first glance seems to be flat: does it extend to infinity or does it have an edge? Neither, in fact. The Earth is spherical and could not be otherwise since real infinity does not exist and the idea of edges is nonsensical. The same may well go for the Universe. It is very hard to believe that it has an edge, and real infinity does not exist: space-time is therefore quite probably spherical in four dimensions, as measurements seem to confirm. That is to say, if we could go completely straight ahead at infinite speed, we would end up back where we started, arriving from behind.[119] Only models that fail to take all elements of reality into account could imagine the contrary, and neither a flat Universe (with no curve) nor a hyperbolic one (with a negative curve) has any meaning in physics.[120]

Nor, most probably, as we've seen above, is there infinite past time. The standard model of the Big Bang, based on Einstein's Theory of Relativity, suggests that time, space, and matter started together, and coherent support for the thesis of an eternal Universe has become extremely difficult to maintain.[121]

---

117. Cf. "Planck evidence for a closed Universe and a possible crisis for cosmology," published in 2018 in the journal *Nature*: https://www.nature.com/articles/s41550-019-0906-9.

118. All imagined models of infinite cycles in the past come up against major theoretical obstacles. If the Universe had existed for an infinitely long time, our satellites should be able to detect an infinite number of observable signals from an infinite number of sources, in the spirit of Olber's paradox. Black holes would have formed in infinite numbers within these cyclical universes, and the shape of the Universe would not be as homogeneous as it is, rather it must have taken on the shape of an infinitely elongated cigar. Clearly, none of this is the case.

119. The latest measurements of the Planck satellite confirm that the Universe is spherical and that it has a spatial curve, one that is very small but slightly positive. See https://en.wikipedia.org/wiki/Shape_of_the_universe.

120. In cosmology, as in other fields of science, we approach the truth when physical models, mathematical equations, experiments, and observations agree. Without all this, theories remain incomplete and often highly speculative.

121. The etymology participates in the idea that the infinite is impossible, for *"infinite"* signifies *"not finite,"* that is *"not determined,"* almost *"not real"*: it is only negatively that the infinite can be defined against the reality of the material, temporal, and spatial Universe.

A brief historical overview of the question and its philosophical significance

For most of history, almost everyone (with the striking exception of readers of the Bible) have thought that the Universe was infinite in time[122] for these two reasons:

- The first is because this intuitive-feeling notion seems like common sense. We humans perceive the Earth as flat, the Sun as revolving around the Earth, and the Universe as infinite in space and time for the simple reason that from our vantage point we can't see any limit, or even how there could be a limit. Unfortunately, sometimes our intuition deceives us. Convincing us it's wrong takes great scientists and major discoveries.
- The second reason is that the very fact of acknowledging the absolute beginning of the Universe amounts to admitting the existence of a creator God because, as was first noted by Parmenides and taken up by Lucretius, *ex nihilo nihil* ("*nothing can come from absolute nothingness*"). It is therefore evident that there has never been absolute nothingness and that something has always existed. There are then only two possibilities: either the Universe has always existed, or a transcendent God has always existed outside of the Universe—its creator, in short.

It is therefore understandable that materialists from all eras have consistently defended the eternity of the Universe because they simply had no other choice. From Parmenides, Heraclitus, Democritus, Epicurus, and Lucretius, to Marx, Engels, Lenin, Mao, and Hitler, via Nietzsche, Schopenhauer, Feuerbach, Hume, Sartre, the atheist philosophers of the nineteenth century, and even Spinoza, Auguste Comte, Mach, Arrhenius, Haeckel, Berthelot, Russell, Crick, etc.... they have all been obliged to

---

122. Historically, the majority of peoples, civilizations, and scientists have believed that the Universe is eternal and cyclical. The idea comes naturally to mind when we observe the stability of the world and the cycle of the seasons. Aristotle, for example, and all the philosophers of his movement, believed in an eternal Universe that did not have a beginning and would not have an end. Einstein and the majority of leading scientists believed the same thing only 100 years ago.

affirm that matter is, in one sense or another, eternal in the past and that the Universe never began.[123]

This historical panorama and the description of the philosophical issues at stake allow us to better understand why the atheist regimes of the 20th century reacted with such incredible violence, for reasons which were exclusively ideological, as we shall see in the next chapter.

---

123. "*An eternal universe seemed to strike a chord with the scientific community, because the theory had a certain elegance, simplicity and completeness. If the universe has existed for eternity, then there was no need to explain how it was created, when it was created, why it was created or Who created it. Scientists were particularly proud that they had developed a theory of the universe that no longer relied on invoking God.*" In Simon Singh, *Big Bang: The Origin of The Universe* (Harper, 2004), 79.

8.

The Big Bang, A Noir Thriller

Theories to be shot down

Who could have ever imagined that the history of certain scientific theories, elaborated in the studious silence of laboratories, would end for some researchers in the icy anonymity of the Gulag or the bitter roads of exile? Landau, Kozyrev, Bronstein, Frederiks, Hausdorff, and Stern—so many broken or maltreated souls. What brought tragedy to these peaceful men, who were content to bury themselves in their cerebral equations, so far from the calculations of political intrigue?

To understand these events, we have to go back to the discovery of the thermal death of the Universe. This theory, derived from the second principle of thermodynamics, had obvious metaphysical implications that, despite the scientific nature of the question, did not escape the attention of philosophers and ideologues.

Heat death implies that the Universe had a beginning—implying the existence of a creator God. But these conclusions undermined the doctrines of contemporary materialist ideologies, especially Marxism. The expansion of the Universe, verified soon after, went further in the same direction. A short time later, the discovery and dating of the Big Bang completed the demonstration.

The ideological persecution of Big Bang supporters is telling: it reveals how strong and dangerous this evidence was considered from the beginning

The materialists' outrage was proportionate to the danger the Big Bang posed to their fundamental beliefs. To save their ideology, they had to silence at all costs the scientists researching and publishing these theories. Herein lies the great interest of the history of these persecutions. The materialists would not have unleashed such extreme violence upon scientists who theorized the expansion of the Universe and the Big Bang had they not been convinced that these theories made a strong case for the existence of God. The use of force is, here more than ever, an admission.

Soviet Russia and Nazi Germany represent two sad and emblematic cases of this scientific "witch hunt," which squeezed leading scientists in a vise between a red and a brown terror. Even after 1945, the Big Bang continued to tear the scientific community apart, and ideological violence against its supporters smoldered on in a more insidious form.

As we turn the sad pages of this little-known history, we see that reality is often more outrageous than fiction. We find denunciations and summary executions, threats and escape attempts, all provoked by nothing more than scientific discoveries and scholarly demonstrations.

We hope the reader will forgive this break from the more academic style of the previous four chapters. Writing the thrilling story of the Big Bang demands a style verging on that of a spy novel. But in the midst of the story's diversion and suspense, we must not forget the real human tragedy that lies behind it. Unfortunately, all these stories are true.

I. Soviet suppression of the Big Bang Theory

1. "God is dead!"

Sharp as the crack of a whip, searing as a cold flame, this cry was first raised by Nietzsche in *The Gay Science* (1882). Of all the thoughts penned by the great German philosopher, this one would have the greatest influence on a certain young student who passed the entrance exam for Saint Petersburg State University in 1891, earning the highest mark in every subject.

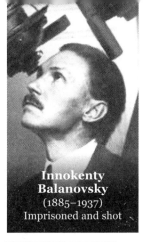

Innokenty Balanovsky
(1885–1937)
Imprisoned and shot

Yevgeny Perepyolkin
(1906–1938)
Imprisoned and shot

Vsevolod Frederiks
(1885–1944)
Dead after six years in the Gulag

Lev Landau
(1908–1968)
Imprisoned, tortured, then rehabilitated. Nobel Prize 1962

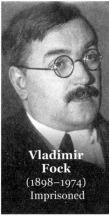

Vladimir Fock
(1898–1974)
Imprisoned

Scientists persecuted or assassinated for their work on the origins of the Universe

Matvei Bronstein
(1906–1938)
Tortured and shot

Maximilian Musselius (1884–1938)
Dmitri Eropkin (1908–1938)
Boris Numerov (1891–1941)
Imprisoned ten years then shot

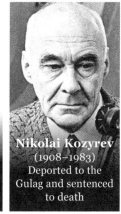

Nikolai Kozyrev
(1908–1983)
Deported to the Gulag and sentenced to death

Leonid Plyushch
(1939–2015)
Jailed in a psychiatric hospital

George Gamow (1904–1968)
Jacob Tamarkin (1888–1945)
Escaped and fled to the USA

Albert Einstein (1879–1955)
Max Born (1882–1970)
Otto Stern (1888–1969)
Forced into exile

His name? Vladimir Ilyich Ulyanov—better known as Lenin

The twenty-one-year-old law student was still a long way off from the smoking barricades of the October Revolution. His ideas were still embryonic, but his voracious intellect had already devoured several of the works that would prepare him for his fateful plunge, especially the thousand pages of Karl Marx's *Capital*. Furthermore, he already had a model to follow. In 1887, his older brother Alexander, an anarchist affiliated with the terrorist group *People's Will*, had been hanged at age twenty-one for his involvement in a plot to assassinate Tsar Alexander III. Haunted by the sacred chants and prayers that accompanied the execution, Lenin was slowly gripped by the conviction that religion is the opiate of the masses, and that a new social order would arise only when the idea of God was annihilated. Unbeknownst to the young agitator, this renunciation would form the core of the Bolshevik ideology that would blaze a trail of iron and blood through all of Russia in the 1930s—dialectical materialism. The famous term "Marxism" was not coined by Marx himself—he never used the phrase—but by Lenin and his followers, most notably Stalin.

Matter alone

The new philosophical trend called for a re-ordering of reality within a strict materialist framework. According to Marxist-Leninist ideology, reality exists in matter alone and not in any other dimension, least of all the "spiritual." More precisely, dialectical materialism uses the dialectical method to analyze reality within the framework of materialism, taking as axiomatic the eternal nature of matter. Materialist thought therefore consists in relying on Hegel's dialectic, stripped of its "idealist" dimension, to represent the eternal transformation of reality without beginning or end. The spirit? A fantasy. God? When asked this question at the start of the revolution—while he was preoccupied with dismantling the tsarist order—Lenin merely shrugged his shoulders.

But by the early 1900s, science had already turned to a completely different understanding of reality. The great Austrian physicist Ludwig Boltzmann had demonstrated that the Universe's entropy—in other words, its loss of heat—increases with the passage of time, as coffee gradually cools in a mug. This observation necessarily leads in the very distant future to the "thermal death" of the Universe. The consequence of this phenomenon is therefore that the Universe cannot be eternal.

In 1917, the revolutionaries of the October Revolution, led by Plekhanov and Lenin, had already recognized the threat posed by this scientific advance. Lenin explained to anyone who would listen that if matter were not eternal, some people—perhaps even nonbelievers—might be tempted to think that it was created by that transcendent cause that some called God. Therefore the churches became the enemy to be destroyed. Emboldened by their new master, the Bolsheviks wasted no time in wiping out their enemies. The State Political Directorate (GPU), a fearsome political police apparatus created in 1922, relentlessly persecuted Orthodox priests and bishops, from one day to the next throwing them into prison, deporting them to Siberia, or roughly shoving them against a wall to die by firing squad. This murderous repression would continue until Lenin's death on January 21, 1924, drowned in his own blood from a stroke.

The death of the father of the October Revolution did nothing to abate the struggle against God and religion. On the contrary, repression took on an ever more terrifying aspect, and its violence was redoubled. Over the next two decades, the man who took the reins of power in 1925 would become one of the bloodiest dictators of the first half of the twentieth century. The byname he gave himself describes him well: *Stalin*, Russian for "man of steel."

For the man of steel, who in his youth, after four years in the Tiflis seminary, had almost become a priest, one thing was clear: God does not exist. Despite searching for four years, he had failed to find him. The human race had to be freed of this pernicious illusion. After taking power in 1925, Stalin brought the persecution of priests to a new level, systematically burning and destroying monasteries and churches. On May 15, 1932, the Soviet government unveiled an atheistic "Five Year

Plan" with the brazen slogan: *"No more God by 1937!"* To accomplish this veritable "theocide," Stalin came up with a radical program to *"bring to completion the liquidation of the reactionary clergy in our country."*[124]

Despite systematically destroying the clergy, church property, and the public expression of religion, the ideologues of Stalin's party soon faced a new enemy. True, the Bolsheviks had torn down the holy places, waded gleefully through the blood of murdered priests, and obliged their "comrades" to take courses in atheism. Nevertheless, beneath the surface still smoldered the idea that someone or something created the Universe. But how? In the 1920s, scientists had begun to publicize discoveries that suggested the Universe is not static, as so many people had come to believe; rather, it was actually growing every minute. In other words, it was "expanding." If true, then at some time in the remote past, the cosmos had a beginning, and this beginning inevitably had a transcendent cause, which some would not hesitate to call God.

The light of Alexander Friedmann

The story begins in 1922 at Saint Petersburg State University, with a then-unknown mathematician named Alexander Friedmann. This modest scholar who worked quietly in the meteorological center in Saint Petersburg had two impressive assets. First, he was a class above his contemporaries in the field of mathematics. At just seventeen years old, he had published his first article on number theory in the prestigious *Annales de Mathématiques,* along with his high-school friend Jacob Tamarkin. Second, he spoke and read several languages including, significantly, German. He was also one of the very first Russian scientists to learn in the autumn of 1917 of a monumental theory that was about to change the world: General Relativity. As he quickly scanned the sea of equations, he noticed almost effortlessly that Albert Einstein had made

---

124. Dimitry V. Pospielovsky, *A History of Soviet Atheism in Theory and Practice, and the Believer,* vol. 2, *Soviet Antireligious Campaigns and Persecutions* (New York: St. Martin's Press, 1988), 88.

a mistake. Worse, Einstein had employed a fudge factor, assigning an ad hoc value to the cosmological constant in order to yield a static Universe, in line with his expectations.

Let's go back to 1916. At the time, Einstein was adamant that the Universe had been fixed in size for all eternity, that it never had a beginning, and that it would exist forever. For this stance, he was welcomed with open arms by the Bolsheviks, who praised him to the skies, though they hadn't read a single line of his work.

Upon closer examination, the equations of Relativity did not at all demonstrate that the size of the Universe is fixed. Quite the contrary. Einstein reworked the equations backwards and forwards, but to no avail. The solution stubbornly indicated that the Universe is expanding.

But in his view, the result was absurd.

In the interest of removing this absurdity and ensuring that the equations yielded a static Universe, Einstein introduced what he called the "cosmological constant" into the equations, to yield a static Universe. He assigned it the precise value necessary to balance the force of gravity, in order to represent the Universe as stable, that is to say, without expansion or contraction. He played this card in 1916.

Until 1922, no one had detected this sleight of hand. No one, that is, except Friedmann, who at first glance noticed that this famous constant had no reason to be there other than forcing the equations to say something they would not otherwise say. Irritated and determined to find out the truth, Friedmann recopied the interminable lines of calculations one by one. Then, with a stroke of his pencil, he crossed out the cosmological constant. At last, he was free to plunge into the equations unencumbered. The calculations were terribly long and complex. But with the help of his old friends, the brilliant mathematician Jacob Tamarkin and the talented Yuri Krutkov, he was able to find that the equations of General Relativity naturally lead to the possibility of a dynamic Universe. Friedmann was the first to discover that unless the cosmological constant were arbitrarily fixed at an extremely specific and improbable value, the field equations would yield solutions that lead to a dynamic Universe—not a static one.[125]

An expanding Universe

The claim was mind-boggling. Ignoring colleagues who urged him to hush up this discovery, which was so contrary to everything the prevailing regime tried to drum into the Russian people, Friedmann decided, as already mentioned, to publish his work on June 29, 1922 in *Zeitschrift für Physik*, the most renowned journal of the era. With its direct style and unapologetic message, the article hit the journal's main readership like a ton of bricks. One morning, Einstein received the famous publication and came across Friedmann's article. This Russian he had never even heard of was suggesting his calculations were faulty. Stung to the quick, he published a scathing response on September 18, 1922, in the same journal: *"The results obtained in [Friedmann's] paper regarding a nonstationary Universe seemed suspect to me. In fact, it turns out that the solution given does not agree with the field equations."*[126]

For Friedmann, Einstein's rebuttal came as a heavy blow. He had already sensed a silent hostility growing between himself and certain colleagues affiliated with the Communist Party. But if Albert Einstein himself, a recent winner of the Nobel Prize, had spoken against his theory, he could not do much to make his voice heard. On the eve-

---

125. As mentioned earlier, the cosmological constant is a natural result of mathematics which made it possible to establish the equations of General Relativity. Einstein had fixed the theory with the precise value necessary to maintain a static Universe. After recognizing that the Universe was expanding, he removed the value from his equations. Later, cosmologists concluded that the cosmological constant was necessary in order to take into account what we today call dark energy. This energy permeates all empty space and is at the origin of the acceleration observed in the expansion of the Universe.

126. Albert Einstein, "Comment on A. Friedmann's Paper: 'On the Curvature of Space,'" *Zeitschrift für Physik* 11 (1922), in *The Collected Papers of Albert Einstein, Volume 13: The Berlin Years: Writings & Correspondence January 1922–March 1923* (English translation supplement), ed. Diana Kormos Buchwald, József Illy, Ze'ev Rosenkranz, and Tilman Sauer, trans. Ann M. Hentschel and Osik Mose (Princeton, NJ: Princeton University Press, 2012), 271–272.

ning of December 6, 1922, Friedmann called upon the aid of one of his students, Vladimir Fock, who had a stake in the fight because his thesis director was none other than Friedmann's faithful supporter, Krutkov. Together, they wrote the following note: *"Considering that the possible existence of a non-stationary world has a certain interest, I will allow myself to present to you here the calculations I have made… for verification and critical assessment… Should you find the calculations presented in my letter correct, please be so kind as to inform the editors of the* Zeitschrift für Physik *about it; perhaps in this case you will publish a correction to your statement or provide an opportunity for a portion of this letter to be published."*[127]

It worked: in early May 1923, Krutkov set out to find Einstein, meeting him in Holland. Walking Einstein through the calculations step by step, on May 7, 1923, Krutkov succeeded in convincing him (with the aid of indispensable theoretician Paul Ehrenfest, an old friend of Einstein's) that Friedmann's approach was sound. As soon as he returned to Berlin on May 21, the father of Relativity sent what was to become a historic note to the *Zeitschrift für Physik* (published June 29, 1923): *"My criticism, as I became convinced by Friedmann's letter communicated to me by Mr. Krutkov, was based on an error in my calculations. I consider that Mr. Friedmann's results are correct and shed new light."*[128]

A calculation error! A thundering admission! The note was published Friday the 29th of June, 1923. With these few lines, modern cosmology was born.

---

127. Friedmann to Einstein, December 6, 1922, in *Relativity: Modern Large-Scale Spacetime Structure of the Cosmos*, ed. Moshe Carmeli (Singapore: World Scientific Publishing, 2008), 321.

128. Remark on the Work of A. Friedmann "On the Curvature of Space," *Zeitschrift für Physik* 10 (1922).

The Big Bang pioneer triumphs

Friedmann could finally savor his victory. Straight away in that same year, with the help of Vsevolod Frederiks, he published a work entitled *The World as Space and Time*. His readers were astounded to discover that the Universe had a beginning billions of years ago. What sort of a cosmos do we live in, then? At the dawn of time, the entire Universe, with its billion stars, was contracted *"within a point (of zero volume) and then, from this point, it increased its radius to a maximum value, then it decreased again to become a point again, and so on."*[129]

This kind of conclusion was simply insupportable for the Bolsheviks, all the more so since Einstein himself now seemed to recognize its validity. The murmurs that had spread slowly throughout the Saint Petersburg University halls had subsequently grown into a roar threatening the tenets of Marxism-Leninism. For although the authorities had up till now tolerated this scholar's writings, every line they contained starkly contradicted dialectical materialism, which held that the Universe is static, eternal, and without beginning. For the new masters of the regime, the solution was simple: end the myth of creation once and for all, along with anyone who dared to propagate it.

Destroying the myth of creation

Thenceforth the order of the day was to eliminate any scientist even vaguely suspected of violating the teachings of dialectical materialism. The regime accused them of anti-Soviet propaganda. These supposed *"enemies of the Bolshevist state"* began to disappear to far-off Siberia or fall before the firing squad.

The merciless hunt commenced in the winter of 1923. Measures were taken to silence Friedmann. Several of his seminars and courses were "inspected" by agents of the political police, who came to the lecture halls armed to the teeth. But Friedmann would not be bullied into

---

129. Alexander A. Friedmann, *Mir kak prostranstvo i vremya* (Moscow: Éd. Nauka, 1965; orig. pub. 1923).

silence. Clashes broke out between the Bolsheviks and the mathematician's students, in particular four talented young men who were seen together so often that they were known by all Leningrad as *"the musketeers."*[130] The youngest was a mathematical genius named Lev Landau, who earned his bachelor's degree at the age of thirteen and was still an adolescent in 1924. His best friend, George Gamow, knew the theory of Relativity better than anyone. Matvei Bronstein had already begun to write a book on the origin of matter.[131] And Dmitri Ivanenko knew all there was to know about what was then coming to be known as "the Friedmann equations."

These brilliant mathematicians enthusiastically spread far and wide the astonishing news that the Universe is not fixed and that, therefore, it had a beginning. In packed lecture halls, Friedmann was greeted with thunderous applause each time he uttered in a hushed voice: *"Gentlemen, we have demonstrated that the Universe has not existed forever. It had a beginning, several billion years ago, in a far-off age when it was no bigger than a speck of dust!"*[132]

He was a resounding success, even among non-scientists. The party leaders were not pleased.

The fearsome chief of the political police, Felix Dzerzhinsky, ordered his men to start tailing Friedmann in 1924. In spring 1925, the university directors who were affiliated with the party tried to prevent him

---

130. George Gamow, interview by Charles Weiner, April 25, 1968, Niels Bohr Library & Archives, American Institute of Physics, College Park, MD, USA, https://www.aip.org/history-programs/niels-bohr-library/oral-histories/4325.

131. Matvei Bronstein, "On the Expanding Universe," *Physikalische Zeitschrift der Sowjetunion* 3 (1933): 73–82.

132. In Alexander Friedmann, *The World as Space and Time* (Montréal: Minkowski Institute Press, 2014), 79. Friedmann crossed the line Einstein had feared to, writing: *"The Universe contracts into a point (into nothing), and then again increases its radius from a point up to a certain value, then again, diminishing its radius of curvature, transforms itself into a point, etc."* Jean-Pierre Luminet explains: *"For the first time in the history of cosmology, the problems of the beginning and the end of the universe were posed in scientific terms, but Friedmann couldn't help but see the metaphysical implications: 'We may also speak of the creation of the world from nothing.'"* Cf. https://www.cairn.info/revue-etudes- 2014-1-page-67.htm.

from teaching his theory of cosmic expansion. But this pioneer of the Big Bang abruptly died the following September at age 37, reportedly of typhoid fever.

The young George Gamow was convinced that his teacher had been the victim of a conspiracy. He confided his suspicion one autumn evening in 1925 to mathematician Jacob Tamarkin, Friedmann's most faithful companion. Tamarkin never forgot those nights spent together when the two friends demonstrated that space-time had been created in a single instant in the distant past. In the seventies, the Nobel Prize-winning Russian Andrei Sakharov would call this instant "the Friedmann singularity."

After his clandestine conversation with Gamow, Tamarkin realized that his "highly suspect" ideas and his proximity to the Friedmann group made him a prime target for the political police. He made up his mind just a few hours later: he would leave the Soviet Union. So it was that in 1925 Russia lost two of its most brilliant scientific minds: Friedmann, who was carried off in September by a suspicious fever contracted upon his return from his honeymoon, and his close friend Tamarkin, who secretly emigrated to the United States, where he began a new life.

Unfortunately, Gamow was not so lucky; nor were Friedmann's other students, least of all the ones most closely associated with him. These faithful disciples would soon begin to feel the weight of the regime's iron fist. But not quite yet.

After Friedmann's death, the campaign of repression and intimidation against scientists who believed in the creation of the Universe was put on hold for several years. First of all, since the death of their leader, his students, now well aware of the danger, began to act more discreetly. For another thing, during this period the Soviet regime had to grapple with two recent scientific events whose impact resounded throughout the whole world. The first took place in 1927: the bombshell publication of Fr. Lemaître's new theory of cosmic expansion.

The scientists of the regime swatted away the Belgian canon's article with a backhand stroke. What a waste of time! Two years later, in 1929, American astronomer Edwin Hubble actually observed the expansion of the Universe using the huge Hooker telescope on Mount Wilson: here was the first observational confirmation that Friedmann had not been mistaken. Stalin and his lieutenants reacted at once, officially disparaging the monumental discovery as American propaganda and silencing any proponents of expansion. Russian astrophysicists, led now by Gamow, could no longer speak out. As Stalin slowly took over the Soviet Union's enormous state apparatus with the aid of the fearsome Molotov, the campaign against ideas threatening Marxist materialism's sacrosanct doctrines became more and more brutal. In 1926, Vyacheslav Menzhinsky succeeded Felix Dzerzhinsky as head of the secret police and at once unleashed his bloodthirsty agents upon the supposed dissidents.

Meanwhile, Stalin had found his bearings. He secretly began to prepare the grand show trials and mass executions that from the early 1930s would successfully reduce to silence those they now called "Lemaître's henchmen."

2. Mass crimes against cosmologists

In the early 1930s, under pressure from the man who now referred to himself as the "Little Father of the People," the machinery for eliminating these scientists was set in motion and quickly spiraled out of control.

One particularly striking example of the new mass repression was "the Pulkovo affair,"[133] a long series of state-sponsored murders organized in the '30s against astronomers and astrophysicists, most of whom had been close to Alexander Friedmann and were firmly convinced that the Universe was not eternal.

---

133. See Robert A. McCutcheon, "The 1936–1937 Purge of Soviet Astronomers," *Slavic Review 50*, no. 1 (1991): 100–117.

It was no accident that this murderous repression was centered in Leningrad. For one thing, Leningrad was Friedmann's native city, the place where he had discovered and circulated his ideas. But there was another reason. In 1934, one of Stalin's devotees, Andrei Zhdanov, was named Secretary General of the Communist Party of Leningrad. A self-proclaimed idealogue of the regime and the methodical organizer behind the great Stalinist purges, Zhdanov summed up the Soviet stance toward the Big Bang in a few vitriolic words, calling Lemaître and his crew *"imposter scientists who seek to bring back to life the fairy tale according to which the world came out from nothing!"* His objective was to savagely hunt down Lemaître's *"reactionary scientists."*[134] Anyone who spoke or wrote about expansion was systematically eliminated.

By 1931, Gamow and his friends in Moscow knew that western scientists had begun to discuss the possibility of matter itself having a beginning. They were resolved to open the conversation in Russia, but the Soviet empire's new masters—Stalin, Molotov, Bukharin, Beria—were now openly hostile to western science. They believed matter was eternal, and no one, especially not Friedmann's former students, was allowed to say otherwise.

Gamow was the first to sense the danger. When his passport was confiscated in 1931, he felt the walls of his country close in around him like a prison. In the spring of 1932, he attempted to cross the Black Sea in a kayak with his wife, hoping to reach Turkey. A few months later, he tried to flee to Norway, but this attempt nearly ended in shipwreck. All seemed lost until 1933, when the regime miraculously granted him a visa requested by Niels Bohr and Paul Langevin to attend the prestigious Solvay conference in Brussels. He would never again set foot in Russia. Immediately declared a *"traitor to the regime,"* he was sentenced to death. But through the help of Marie Curie, Gamow and his wife left Europe for good and moved to the United States in 1934.

---

134. Andrei Zhdanov, "Against Idealism in Modern Physics" (1948): https://www.balcanicaucaso.org/eng/Areas/Russia/Stalin-the-big-bang-and-quantum-physics-176560.

Unfortunately, the other musketeers were not so lucky, as we shall soon see. Landau, Ivanenko, and Bronstein followed a different path, falling into the clutches of the repressive regime.

The tragic Pulkovo affair

It is March 1935, the end of a long winter.

It is night, but a man has been at work for a long time. His name is Vladimir Fock. This brilliant mathematician and physicist is a chip off the old block; he is the son of Alexander Fock, a surveyor and the USSR's Chief Engineer of Waters and Forests. But that's not all. While studying physics at the University of Leningrad, he was a loyal student of Alexander Friedmann.

Though his relation to Friedmann was a maddening thorn in the Bolsheviks' side, Fock skillfully navigated the pitfalls of Leningrad university life and became Chair of the Department of Theoretical Physics in 1934. That year, he gave a series of conferences at the celebrated astronomical observatory of Pulkovo, well known as the premier center for astronomical research of the Academy of Sciences of the Soviet Union.

His arrest by police officers takes place in the middle of the night: Fock is accused of high treason and of plotting against the ideology of the State. After three extremely trying days of isolation, during which he was denied food, Fock learns that he has been linked to the Pulkovo conspirators, traitors to the State who claim that the Universe had a beginning billions of years ago. However, for his torturers, dialectical materialism epitomizes the triumph of reason. The Universe has always been just as it is. And so it will be forever.

If the Universe did indeed have a beginning, who then set it all in motion? Fock senses that they want to extort a declaration of faith from him, to make him say that God created all that exists. But the physicist knows the price for such an admission: the penalty for such a crime would be death.

At the end of this kangaroo court, the Bolsheviks told him that he would be shot. His crime was to have been Friedmann's student and to think, as he did, that the Universe was in expansion— to believe that in the distant past, the Universe was no larger than a speck of dust. He is accused, like Friedmann, of having fudged his calculations to yield a false result for the field equations of General Relativity. Faced with a situation simultaneously horrible and absurd, disbelief weighs heavily on Fock's shoulders. But even as the execution day approaches, a miracle happens. The miracle worker's name is Pyotr Kapitsa. An extraordinary scientist in the USSR, Kapitsa would go on to win the Nobel Prize in Physics in 1978, the same year as Arno Penzias and Robert Wilson, who discovered the famous trace left in space by the Big Bang: fossil radiation. Kapista was also the only scientist outside of the Communist Party to be admitted to the Russian Academy of Sciences, and thus the only Russian scientist of note opposed to Stalin. Having already saved the life of Lev Landau, one of Friedmann's four musketeers, he also saves Fock in 1935. Once freed, the physicist goes right back to work.

The nightmare begins again

On a frigid night in 1937, Fock is again arrested and thrown into prison. He screams that he has already been tried and cleared, but to no avail. More "charges" have been brought against him. First of all, he had been seen revisiting the Pulkovo observatory, that astronomical sanctuary where Friedmann and his students had passionately worked in the early 1920s—a place literally cursed by the regime. And then, wasn't it a spy at Leningrad University who discovered that he had delivered a scandalous lecture on the alleged origin of the Universe? Kapitsa moves heaven and earth to secure an audience with Stalin and once more obtains an order for Fock's release.

This time, the nightmare is finally over for good.

Yevgeny Perepyolkin: sent to the Gulag and shot

Most Russian astronomers weren't so lucky. The first to fall was a young physicist named Yevgeny Perepyolkin, who had followed Friedmann since 1922 and was a passionate participant in the debates sparked by the controversy between his teacher and Einstein over the origin of the Universe. In 1934, when he was only twenty-eight, he was named professor of astronomy and then chief of the astrophysics lab at the Pulkovo observatory. He became a champion of Friedmann's teachings. Reworking his teacher's calculations step by step, the young Perepyolkin again demonstrated, beyond a shadow of a doubt, that the field equations of General Relativity point necessarily to a *dynamic* Universe rather than the static one conceived of by Einstein. In other words, the cosmos is not set and eternal—it is growing with every passing instant. The conclusion is inevitable: the Universe had a beginning long ago. In his courses, the young astrophysicist hammered this great discovery into his students. But he did not recognize the danger that hung over his own person. In the middle of the night on May 11, 1937, he was suddenly snatched from his bed and arrested. He was accused of being an accomplice to Leon Trotsky's anti-Soviet activities, and, after a rushed trial, was sentenced to five years in prison. But this was not enough. One morning while he was still serving his sentence, the terrible NKVD—predecessor of the KGB—decided to retry him, this time for much graver crimes: undermining the regime's ideological basis by denying the eternity of the Universe and refusing to admit that the idea of God is completely without merit. He was shot the next day, on the frigid morning of January 13, 1938;[135] he was barely thirty years old. Soon to meet the same fate was a school friend he knew well and deeply admired, Matvei Bronstein.

Bronstein: tortured and shot

The young man was one of the four musketeers of Leningrad, a disciple of Friedmann alongside Gamow, Landau, and Ivanenko. A mathemati-

---

135. https://en.wikipedia.org/wiki/Yevgeny_Perepyolkin.

cian more than half a century ahead of his time, he had tried to develop a physics that united the infinitely large and infinitesimally small in a single theoretical body—in other words, combining General Relativity and quantum mechanics. Bronstein's "quantum gravity" remains mysterious to this day. As a youth Bronstein befriended Nobel laureate Andrei Sakharov. A few years later, he wrote wonderful books of popular science for children. Alongside his friends Gamow and Landau, he broadcast the news far and wide that the Universe had a beginning billions of years ago.

In 1937, as he was working peacefully at home, he was dragged away by officers right in front of his wife. She never saw him again. Sentenced to death by the Military Collegium of the Supreme Court of the Soviet Union, Bronstein was shot in the back of the neck on February 18, 1938. He was only thirty-one years old.[136]

But the regime's hit list had many more entries.

Dmitri Eropkin: condemned and shot

On September 4, 1936, astrophysicist Dmitri Eropkin, who also found Friedmann's revolutionary ideas intriguing, was arrested in the middle of the night, just like the others. On May 25, 1937, he was sentenced to ten years in prison without right of correspondence or visitation. But the NKVD thought the sentence too light: only death could put an end to the contagious heresy that everything in existence originated with a God more powerful than the Soviet Union. Accused of spreading anti-Marxist propaganda among his fellow detainees, Eropkin was shot on January 20, 1938, at just twenty-nine years old.[137]

---

136. Gennady Gorelik and Victor Frenkel, *Matvei Petrovich Bronstein and Soviet Theoretical Physics in the Thirties* (Basel: Birkhäuser Verlag, 1994).

137. See https://fr.m.wikipedia.org/wiki/Dmitri_Eropkine; On the Pulkovo affair, see: http://www.owlapps.net/owlapps_apps/articles?id=9928572.

The Kresty Prison still stands today in St. Petersburg, formerly Leningrad

Boris Numerov: deported, tortured, and shot

A month after Eropkin's arrest, it was the turn of one of his best friends, Boris Numerov. But this man was not just any scientist. First of all, he was a brilliant mathematician—a quality he shared with Friedmann, whom he had known for many years and with whom he spent long hours working out equations on the blackboard. Since 1920, the whole university had used the famous "Numerov method" to resolve difficult second-order differential equations in the blink of an eye. Always in good spirits and ready to help others, he climbed swiftly at the university; at age twenty-one, he was already the president of the Russian Astronomical Society. Two years later, he was made head of the Astronomical Institute at Leningrad, where he immediately invited Friedmann to give a series of conferences. But Friedmann's success had reached Stalin's ears. Although the man of iron was not yet in the grip of that murderous folly which set off a terrifying wave of purges in the 1930s, the "little father of the peoples" had a long memory. Numerov's friendship with Friedmann left faint traces of suspicion in Stalin's mind. On the night of October 22, 1936, Numerov was arrested and then placed into solitary confinement in the terrorist wing of Kresty Prison, northeast of Leningrad. He did not see another human being, nor did he leave his cell, until the day of his trial, May 25, 1937. On that day, he was sentenced to ten years of forced labor for espionage and anti-Marxist propaganda. But once again, the regime found the punishment too light. In the autumn of 1941, Numerov was blindfolded and led into the forest of Medvedev. On Stalin's personal orders, Numerov's torturers broke his legs with the butt of a rifle before beating him to death like a dog.[138]

Maximilian Musselius: condemned, imprisoned, and shot

From 1937 onward, the repression intensified against the adversaries of the theory of an eternal Universe. On February 11, 1937, Professor Maximilian Musselius was arrested and sentenced to the usual ten years of prison. Again as usual, this "light" sentence attracted the NKVD's bloodhounds, who decided to "review" the earlier trial. Condemned a second time for anti-Soviet propaganda, the unfortunate astronomer fell before the firing squad on January 20, 1938.[139]

Vsevolod Frederiks: arrested, condemned, and dead after six years in the Gulag

The same treatment was reserved for Vsevolod Frederiks, one of Friedmann's most faithful intellectual companions. In the '20s, his seminar on General Relativity at Saint Petersburg University had packed the lecture halls. Frederiks had also contributed to the calculations for the dynamic solution of Einstein's equations. Thus he was absolutely convinced—and had the math to prove it—that the Universe had a beginning.

---

138. Robert A. McCutcheon, "The 1936–1937 Purge of Soviet Astronomers," *Slavic Review* 50, no. 1 (1991): 100–117. For further information on the fate of the Soviet scientists described in the above pages, see also Igor and Grichka Bogdanov, *Trois Minutes pour Comprendre le Big Bang* (Paris: Courrier du Livre, 2014).

139. https://fr.m.wikipedia.org/wiki/Maximilian_Musselius.

He continued to defend this position long after Friedmann's death. Accused of "creationism" in 1937, he was similarly sentenced to ten years in chains. Miraculously, he was spared the firing squad, but hunger and the ill-treatment he received in prison left him gravely weakened. He died in 1944 after six bitter years of forced labor.[140]

Innokenty Balanovsky: denounced and shot

Another of Friedmann's colleagues, Innokenty Balanovsky, was denounced to the police in the same period for defending Friedmann's ideas. At first he was sentenced to 10 years in a labor camp.[141] According to some, his life ended with an execution by firing squad in 1937.[142]

Nikolai Kozyrev, sent to the gulag and sentenced to death

Nikolai Kozyrev, initially sent to a forced labor camp in 1937, was also sentenced to death, accused by Zhdanov's men of teaching the detainees that the Universe was born in an explosion more than ten billion years ago. Miraculously, he was not shot for the simple reason that on the morning of the fatal day there were not enough executioners to form the firing squad.[143]

The Stalinist regime had intended to make a chilling example out of these towering intellectuals. But in April 1948, the Big Bang theory took a giant leap forward. George Gamow—Friedmann's closest disciple—burst back onto the scientific scene from across the Atlantic, through the joint publication (along with Ralph Alpher and Hans Bethe) of an explosive

---

140. N.V. Tsvetkov, "Vsevolod Konstantinovich Freedericksz: The Founder of the Russian School of Physics of Liquid Crystals," *ResearchGate* (June 2018).

141. See https://fr.wikipedia.org/wiki/Innokenti_Balanovski.

142. L. J. Reinders, "Repression at the Leningrad Physico-Technical Institute," in *The Life, Science, and Times of Lev Vasilevich Shubnikov* (Cham: Springer, 2018), 139–144. See https://fr.m.wikipedia.org/wiki/Innokenti_Balanovski.

143. See: https://www.britannica.com/biography/Nikolay-Aleksandrovich-Kozyrev.

article in *Physical Review*. Its provocative claims quickly spread across the globe like wildfire. What do the authors assert? They propose that the light elements of matter—protons and neutrons—could only have formed during an extremely hot phase of the Universe, with temperatures soaring to billions of degrees, far exceeding those of stars. And they calculated the relative abundance of the lightest elements–75% hydrogen, 25% helium along with traces of lithium and beryllium–which in itself constituted an extraordinary finding and which was a compelling new way to prove that the Big Bang took place.[144]

Just as they had suppressed Hubble's discovery in 1929, for a time the Soviets were able to stifle Gamow's article, along with all serious research in cosmology. In fact, in the '40s and '50s nearly all astronomers scrupulously toed the party line, led by physicists like Boris Vorontsov-Velyaminov of Moscow, who shouted down every hall in the Academy of Sciences that Gamow was an *"Americanized apostate."*[145] Then there was V. E. Lov, who supposedly referred to the Big Bang as a *"cancerous tumour that corrodes modern astronomical theory and is the main ideological enemy of materialist science."*[146]

Censure and condemnation lasted into the second half of the twentieth century, a dark period in which Bolshevik repression would assume its final form.

3. The fall of dialectical materialism

Spring 1963

The early 1960s marked a sea change for scientists, especially those who studied the Universe. Ten years earlier, the Great Purge had ended

---

144. R. A. Alpher, H. Bethe, and G. Gamow, "The Origin of Chemical Elements," *Physical Review* 73, no. 7 (April 1, 1948): 803–804.

145. Helge Kragh, "Science and Ideology: The Case of Cosmology in the Soviet Union, 1947–1963," *Acta Baltica*, 1, no. 1 (2013).

146. As cited in Simon Singh, *Big Bang: The Most Important Scientific Discovery of All Time and Why You Need to Know About It* (Notting Hill: Fourth Estate, 2004), 364.

with Stalin's death, and the new leaders of the USSR were committed to radical "de-Stalinization." A new generation of physicists emerged in the more tolerant atmosphere that now prevailed, and this fresh crop of scientists was delighted to rediscover the research of Friedmann and Gamow on the origin of the Universe. They were led by Igor Novikov and Andrei Doroshkevich, two young researchers from the University of Moscow. Both were students of the legendary theoretician Yakov Zeldovich, one of the fathers—along with Sakharov—of the hydrogen bomb. In 1964, Novikov and his colleague made a prediction that Zeldovich gladly supported. Following the main points of Gamow's theory, they advanced the thesis that though the wave of intense heat from the Big Bang (100,000 billion billion billion degrees) has cooled down, a trace of it remains in the sky. The two young researchers were also the first to propose that this ultra-cold relic is black-body radiation. In other words, it is in thermal equilibrium and therefore extremely uniform at every point in the cosmos. Quite sure of themselves, they added that it is possible to detect this "fossil radiation" in the microwave range of the spectrum.

These two extraordinary predictions could have propelled the USSR to the front of the Big Bang race, but the regime's passive hostility to cosmology meant that the field was ceded entirely to American scientists. In May of the same year (1964), two young American astronomers, Arno Penzias and Robert Wilson, using an immense radio telescope in their work for Bell Telephone Company, made what is often considered the greatest discovery in history: fossil radiation, a faint "echo" of the Big Bang. The pair would go on to become Nobel laureates in Physics in 1978, alongside the courageous Russian scientist Pyotr Kapitsa, who had stood up to Stalin in the '30s and snatched Landau and Fock from certain death.

Deeply impressed by the discovery of fossil radiation, the leading lights of nuclear physics turned toward cosmology in the 1960s, more or less openly adopting the Big Bang model. But the game was still far from over. Despite its crucial importance, Penzias and Wilson's discovery did not radically change the doctrine of the Soviet regime, which obstinately clung to the belief that the Universe is composed of eternal matter. This ideological blindspot meant that Soviet scientists were practically forced to keep silent about the Big Bang until the collapse of the USSR. While death sentences and deportations to the Gulag stopped with Stalin's death in 1953, the regime devised other methods such as house arrest—or worse, commitment to a psychiatric hospital.

Psychiatric prison

The year is 1968. A fresh breeze of liberated thought is blowing across France, and certain gusts travel as far as student groups in Russia. In that year, a young mathematician, Leonid Plyushch (a researcher at the Moscow Institute of Cybernetics), became a dissident. The "Plyushch affair" caused a stir throughout Europe, especially in France, and was typical of dialectical materialism in its death throes.

The affair began with four students receiving the harsh sentence of forced labor for the crime of "anti-Soviet agitation and propaganda." Disgusted by this, the young mathematician joined a group of intellectuals who, like him, were determined to put an end to indoctrination. Keen on cosmology and convinced that fossil radiation definitively resolved the question of the Big Bang, Plyushch could not stomach the regime's refusal to let him organize scientific conferences or speak the truth about a subject he loved so much. Heedless of the consequences, he sent a letter to the United Nations Commission on Human Rights, demanding an investigation into the USSR's violations of the rights to freedom of conviction and to propagation of ideas through legal means. The regime's reaction was swift. He was immediately dismissed from the Institute of Cybernetics and interrogated by the KGB, who also confiscated his manuscripts. After a period of increasingly strict surveillance, accused of anti-Soviet activities, he ended up being arrested in 1972 and thrown in prison. One year later, after a secret trial he was not even allowed to attend, the hammer fell: Plyushch was declared insane and sentenced to "treatment" at the Special Psychiatric Hospital at Dnipropetrovsk, where patients with the most serious cases of psychosis were confined. Overnight, Plyushch was locked in a windowless padded cell, which he left only to undergo frightful sessions of electroshock therapy. Little by little his mind began to falter, and for a time he lost the power of speech. As he explained after his liberation in 1976 in his book *History's Carnival*: "*Although I tried to spit out the drugs when I could, they were killing my desire to read or think...My memory was slipping away, and my speech became jerky and abrupt. I was overcome by autism and tried to sleep...I was increasingly afraid that my deterioration was irreversible...I, too, was losing my will to live. I maintained a grip on myself only by saying over and over: I must not become embittered; I must not forget; I must not give up!*"[147]

After a year of detention, the psychiatrists commissioned by the regime were told to examine the patient and provide a diagnosis. The special commission concluded that Plyushch was suffering from "reformist delirium" exacerbated by "Messianic elements." Like the scientist-martyrs who came before him, his case centered around the question of God. This is what he confided in 1992 to Alexandre Guinzburg, the young student for whom he fought in 1968 and who was the origin of his condemnation: "*When I was eleven years old, and a devout believer, one of my teachers managed to prove to me that God did not exist. With this denial, I became an 'educated' man.*"[148]

In the early 1970s: clearly still no room for God

The imprisonment of Plyushch set off international waves of protest. Six hundred and fifty American mathematicians wrote a letter to the Soviet ambassador, and Henri Cartan brought the matter before the 1974 International Congress of Mathematicians, held in Vancouver. But

---

147. Leonid Plyushch, *History's Carnival: A Dissident's Autobiography* (New York: Harcout Brace, 1979).

148. Translated from A. Guinzburg, "Quand l'Est était rouge: l'Utopie meurtrière," *L'Express*, February 1, 1992, https://www.lexpress.fr/informations/quand-l-est-etait-rouge-l-utopie-meurtriere_591530.html.

the most notable intervention was that of physicist Andrei Sakharov, who had just completed his model of the Universe based on the Big Bang. He even dared to call the origin of the Universe the "*Friedmann singularity*"—the ultimate provocation!

Sakharov denied that matter was eternal and held that the Universe was finite in both its past—the Friedmann singularity—and its future. In particular, he was the first to postulate the disintegration of the proton, the particle of matter previously assumed to be eternal. This claim earned him the wrath of the Academy of Sciences.

As his ideas continued to find confirmation, the great physicist increasingly chose the path of dissidence, undeterred by repeated calls to order from the regime's tribunals. In the West, his actions met with praise and the highest recognition of the Nobel Peace Prize. As might be expected, the Soviets forbade Sakharov from crossing the Iron Curtain, and his wife had to go accept the prize in his name. He kept up his activism. The day of the prize ceremony, Sakharov traveled instead to Vilnius to advocate for Leonid Plyushch at his retrial, an act that proved decisive for Plyushch's liberation and expulsion to France.

In the same period, the father of the H-bomb increasingly turned away from hard physics and towards cosmology. He grew close to researchers committed to the Big Bang and organized conferences and gave seminars on the origin of the Universe. But the subject remained dangerous. In the end, his outspoken defense of Plyushch, his widely publicized research on the Big Bang, and his other activities would cost him dearly.

On the morning of January 22, 1980, Sakharov was arrested in the middle of the street. A few hours later, without a trial, he was deported to Gorky (today Nizhny Novgorod), a city off-limits to foreigners. There he was put under house arrest, with armed police at his door night and day. He was forbidden to correspond or meet with anyone and was effectively reduced to silence. His rooms were searched each week: anything he wrote or hid from his guards was inevitably confiscated. In May 1984, he began a hunger fast that left him hospitalized. At the hospital, he was force-fed and kept for four months in total isolation. Upon returning to Gorky, he was still denied the right to any contact with the outside world. In April 1985 he began a second hunger strike and was again brought to the hospital to be force-fed.

The ordeal ended on December 19, 1986, when Mikhail Gorbachev called Sakharov to tell him that he was free to return to Moscow. He was the last on the long blacklist of scientists persecuted by the regime for their conviction that matter is not eternal and that it does not constitute the whole of reality.

In Soviet Russia, most scientists who believed in the Big Bang suffered physically for spreading a revolutionary theory that challenged the eternity of matter. This noir thriller has had its heroes and hangmen, its reversals and tragic twists. Between the lines we can read the doubts and inner conflicts these scientists endured for their faith in this scientific phenomenon, a drama reminiscent of the novels of Dostoevsky.

In Nazi Germany, the violence exercised against partisans of the Big Bang took on an administrative and procedural guise more reminiscent of a Kafka novel, until the mask was dropped to reveal a face twisted by the fanatical brutality of the Nazi regime.

II. Nazi opposition to the Big Bang[149]

1. Hitler declares war on God

As we shall now see, while persecution was less widespread, the fate reserved for supporters of the Big Bang in Nazi Germany was hardly more enviable. Our story begins with certain German scientists, oppressed

---

149. We should not make the mistake of thinking that cosmology was only a minor point for Hitler, Himmler, or the Nazis in general. On the contrary, the racist theories underpinning National Socialism were constructed out of mythological material and cosmological fantasies that their ideology and ignorance led them to swallow hook, line, and sinker, as we shall see later.

since the early 1920s by the rising tide of a then-obscure movement. No one suspected the future danger it posed to society, nor how it would expand and organize itself into the second most bloody totalitarian regime of the twentieth century. Right from the start, it was also willing to use its iron boots to stamp out all science supporting the idea that matter was not eternal.

———

February 1920. It is the second time the tiny German Workers' Party, founded by the ultranationalist Anton Drexler, has met this year. Perched on a crate, the orator unleashes a torrent of black oaths against the villains who have banded together to bring the great nation of Germany to its knees. The man's name is Hitler. A few sharp cries hammer his point home: the time has come to replace the German Workers' Party with the National Socialist German Workers' Party. The galvanized audience roars its approval.

On this night, the Nazi party is born.

———

November 1922, at the Friedrich Wilhelm University in Berlin. The most popular physics course in the university has just ended. The students leave the lecture hall in small groups. They know they are fortunate: their professor is none other than Albert Einstein. Two electrifying discoveries in less than three years have made him a household name the world over. The first came in 1919, when Sir Arthur Eddington experimentally verified the famous theory of General Relativity. Then in 1921, Einstein won the Nobel Prize, making him an intellectual celebrity.

That evening, Esther Salaman, a student studying theoretical physics, approaches Einstein's desk. After a moment's hesitation, the burning question escapes her lips, and she dares to ask the great scientist what it is exactly that he is looking for in his equations.

Einstein answers the student in little more than a whisper: "*I want to know how God created this world. I am not interested in this or that phenomenon, in the spectrum of this or that element. I want to know His thoughts; the rest are details."*[150]

God's thoughts

The phrase caught on. It traveled around Germany, enraging some members of the newly created National Socialist Party. In 1922, the Nazi terror was still a ways off. But that year, Einstein began receiving anonymous death threats by post. His fate, and that of a great number of scientists and experts on the evolution of the Universe, was already sealed.

Only when he came upon Soviet scientist Alexander Friedmann's famous article did Einstein at last began to suspect—though he did not yet admit it—that his own equations of General Relativity might describe an expanding Universe. They might mean that the Universe had a beginning! In the early 1920s, this foolish idea was nothing more than a rumor. But as we shall see, this faint spark would ignite into a fire greater than anyone could have imagined.

In December 1923, Einstein's shocking declaration about God made its rounds at the university. Privately, the students mocked their professor for mixing God and science. But it also ruffled the feathers of the Prussian professorial elite, backs ramrod straight under their wing collars. The majority had been gradually won over by the theories of the "völkisch" thinkers – that is, by German far-right populism—and so had become proponents of a pantheist or animist conception of the Universe, hostile to the idea of an intelligible God. This new mystical and anti-scientific view of the German race sought to return to pre-Christian Germanic sources in which nature is eternal. In this context, any ideas that placed limits on the power of the gods in German mythology—represented, for example, by Wagner's Wotan—became increasingly suspect.

---

150. In Esther Salaman, "A Talk with Einstein," *The Listener* 54 (1955): 370–371. Einstein's thinking on God is a complex question to which we devote an entire chapter later on (see Chapter 14).

The worst was yet to come

As we have already seen, on June 29, 1923, Einstein published a note in the famous scientific journal *Zeitschrift für Physik* in which he confirmed Friedmann's hypothesis that the Universe is not fixed. Even worse, he recognized that the equations of Relativity describe an expanding Universe.

The publication had made waves in the Soviet Union, but its reverberations were also felt in Germany, making their way to the ears of the leaders of the newly-formed Nazi party. Einstein saw threatening clouds gathering on the horizon.[151] Who was his main enemy? The agitator we met earlier, Adolf Hitler.

Hitler!

Although he didn't understand a word of it, he found its suggestion that matter is perishable disgusting. According to his close friends—Bormann, Heydrich, Himmler, Goebbels, and Speer—Hitler, seething with rage over the bad luck that had landed him in prison and profoundly influenced by party ideologues Alfred Rosenberg and Gottfried Feder, became increasingly viscerally materialistic. Despite his facade of Christianity, Hitler began to distance himself from both the Church[152] and Christianity[153] in 1925, as reflected in his book *Mein Kampf* and his public speeches. From the 1930s onward, under the influence of the violently anti-clerical Himmler, he rejected any idea of an immaterial God outside nature. Matter, he was now convinced, is eternal and any attempt by

---

151. Alexander Churkin, "The Life and the Memory": https://www.academia.edu/29326698/Esther_Salaman_The_Life_and_The_Memory.

152. "*It would correspond to the meaning of the most noble in this world if our two Christian churches, instead of annoying the negroes with missions which they neither wish nor understand, would teach our European mankind with kindness.*" Adolf Hitler, *Mein Kampf: Complete and Unabridged, Fully Annotated*, ed. John Chamberlain (New York: Reynal & Hitchcock, 1939), 607–608.

153. "The destruction of Christianity was explicitly recognized as a purpose of the National Socialist movement," Nazi leader Baldur von Schirach explained. Baldur von Schirach, quoted in Joe Sharkey, "Word for Word: The Case Against the Nazis; How Hitler's Forces Planned to Destroy German Christianity," *New York Times*, January 13, 2002, 899.

science to question this principle should be, in his own words, "*killed in the cradle.*"[154] It comes as no surprise, then, that the Jewish Einstein would become a target of the Nazi regime.

Einstein was more or less aware of all this. He had already been heckled by a group of nationalists for presiding over the third congress of the World Anational Association (SAT), held in the university town of Kassel, Germany, in 1923, with the aim of promoting universal, supranational bonds between peoples. The ultranationalists took this decision as a direct provocation. That same year, Einstein's difficulties began to mount. He had already weathered the furious attacks of experimentalist Johannes Stark, winner of the 1919 Nobel Prize in Physics. His goal, which he shared with many of his colleagues, was to utterly destroy General Relativity, the product of what they called "*Jewish physics.*" This group vaunted themselves as the only representatives of "*pure Aryan physics,*" and they harshly criticized Relativity for opening the door to the possibility of a non-eternal Universe. They remembered how the Russian mathematician Alexander Friedmann had burst through that door in 1922 by publishing his famous article on the Big Bang theory in the German review *Zeitschrift für Physik*. Stark and most of his colleagues were part of the National Socialist movement.

One Thursday evening, the professor has just finished his lecture. With his notes tucked under his arm, he hastens in short steps down the interminable corridors of the university. He is soon joined by his long-time friend, physicist Paul Ehrenfest, who is a Jew like himself. Ehrenfest speaks Russian fluently and for years gave courses at Saint Petersburg State University. He often met with the young Friedmann—one of his best students—and knows his model of an expanding Universe like the back of his hand. He is ready to defend it against all comers.

Soon, the two scholars hear patriotic chants coming from the inner courtyard. Next, far-off cries, as if a brawl has broken out in the darkness.

---

154. *The Speeches of Adolf Hitler: April 1922–August 1939*, ed. Norman H. Baynes (New York: Howard Fertig, 1969).

Slightly worried, Ehrenfest takes Einstein by the arm and leads him toward the exit. But Einstein is nervous. Something has to be done, he says, about the rise of anti-Semitism, which is growing stonger by the day. Ehrenfest nods his approval. But something else is bothering him: he confides to Einstein that a famous Russian mathematician named Friedmann might be right about his new solution to the equations of General Relativity. A dynamic solution demonstrating that the Universe is not fixed and that it had a *beginning.* Ehrenfest is completely convinced that Friedmann is right and had already broadcast this fact loud and clear a few weeks ago during a conference at the university in Berlin. It had earned him a flood of insults from a band of brown-shirted youths huddled in the back of the lecture hall. Even worse, he had been reprimanded by the vice-president of the university, who advised him to moderate his opinions. But what can be done? How can we not recognize that Friedmann's solution is the most logical and that we have to rethink the equations of Relativity? When he hears the word "*relativity,*" Einstein raises an eyebrow but is only half listening. To hell with calculations! What worries him is the violence against Jews. Against science.

Autumn 1925. Einstein learns that Alexander Friedmann—who two years earlier had verified the calculations demonstrating that the Universe had an origin—has just disappeared. Everyone at the university knows that he has been entertaining Friedmann's notion that the Universe is not eternal. One evening, Rosenberg told his Nazi comrades that "*that renegade Einstein*" is giving credence to the mad theories of these Russian scientists who believe matter came to be at some time in the distant past.

In fact, this same year sees the first serious rifts appear between Einstein and the German scientific community. Every day—at the university and elsewhere—he is forced to defend his ideas against a stream of vicious attacks. He ignores none of the increasingly bitter critiques raised against the theory of Relativity even after he won the 1922 Nobel Prize. Day by day, a growing number of extremist physicists view his scientific constructs as an insult to nature, as the regrettable result of "*typical Jewish*" thinking. Gradually the opposition organizes and begins to operate openly against Einstein. First, we have the particularly caustic assaults from the two German physicists mentioned earlier: Philipp Lenard, whose work on light quanta had inspired Einstein in 1905, and especially Johannes Stark, already mentioned, who had discovered the shifting or splitting of spectral lines in the presence of an electric field. These two well-respected Nobel laureates had become staunchly anti-Semitic. Further, since their studies focused entirely on matter, they were adamantly convinced that matter had neither beginning nor end.[155] Together they swore to "*purge science*"[156] of the very idea that atoms and other elementary particles were not eternal. And of course their first target was Einstein, who had dared to commit to writing that Friedmann was correct. For the adversaries of a finite Universe, this was a stunning blow. Relativity had done nothing less than usher in a new cosmology based on the concept of an expanding Universe, a cosmos finite in space and time.

Faced with this new danger, the cohort of Nazi ideologues was forced to look for an appropriate response. All the more so, as in 1927 Einstein would give a new provocation that would further radicalize the movement against him and the science he represented.

On October 29, 1927, the finest minds in theoretical physics gathered at the Hôtel Métropole in Brussels, an elegant nineteenth-century palace smelling of leather and wax. That year, eighteen Nobel laureates had met for the famous Solvay Conference, a meeting whose grand goal was to assist in the birth of a new physics. Einstein met a physicist in a cassock, the Belgian canon Fr. Lemaître, who had just put the finishing touches on a strange theory that suggested, like Friedmann's, that the Universe was expanding.

---

155. Klaus Hentschel, "Philipp Lenard and Johannes Stark: The Hitler Spirit of Science (May 8, 1924)," in *Physics and National Socialism*, ed. Klaus Hentschel and Ann Hentschel (Basel: Birkhäuser Verlag, 1996).

156. "*Lenard believed deeply that space was filled with a material, an 'ether' through which electromagnetic waves traveled,*" from the publisher's review of Bruce J. Hillman, *The Man Who Stalked Einstein: How Nazi Scientist Philipp Lenard Changed the Course of History* (Guilford, CT: Rowman and Littlefield, 2015, https://jewishbookworld.org/2015/07/the-man-who-stalked-einstein-how-nazi-scientist-philipp-lenard-changed-the-course-of-history-by-bruce-j-hillman/.

Einstein was strangely moved by this encounter. It was the second time that a physicist besides Friedmann had spoken to him about the possibility of a cosmic origin. But he was not in the mood to listen today. Earlier that morning he'd had a furious discussion with the formidable theoretician Niels Bohr, who that very year founded a brand new mechanics of the infinitely small: quantum theory. Bohr had become his worst enemy. As the tone of the debate between the two giants of physics became more heated, Einstein had abruptly concluded: *"God does not play dice!"*[157]

Niels Bohr (1885–1962)

Once more, the catchphrase made the rounds and proved impossible for partisans of the Nazi regime to bear. Stark and Lenard knew that Einstein had an enormous influence on the whole world. To permit him to say that God governs the Universe was therefore unthinkable. But what could they do? Yet upon reflection, the father of Relativity did have one vulnerable point: he was Jewish. This fact made their job easier.

Above all, they had to retake the upper hand. As good physicists, Stark and Lenard knew it would be difficult to eliminate Einstein and his ideas. To assure the triumph of their "Aryan Physics," the two agents of the regime launched their anti-Einstein campaign by founding the "German Scientific Work Group." Its sole aim: to destroy "Jewish physics." On the evening of December 11, 1927, the famous working group opened their inaugural conference with great fanfare in the hall of the Philharmonie

---

157. Letter to Max Born, one of the fathers of quantum mechanics, 4 December 1926. Published in *The Collected Papers of Albert Einstein Volume 15: The Berlin Years: Writings & Correspondence, June 1925-May 1927 (English Translation Supplement)*.

Berlin, intending to purge physics once and for all of all Jewish theories like those of Einstein. A serious attack to come from two Nobel laureates. Curious to understand his enemy's motives, and, more importantly, their arguments, Einstein decided to attend the famous conference in the company of physicist and chemist Walther Nernst (himself a 1920 Nobel laureate). Concealed in the back of a private box, the Professor listened attentively to every word as colleagues in the first throes of Nazi delirium dragged his reputation through the mud. These fanatics accused him of deceiving the human race with false beliefs.

Concerned for his personal safety, Einstein asked Nernst to accompany him back to Haberland Strasse. He spent the walk back plunged deep in thought, uneasily recalling how in 1922 his friend Walther Rathenau, then a minister in the Weimar Republic, had been murdered in his car, merely because he was a Jew.

America, January 29, 1931. Einstein has just arrived at the peak of Mount Wilson in California, more than 2,000 meters above sea level

Einstein is exhilarated by the fantastic photos that the American astronomer Edwin Hubble has spread out before him. They were taken using the giant Mount Wilson telescope with its 100-inch lens. He has even glued his eye to the eyepiece of the colossal instrument. And the verdict was clear: the Universe is indeed expanding.

But far from convincing the Nazi ideologues, Hubble's fantastic discovery had radicalized them even further. Their only objective was now to convince the German people at all costs that General Relativity was a deception based on a jumble of errors. A dozen determined physicists got to work. Once again, Stark and Lenard proposed an ideal solution to counter Einstein—a lethal blow from which they hoped he would never recover.

October 1931, Berlin. One morning, while Einstein was at his home at 5 Haberland Strasse, the faithful Ehrenfest knocked on his door. Visibly agitated, he waved under the Professor's nose a fat volume entitled *Hundert Autoren gegen Einstein—One Hundred Authors against Einstein*.[158]

Einstein leafed through the publication and furrowed his brow. Not a single well-known scientist had contributed to the volume. But the pamphlet's tone was inversely proportional to the scientific quality of its authors. Riddled with vitriolic personal attacks against Einstein, the work exuded breathtaking aggression. But it was also careless. Once more the father of Relativity was treated as an ignoramus and a traitor to German science, his theory consigned to the flames. Among other choice obscenities, Einstein found himself portrayed as an *"ass"* and a *"weathercock"* who had once defended the idea of an eternal Universe and now had reversed course, pretending to believe the opposite. Shrugging his shoulders, the Nobel laureate returned the volume to Ehrenfest and replied gravely, *"To defeat relativity one does not need the word of one hundred scientists, just one fact!"*[159]

Einstein may have dismissed the work with a wave of his hand, but the publication's effects spread like wildfire. A few months later, he was removed from the Bavarian Academy of Sciences, under the excuse that *"his erroneous teaching had perverted the German people."* Next, Friedrich Wilhelm University closed its door on him definitively.

Pressure against those who defended relativistic cosmology began to mount outside the sacred groves of academe. Realizing that simply writing a book would not be enough to eliminate their sworn enemy, the advocates of Aryan Physics decided to shift their campaign into a higher gear.

On an autumn night in 1932, Einstein and Ehrenfest met as usual in the Kaffeehaus. Suddenly, two baton-wielding figures appeared out of the darkness, banded and black-booted men dressed in the brown

---

158. *Hundert Autoren gegen Einstein*, ed. Hans Israel (Innsbruck, Germany: University of Innsbruck, 2012, reprint).

159. Cited in https://www.britannica.com/biography/Albert-Einstein/Nazi-backlash-and-coming-to-America.

uniforms of the SA. Swiftly, one of them picked up a paver and hurled it through the window, shattering it and breaking the crockery on the scientists' table. Tied to the rock was a message that Einstein knew was meant for him: "*We'll have your hide, you dirty Jew!*"

But these were only the first dark clouds presaging a terrible storm. It would break one year later.

January 30, 1933, midnight. After brutally seizing power (supported by Ernst Röhm's SA street fighters), Hitler is elected chancellor of the Weimar Republic. His iron fist swiftly fell on Germany. Within a few weeks, he dissolves and burns the Reichstag, Germany's parliament building, and unleashes his Stormtroopers to murder or remove all his political opponents, in particular the previous chancellor, Kurt von Schleicher. Hitler's dream of making all Germany march together in goose step was about to become a reality. At the same time, the police apparatus was set to work crushing dissident scientists. There was nothing to stop him.

First stage: sweeping away deceptive theories about a cosmos limited in the past and in the future

Hitler already had a pretext close at hand: freeing Germany once and for all from the infection of "*Jewish science*," a science dominated by "*perverse manipulators*"[160] who dared to suggest that no race was superior to any other and that the eternity of nature is a myth.

Results were not long in coming. In January 1933, while Einstein was travelling in the United States, a bounty was put on his head. In Germany, his books were thrown in the street and burned alongside his effigy. Hitler would not budge: "*this filthy Jew*" had to be eliminated, together with theories that supported the idea that the Universe had an origin.

---

160. Adolf Hitler, *The Speeches of Adolf Hitler: April 1922–August 1939*, ed. Norman H. Baynes (New York: Howard Fertig, 1969).

But what did he propose to take its place?

Hitler ordered Alfred Rosenberg, the Party's ideological leader at the time, to find a solution. Like his master, Rosenberg was violently opposed to Christianity and any appeal to God. He believed only in the immutable superiority of the Aryan race and in the eternity of matter,[161] and it was with this harsh creed in mind that he turned to Philipp Lenard. Immediately Einstein's tireless adversary once more took center stage. With physicist Hermann Oberth, he fabricated a bizarre theory to please Rosenberg and the Nazi theorists, called the "World Ice Theory" or "*Glacial Cosmogony.*"[162]

Philipp Lenard (1862–1947)

Hans Hörbinger (1860–1931)

This pseudo-scientific cosmogony of "eternal ice" was forged in the beginning of the 20th century by an obscure engineer named Hans Hörbiger, with the support of amateur astronomer, Philipp Fauth. Hörbiger proposed that the Universe has existed for all eternity and will exist for all eternity, as it rests on the beginningless and endless rule of the eternal ice. An ever-renewed struggle of fire against cold that finds, with every cycle, ice the victor. Of course, this insane theory would have remained in the shadows—if it were not for Philip Lenard. As it happened Hörbiger made his task easier since, in his cosmogonic delirium, the engineer insisted

---

161. Alan Bullock, *Hitler: A Study in Tyranny* (London: Odhams Press Limited, 1955).

162. Ibid.

that from the heart of the cosmic ice, a superior race emerged—giants with blond hair and blue eyes, the seeds of a new humanity. Thus in 1932, turning a blind eye to the profound absurdity of the model, Lenard presented Hans Hörbiger's "sound German science" to Heinrich Himmler, a top official of the Third Reich and Hitler's right-hand man. Himmler, completely ignorant of science, immediately fell in love. As a result, in March 1933, the preposterous Glacial Cosmogony became the official doctrine of the Third Reich.[163] In the wake of this decision, Himmler decided to issue a new uniform to the SS soldiers under his command. His SS would wear—what else?—a "glacial" uniform! A sinister black outfit meant to evoke the eternal reign of the Third Reich at the heart of a Universe with neither end nor beginning.[164]

Meanwhile, upon returning from their voyage to America, Einstein and his wife Elsa docked at Antwerp on March 28, 1933, only to receive two pieces of bad news. First, a bounty had been put on Einstein's head, and there was real fear he would be assassinated. Second, the German Reichstag had just adopted a resolution that transformed Hitler's government into a de facto legal dictatorship. Einstein could never again set foot in Germany.

To make matters worse, the professor's summer retreat, a small house in the Brandenburg countryside, had been pillaged and sacked by the Nazis. This was too much! Einstein went straight to the German consulate to give up his passport, officially renouncing his German nationality. Three days later, he sent a scathing letter of resignation to the Prussian Academy of Sciences. When, on the following day, the Nazi authorities initiated a formal exclusion procedure against the scientist, they found he had anticipated them with his letter of renunciation. The physicist Max von Laue, who was present that day at the Ministry of Research, noted in his journal that the fury of the directors at having been outmaneuvered was *"indescribable."*

---

163. Peter Levenda, *Unholy Allliance: A History of Nazi Involvement with the Occult*, 2nd ed. (New York & London: Continuum, 2002), 197–200.

164. For the design of the new SS uniform, see Chris McNab, *Hitler's Elite: The SS 1939–45* (Oxford: Osprey, 2013), 90.

Will the escalation stop there? Of course not!

In April 1933, while he was living for several months in the small Belgian village of Coq-sur-Mer, Einstein was alarmed to discover that the new German government had adopted laws forbidding Jews—and also certain "suspect" non-Jews—from occupying official posts, particularly from teaching in the universities. Overnight, thousands of Jewish scientists were forcibly obliged to renounce their university positions.

On May 6, 1933, hoping to staunch this terrible hemorrhage, Max Planck met with Hitler in person. After shouting that "a Jew is a Jew," the Führer went on: *"Rather than arguing with me, why don't you go see Stalin? Jewish science is perverting thinking about the Universe and is trying to convince us that it has not existed forever!"* Planck tried to argue further, but Hitler roughly cut him off: *"People say I suffer occasionally from nervous debility. This is slander. I have nerves of steel."*[165] Planck was forced to beat a hasty retreat without achieving his goal. Science met with another disaster that day.[166]

But repression of those who refused to stomach the indigestible Cosmic Ice Theory was only beginning. At the end of March 1933, German mathematician Ludwig Bieberbach, fiercely anti-semitic and an extreme Nazi zealot, took matters into his own hands. First, he chased out all his Jewish mathematical colleagues at Friedrich Wilhelm University *"in the name of the party."* Next it was the physicists' turn. In April 1933, Einstein's works were among those targeted by the German Student Union. On April 27, in Berlin, without the slightest qualm, the SA and other henchmen of the regime threw thousands of books written by Einstein and other relativist astronomers onto a huge bonfire. Intoxicated by this spectacle, the arsonists under the orders of Röhm's men set fire to the houses of two astronomers

---

165. Translation taken from "Max Planck: My Audience with Adolf Hitler [May 6, 1947]," in *Physics and National Socialism: An Anthology of Primary Sources*, ed. Klaus and Ann Hentschel (Basel: Birkhäuser Verlag, 1996), 359–361.

166. James C. O'Flaherty, "Max Planck and Adolf Hitler." *AAUP Bulletin* 42, no. 3 (1956): 437–44. https://doi.org/10.2307/40222051.

accused of betraying the German workers. Applauding these crimes, the Nazi Minister of Propoganda, Joseph Goebbels, took the stage and proudly proclaimed: *"Jewish intellectualism is dead!"* As for Einstein, even though there was already a bounty on his head, a German magazine put him on a blacklist of criminals most dangerous to the regime, with a subtitle: *"Not yet hanged."* They offered an additional prize to the person who turned him in to the authorities, dead or alive. Shaken to the core, Einstein wrote to his friend Max Born, who had already prudently fled to England some months earlier: *"I must admit, however, that they have surprised me not a little by the degree of their brutality and cowardice."*[167] Day by day, the air around his Belgian refuge became more tense. Strange figures lurked on his street at night, and the death threats accumulated to the point that King Albert decided to place a cohort of policemen and bodyguards at his guest's disposal. Einstein found their constant presence oppressive. He could not take it any longer, and so he made a radical decision: to go to America and never again set foot in Europe.

October 7, 1933. The American liner *Westernland* weighs anchor, and the port of Southampton slowly disappears on the horizon. On board, Einstein turns his back on Europe for good, like many other scientists fleeing the barbaric folly that has taken over Germany and threatens to push the whole world over the brink.

2. The Nazi war machine targets the Big Bang

Nuremberg, September 5, 1934. For more than an hour, Hitler has been unleashing a unrelenting flood of anger upon the immense crowd, which stretches as far as the eye can see in the courtyard of the colossal monument designed by regime architect Albert Speer. The Führer would not have missed this grand spectacle for anything in the world. He had designed it as a pagan cult celebration at which, each year, he would speak directly to the people—to all Germany—in this grandiose style, like a god addressing his creatures at his feet.

---

167. Albert Einstein, "Letter to Max Born, May 30, 1933," quoted in Ute Deichmann, *Biologists Under Hitler* (Cambridge, MA: Harvard University Press, 1996), 22.

Between two flights of soaring rhetoric, he returns with a mix of delight and rage to what he considers one of his sweetest victories: the expulsion of the Jew Einstein from the Reich "*with a kick in the gut.*" A roar of enthusiasm rises from the crowd, rolling in waves to the immense balcony on which the orator is perched. The time for purification had arrived: Germany had freed itself of one of its worst enemies. The Führer launched into a mystical tirade, exclaiming, "*From this day forth, we will rule the Universe forever!*"[168]

With this astonishing statement began the second phase of the regime's offensive against those who denied that matter and eternity are one.

In contrast to the years before the Nazis came to power, the elimination of scientists now proceeded efficiently and methodically, starting with notes and pamphlets targeting Einstein's close associates. One of the first victims of this reign of terror was Otto Stern, an old travelling companion of Einstein. Enchanted by the theory of Special Relativity, he was only twenty-three years old when, in 1912, he decided to go to Prague to become Einstein's assistant at the university. When Einstein became the Chair of Physics at the prestigious Swiss Federal Institute of Technology in Zurich, he accompanied him as his assistant. He spent hours in discussions with Einstein, witnessed his progress towards the theory of General Relativity, and shared all of his ideas with him.

Stern's advice would prove invaluable to Einstein. While the brilliant scientist could navigate the currents of theoretical physics with ease, Stern had a practical mind. He was an unmatched experimenter who, using his equipment bristling with indicator needles and electroluminescent tubes, had succeeded in penetrating the most intimate secrets of matter. Among other things, as early as 1922 he had succeeded in quantifying "spin"—the rotation of elementary particles—and had also provided proof

---

168. Translated from Fréderic Rouvillois, *Crime et Utopie: Une nouvelle Enquête sur le Nazisme* (Paris: Flammarion, 2014). https://www.google.fr/books/edition/Crime_et_utopie_Une_nouvelle_enquête_ su/5-rjAgAAQBAJ?hl=fr&gbpv=1&dq=discours+de+nuremberg+5+septembre+1934&pg=P- T139&printsec=frontcover

of the wave nature of atoms. However, Stern is best known for measuring the "*magnetic moment*" of the proton, a feat that earned him the Nobel Prize in 1943. In 1923, he was named titular professor at the University of Hamburg and director of the Physics Institute; his performance in both positions was widely lauded in Germany and beyond.

But the Nazis cared nothing for his stellar career. One morning in April 1933, the university's vice-rector paid Stern a visit, flanked by two heavily armed SA officers. When he did not immediately open the door, one of the SA began beating it down. Surprised by the violence of the intrusion, the physicist hurried toward his visitors. Without one word of apology, the grave vice-rector handed Stern a piece of paper and told him that he was no longer director of the Physics Institute, effective immediately. As Stern tried to reply, the SA officer cut him off, warning him "*it's in your interest not to resist.*" Ill at ease, the vice-rector clarified: "*Nevertheless, this measure applies only to the Institute. Until further notice, you will remain a professor at the university.*"[169]

Until further notice! This made Stern's blood boil. He submitted his letter of resignation to the rector the following morning. Fifteen days later, he fled to America, never to return.

Many others share Stern's fate

The first was 1954 physics Nobel laureate Max Born, whose influence in the early 1930s was considerable. A brilliant teacher, he had directed the theses of the famous Robert Oppenheimer, Victor Frederick Weisskopf, Max Delbrück, and even Pascual Jordan, all leading physicists who contributed decisively to the establishment of quantum mechanics, a field concerned with the infinitely small. His assistants included the renowned researchers Werner Heisenberg, Wolfgang Pauli, Enrico Fermi, Eugene

---

169. Translated from Pierre Ayçoberry, *La Science sous le Troisième Reich* (Paris: Le Seuil, 1993).

Wigner, and Edward Teller, among many others. And yet as we shall see, Max Born had all the qualities necessary to displease the Nazi authorities.

He had begun his meteoric ascent in 1904, at the University of Göttingen in Lower Saxony—one of the foremost universities in the world at the time. There he met the teachers who became his mentors: the mathematicians David Hilbert, Felix Klein, and Hermann Minkowski (Einstein's former professor of mathematics at Zurich, whom he helped in the formulation of the geometrical structures of space-time). Gradually he became one of the most indispensable supporting actors on the Göttingen scene. Later, he met Einstein in Berlin. Dismissed due to respiratory failure in 1914, he was available to help Max Planck and Einstein with the difficult calculations of General Relativity. By 1925, Born had, like Stern, entered the inner circle of Einstein's immediate confidants.

It is no surprise that Born and Einstein were interested in the same subjects and had the same worldview decried by the Nazis as "*mystical.*" In 1922, Born was researching the physical constants of the Universe, the pure numbers at the basis of physical reality. In the long list of dimensionless quantities (constants that operate independently of the system of units employed), his attention was particularly drawn to the mysterious fine-structure constant, $1/137$, which quantifies the strength of electromagnetic force (responsible, among other things, for the good old light that illuminates the pages you are reading right now). The great German physicist was the first to discover that this irrational number with its incredibly fine-tuned series of decimals has a close relationship with the speed of light, the charge of an electron, Planck's constant (at the heart of the infinitely small), and even, strangely, with the non-physical constant π. How can this be?

After extensive discussions with his colleagues and countless calculations, Max Born reached a troubling conclusion: "*If alpha [the fine-structure constant] were bigger than it really is, we should not be able to distinguish matter from ether [the vacuum, nothingness], and our task to disentangle the natural laws would be hopelessly difficult.*

The fact however that alpha has just its value 1/137 is certainly no chance but itself a law of nature. It is clear that the explanation of this number must be the central problem of natural philosophy."[170]

He had said it all: the value of this constant is "*certainly no chance.*" Born drove the point home: to his mind, the number could be "*associated with a power of selection and organization*" operative in the Universe.

A "*power of organization!*" Born's phrase sounds quite similar to the "*mind of God*" evoked by Einstein in Berlin, 1922. Thus it also raised the eyebrows of Rosenberg, Röhn, and their uniformed minions.

Nevertheless, Born made a great comeback in 1923 when he was named titular professor of theoretical physics at the extremely influential University of Göttingen. Protected by his mentors, especially by the friendly mathematician David Hilbert, he advanced the physics of his age by leaps and bounds. And yet in May 1934, as he hurried down a university corridor, two members of the SA arrested him and dragged him unceremoniously before the university's new vice president. Dispensing with all courtesy, the stern functionary announced that all Born's activities were suspended and that he had two hours to vacate the premises. A few days later, under the strict surveillance of the SA, he fled to the United Kingdom—first to the University of Cambridge, and then to Edinburgh, where he succeeded Charles Galton Darwin (the grandson of the great naturalist Charles Darwin) as the chair of natural philosophy. He would receive the 1954 Nobel Prize there, rather than at home in Germany, where he had made most of his discoveries.

1934 marked a turning point in this history. It was the end of an age and the start of a darker and more worrisome period, of which the affairs at

---

170. Arthur I. Miller, *Deciphering the Cosmic Number: The Strange Friendship of Wolfgang Pauli and Carl Jung* (New York: W.W. Norton, 2009), 253.

Göttingen were emblematic. For half a century (1880–1930), this town in Lower Saxony had dominated the world of mathematics under the unchallenged reign of mathematicians Felix Klein and David Hilbert.

Hilbert

Thanks to his essential contributions to a broad range of mathematical disciplines, David Hilbert is one of the most influential mathematicians in history. He wrote the celebrated "Hilbert's problems," a list of twenty-three vexing mathematical questions, many of which remain unsolved to this day. In quantum mechanics, he lends his name to "Hilbert space." Rigorous to the point of obsession, but also abounding in imagination, he was limitlessly bold. One evening in 1934, he was invited to an official banquet in the presence of Bernhard Rust, the Reichsminister of Education, who turned to him with a friendly smirk and asked: *"So, Professor, how is mathematics going at Göttingen, now that it is free of Jewish influence?"* Hilbert did not respond immediately. Finally, shrugging his shoulders, he muttered: *"Mathematics at Göttingen? What mathematics?"*[171] Then he left the table.

The purge of intellectuals only intensified. Thus a number of eminent physicists who supported the theory of Relativity and a finite Universe found themselves—like Einstein, Stern, and Born—expelled from German territory. Such was the case of James Franck, Victor Francis Hess, Lise Meitner, Carl Gustav Hempel, and many others, all expelled at the end of 1933. Others preferred to flee like Einstein, including the legendary physicist and mathematician Hermann Weyl, who had succeeded Hilbert in 1930 as the Chair of Mathematics at Göttingen. While he was not a Jew, he was close to Einstein. Predictably, therefore, Rosenberg found his conception of a Universe finite in space and time intolerable and proclaimed to all Berlin that he *"considered it his personal duty to bring the renegade Weyl to account."* His vulnerable point was his wife, Helen. She was Jewish.

---

171. Igor and Grichka Bogdanov, *La Pensée de Dieu* (Paris: Grasset, 2012).

Soon Helen received the first anonymous letters with insidious threats. It was too much. Frightened by this incomprehensible hate and feeling the danger mount, Weyl took Helen by the hand and decided to leave the university under cover of night. They immediately embarked for America, where they joined Einstein at Princeton's prestigious Institute of Advanced Studies.

1936: Nazi anti-scientific repression takes a sudden, more dangerous turn

Until now, the regime's pariahs had been given the chance to flee. But not all would be so lucky. As the '30s wore on, it became increasingly difficult to leave the Reich. Passports were confiscated and fugitives arrested at the border. Soon, Himmler and Hitler (followed by Bormann, Eichmann, Rosenberg, Heydrich, and others) took a new line. From now on they would eliminate the Reich's enemies by sending them to concentration camps or—even worse—slating them for execution.

The mathematician Felix Hausdorff's tragic fate clearly illustrates this downward slide into darkness. Hausdorff was a brilliant mathematician, regarded as the father of modern topology; today's students of mathematics will recognize his name in the "Hausdorff spaces."[172] However, after first considering his ideas a nuisance, the Nazis later set them down as "disgusting." Deep in his calculations, Hausdorff took no notice of the impending danger and continued his scientific work as if nothing had changed. The first warning signs came in 1935, when he was suddenly removed from his position. His works were progressively condemned as "*un-German*," then "*harmful*," and ultimately "*vile*." But again he chose to close his eyes to these portents. Soon, his family's passports were confiscated and it became impossible for him to leave the country. One night in 1942, as he was working late in his office at the University of Bonn, police entered, threw him to the

---

172. In mathematics, an Hausdorff space is a space in which for any two distinct points there exist neighborhoods for each that are disjointed.

ground, and tied his hands behind his back. His wife and sister-in-law received the same treatment. That same night, they were taken to the Bonn transit camp.

They would not survive the night. On January 26, 1942, Hausdorff and the two people most dear to him swallowed barbiturates and died under the guard of their tormenters.

These tragic events were not isolated incidents. Dozens of brilliant scientists were ruthlessly eliminated by the barbaric regime. This systematic campaign of violence aimed to eradicate not just men but their ideas. The detested proposition that neither the Universe nor nature had existed forever, so ruthlessly repressed in the Soviet Union, was among the chief targets of Nazi ire, since it meant that the Germanic race could no longer appeal to an eternal destiny. The fate of mankind was, in the final analysis, inscribed within something much greater than itself.

But the fall of the twentieth century's two great totalitarian regimes did not end this violent and scarcely comprehensible struggle against the idea of a cosmic origin. The following pages will tell the surprising story of repression that continued even after the end of the Second World War.

III. The Big Bang in the West after 1945

Autumn 1945. A fresh wind now blows across the world, dissipating the haze of murderous chaos that wracked Nazi Germany and snuffed out one hundred thousand Japanese lives in the blink of an eye. Slowly, humanity recovered from the deluge of fire and iron that had washed over it during wartime. Ideas gradually began to flow and circulate once again in their proper channels.

And yet, the hypotheses about the origin of the Universe continued to struggle to make headway. There were, of course, no more efforts to eliminate, imprison, torture, or execute the adherents of cosmic origin theories, as had occurred under the totalitarian regimes. Efforts to limit the diffusion of ideas about the beginning of the Universe set off by a transcendent cause became much more insidious and invisible. But they were no less ferocious in their case against its originators.

And Friedmann? Missing and forgotten. Lemaître? Nearly no one believed his theory of the primitive atom. In 1948, George Gamow (Friedmann's student, you will recall), tried to clarify matters, but he met with little success. Gamow and his doctoral student Ralph Alpher published the famous article in the *Physical Review,* mentioned above, "The Origin of Chemical Elements." They claimed that the Universe began billions of years ago when it was very small, very dense, and very hot. More precisely, they told their audience (most of them skeptical) that the light elements, especially hydrogen nuclei, emerged in the first few minutes after the Big Bang. This article caused a sensation that hit the newsrooms of all the leading papers. For example, on April 1 1948, stunned Americans read the following arresting headline in the *Washington Post*: "*World Began in Five Minutes!*"

Would the faithful proponents of the Big Bang now come out of the woodwork? Not really. The sensational article on the nucleosynthesis of the light elements had the great misfortune of appearing on April 1, 1948. The idea that the article was nothing more than an April Fools' joke spread quickly, deeply coloring its reception.

The Big Bang theory tottered. What came next was the coup de grâce. On March 28, 1949, Sir Fred Hoyle, the influential astronomer of His Majesty King George VI at St. John's College, Cambridge, was once more in his London studio at the BBC. Unlike Gamow, who was ignored by journalists, the witty and humorous Hoyle was a media darling. When asked whether he agreed with Gamow that the Universe could have begun in a distant past, he was at no loss for words. A beginning? Hoyle threw back his head and heaved a long laugh that echoed in the studio and broadcast to antennas around the world. Catching his breath, he shifted to a confident tone, as if to assure his audience that Gamow was spinning an enormous hoax. When pressed with questions, he doubled down, swelled his chest, and laughed sarcastically: "*The cosmos didn't begin with some sort of... 'Big Bang'!*"

A Big Bang!

Once launched, the expression shot like a firework all across England and soon around the whole world. The catchy phrase left a permanent mark on scientific vocabulary, since it was ironic and came from a sworn adversary of the idea of an origin to the Universe.

The royal astronomer did not back down. "The Universe is eternal!" Debonair and smiling, he repeated loud and clear that everything Gamow and his students were saying to the poor Americans who listened to them was so much smoke and mirrors, an elaborate ruse calculated to fascinate mere dilettantes in the grove of academe. Firmly convinced he was right, Hoyle rolled up his sleeves and went to work. Through a slew of conferences and articles, he managed to impose his theory of a static Universe—one that is fixed and has existed always, without beginning or end—on England and the rest of the world.

This was a grievous blow to Gamow and his disciples, but it affected many others as well. Down the gentle slope of the 1950s, the idea of a cosmic origin receded into the distance, drowned in the same wave that had swamped the Big Bang.

The Big Bang notion slowly faded from public discourse and even private conversation. Soon, almost no one was talking about it. In peacetime America, as much as in the rest of the war-torn world, people were settling into new rhythms of peace and prosperity. They wanted nothing to do with a band of crackpot intellectuals and their farfetched musings on the possible creation of the Universe. At any rate, the proof for the so-called "Big Bang" was not convincing enough. Thus schools and learned societies everywhere kept drilling it in: matter is eternal, and the Big Bang is a farce. Pushing the rejection of a beginning to the point of caricature, David Bohm went so far as to claim that the partisans of the Big Bang *"effectively turn traitor to science, and distort scientific facts to reach conclusions that are convenient to the Catholic Church."*[173] The British

---

173. Letter 60 to Miriam Yevick (Folder C116), November 30, 1951, in *David Bohm: Causality and*

physicist William Bonnor did not mince words: "*The underlying motive is, of course, to bring in God as creator. It seems like the opportunity Christian theology has been waiting for ever since science began to depose religion from the minds of rational men in the seventeenth century.*"[174] As we have already seen, Sir Arthur Eddington, one of the greatest astronomers of the first half of the twentieth century, was equally insistent and seemed to come unhinged when he heard the term Big Bang: "*Philosophically, the notion of a beginning of the present order of Nature is repugnant to me.*"[175]

David Bohm (1917–1992) denounced the "traitors to science who distort scientific truth to reach conclusions that are convenient to the Catholic Church"

The violence of these counter-attacks will mark a turning point in the lives of these researchers who, at the time, teetered under the blows of their adversaries

Deeply shaken by the barrage of criticism, George Gamow began to disengage from physics and move towards biology. And Ralph Alpher, his best student? Universities shut their doors in his face. After skimming the classified ads, he ended up working for General Electric. As for their companion Robert Herman, he turned his back forever on research to pursue a more prosaic career at General Motors. The man who predicted the Big Bang spent his days designing smaller cars to reduce traffic congestion. Barely a decade later, however, the famous proof was about

---

Chance, Letters to Three Women, ed. Chris Talbot (Heildelburg, Germany: Springer International, 2017).

174. W. B. Bonnor, *The Mystery of the Expanding Universe* (New York: Macmillan, 1964), 117.

175. See J. Stachel, "Eddington and Einstein," in *The Prism of Science*, ed. E. Ullmann-Margalit (Dordrecht: D. Reidel, 1986), 2:189. Eddington made this comment in 1931.

During the 1960s in the United States, George Gamow (1904–1968), Ralph Alpher (1921–2007), and Robert Herman (1914–1997) are forced to abandon physics when faced with the rejection of the Big Bang theory

to explode: "*something*" had created the Universe in the very distant past. In 1964, Penzias and Wilson discovered the "fossil radiation" of the CMB, the mysterious echo of creation that would become the first major evidence for the Big Bang.

And today?

On January 25, 2018, the prestigious *Journal for the Scientific Study of Religion* published a compelling article entitled "Perceptions of Religious Discrimination among U.S. Scientists." In the article, Dr. Elaine H. Ecklung and Dr. Christopher P. Scheitle, both university professors, show that American scientists who identify as religious "*report higher rates of religious discrimination relative to those who do not identify with a religion.*" Even today, scientists who identify as religious are at greater risk of encountering resistance and mistrust in the workplace than are scientists who profess no religious affiliation.

This robust study was conducted with a sample of 879 biologists and 903 physicists belonging to American research institutions, as classified by the National Research Council. It found that 33.8% of biologists and physicists who identify as Roman Catholic say they been subjected to more or less explicit discriminatory practices in the context of their labs or research teams, and 40.3% of researchers who identify as Protestant claimed to have regularly felt pressured or even been sidelined in their work.

The violent, dark story of the reception of the Big Bang theory and the weakness of recent alternative theories only serve to underscore the strength of evidence for the expansion of the Universe, its heat death, and its beginning. At this point in our argument, it is clear that the first two implications of materialist theory are both false. The Universe did indeed have a beginning, and it will end. But a creator God's nonexistence has a third implication, namely, that we should not expect the Universe to be fine-tuned with conditions hospitable to the emergence of life. In the following pages, we will show that this third implication is also contradicted by reality.

9.

The Fine-Tuning of the Universe

You sit down at the breakfast table to read your favorite newspaper over a cup of coffee. As usual, there is plenty of bad news. Global temperatures are projected to climb another deadly degree or two over the next few years, accelerating the melting of the polar ice caps, raising sea levels and causing climate disasters all over the world. On the next page, there is a worrisome report on just how fragile the thin layer of ozone protecting us from the deadly radiation of the Sun has become. On the last page of today's issue, you find devastating statistics about how herbicides are decimating the bee population. If the bees disappear, then plants won't get pollinated, and most of the world's vegetation will go extinct. During your commute to work, you reflect on the extreme fragility and sensitivity of all these precisely-regulated processes, each one so essential to our life on Earth. Isn't it incredible that our existence requires such a specific temperature? Or that terrestrial life requires such a specific proportion of oxygen in the atmosphere, with an ozone layer precisely this thick? With even a slightly lower percentage of oxygen in the atmosphere, life would be impossible. And at a slightly higher ratio, everything on the face of the earth would burn up!

Perhaps a number of other vital parameters come to mind, such as the Earth's protective magnetic field or the perfect inclination of its axis of rotation. All at once the question occurs to you: could such precise fine-tuning be proof of the existence of a creator God? You might think that while it's true that the conditions necessary for life on Earth are very numerous and very precisely regulated, the observable Universe probably has millions of billions of billions of other planets. Isn't it statistically likely that some of these other planets have conditions just as precisely optimized for life as those of our planet? But what about the Universe as a whole—does its existence also depend on such precisely regulated conditions? That would be very interesting indeed. Since there is only one Universe, maybe chance isn't such a satisfactory explanation. In fact, many great thinkers have asked the same question you have.

If the Earth's fine-tunings surprise you, the fine-tuning of the Universe will blow you away.

To dive into the fantastic fine-tuning that made possible the evolution of the Universe and eventual emergence of life is to discover that chance is not a credible explanation. This revolutionary new realization underlies what is known as the problem of the fine-tuning of the Universe

> "*I cannot imagine how the clockwork of the universe can exist without a clockmaker.*" Voltaire is thought to have said these words in the eighteenth century, when scientific understanding about the world and its origin was still embryonic. Extraordinary discoveries over the past few decades have sharpened our understanding of cosmic evolution, making his analogy of the clockmaker even more timely. The Universe now looks like a big "*setup*,"[176] an incredibly precise piece of machinery whose every part shows stunning fine-tunings of design—complex cogs that mesh together miraculously, creating the conditions necessary for the existence and functioning of the whole.
>
> Someone who takes only a quick glance at his watch may feel no wonder at how perfectly its hands turn and mark the hour. But if he cares to open the case to inspect the mechanisms and complicated machinery

---

176. Fred Hoyle's expression, cited in Paul Davies, *Superforce: The Search for a Grand Unified Theory of Nature* (New York: Simon and Schuster, 1987).

that make up this delicate instrument, then the evidence stares him right in the face: the watch must have had an intelligent designer.[177]

In this chapter, join us as we pry open the case of the massive watch that is the Universe to marvel at the delicate fine-tuning that keeps it running. Our inspection will reveal how the initial data and physical constants that have regulated the Universe since its origin are unlikely, ultraprecise, and extremely sensitive.

In the words of George Smoot, 2006 Nobel Laureate in Physics: *"The big bang, the most cataclysmic event we can imagine, on closer inspection appears finely orchestrated."*[178]

What is fine-tuning? The origin, evolution, and continued functioning of the Universe depend on around twenty parameters or constants. The values of these numbers were fixed at the moment the Universe came into being and are invariable in time and space:

Here are the main ones:

- **The force of gravity**, defined by the constant
  $G = 6.67418 \times 10^{-11}$ m³ kg⁻¹ s⁻² with the coupling constant $\alpha_g = 10^{-39}$
- **The electromagnetic force**, defined by the fine-structure constant
  $\alpha = 0.0072973525376$
- **The strong force (or strong interaction)**, which assures the cohesion of atomic nuclei and of nucleons, defined by the coupling constant
  $\alpha_s = 1$
- **The weak force (or weak interaction)** within the atomic nucleus, defined by the coupling constant $\alpha_w = 10^{-6}$
- **The speed of light:** $c = 299{,}792{,}458$ m/s⁻¹
- **The Planck constant:** $h = 6.626070040 \times 10^{-34}$ J/s
- **The Boltzmann constant:** $k = 1.380649 \times 10^{-23}$ J/K⁻¹

---

177. A Darwinian objection doesn't work in this case, because the parameters that determine the Universe have been fixed and unchanging from the moment of its origin.

178. In George Smoot and Keay Davidson, *Wrinkles in Time: The Imprint of Creation* (London: Abacus, 1995), 135.

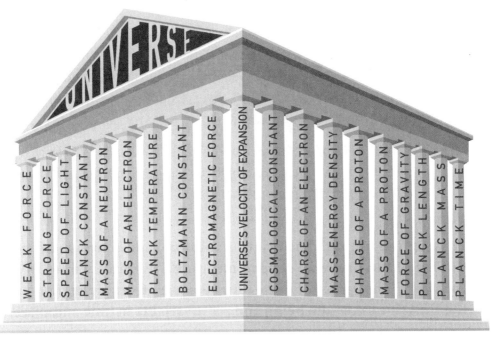

As the Parthenon stands with its columns, the Universe is based on at least twenty numbers, or at most thirty

- **The charge of the proton (+) and the electron (-):**
  $e = 1.6021766208 \times 10^{-19}$ C
- **The mass of the proton:** $1.6726219 \times 10^{-27}$ kg
- **The mass of the neutron:** $1.674927471 \times 10^{-27}$ kg, or 0.14% greater than the mass of the proton
- **The mass of the electron:** $9.10938356 \times 10^{-31}$ kg
- **The mass-energy density** of the Universe at its origin
- **The velocity of expansion** of the Universe at its origin
- **The cosmological constant** that contributes to the acceleration of the Universe: 1.289×10^{-52} m^{-2} or 1.38×10^{-122} Planck units^{-2}
- **The density and mass** of the Universe just after the Big Bang
- **The speed of expansion** of the Universe just after the Big Bang

Where do these numbers come from? What is their purpose? What would the Universe have been like if they had been even slightly different?

Three questions. Three answers.

Where do these numbers come from? There are only two possible responses: either they are the result of chance,[179] or they proceed from the complex calculations of a highly intelligent creator God.

What do they do? The laws, constants, and initial conditions that these numbers describe shape every aspect of the material universe, including its evolution, from the very beginning.

What would the Universe have been like if they had been even slightly different? As surprising as it might sound, we can actually answer this question. For some of these numbers, a very slight variation by even a distant decimal point would have yielded an unrecognizable Universe, and we would not be here to talk about it. This is the essence of the "fine tuning" principle.

We should note in passing just what fantastic scientific progress we have made. Today, when we ask such crazy questions, we can expect to get good answers. With the help of powerful computers, scientists have been able to mathematically model the Universe and say what would happen if certain parameters had been different—if, for example, the twentieth decimal place of the initial density of the Universe were changed.

Let's take the force of gravitation as a familiar example

This force acts between the masses of every object in our Universe according to the formula discovered by Newton: $F = Gm_1m_2/d^2$. Its

---

[179]. For the thesis of chance to hold, one is forced to resort to the highly speculative hypothesis of an enormous number of parallel universes (over 10^{120}, a number that greatly exceeds the number of particles in our Universe), produced by a "universe generator" entirely unknown to us. In the next chapter, we will examine the plausibility of this "desperate" thesis. As Neil Manson aptly puts it: *"The multiverse hypothesis is alleged to be the last resort for the desperate atheist."* Neil A. Manson, *God and Design: The Teleological Argument and Modern Science* (New York: Routledge, 2003), 18.

value depends on the value of the constant G, which we can measure up to the fourth decimal:[180]

$G = 6.6743 \times 10^{-11}$ m^3 kg^{-1} s^{-2}

But why does G have this exact value? "*Its value cannot be explained by any existing theory*,"[181] the astronomer Jacques Demaret declares. Like all the other constants we have seen, it has an arbitrary and inexplicable value. Further, the force of gravity is fine tuned to 1 part in 10^{35} of a range extending from an upper value set as the strength of the strong nuclear force (the strongest of the forces), to a lower range of 0. If its ratio with the strong nuclear force were not close to 10^{39}, all life in the Universe would be impossible. The same holds for the twenty or so other constants that form the structure of the Universe, which are also incredibly fine tuned, as we shall see.[182]

This astonishing fact, verified by numerous computer modelings, is universally recognized today

The "*fine-tuning*" (to use Fred Hoyle's 1953 coinage) of the initial data, laws, constants, and structures of the Universe is an inescapable fact that practically no one contests today, as it is recognized by the greatest thinkers, often themselves agnostic, whose opinion therefore holds particular value.

- **Lee Smolin**, a materialist physicist, recognizes this strangeness and is just as astonished as everyone else: "*We must understand how it came to be that the parameters that govern the elementary particles and their interactions are tuned and balanced in such a way that a universe of such variety and complexity arises…*

---

180. According to the US National Institute of Standards and Technology: cf. https://physics.nist.gov/cgi-bin/cuu/Value?bg.
181. Translated from Jacques Demaret and Dominique Lambert, *Le Principe Anthropique* (Paris: Armand Colin, 1994), 2.
182. Geraint F. Lewis and Luke A. Barnes, *A Fortunate Universe* (Cambridge: Cambridge University Press, 2016), 108-109.

> *We should ask just how probable is it that a universe created by randomly choosing the parameters will contain stars... The answer, in round numbers, comes to about one chance in 10^{229}."*[183]

- **Richard Feynman**, 1965 Nobel Laureate in Physics, adds: *"We use the numbers in all our theories, but we don't understand them— what they are, or where they come from."*[184]

- Physicist **Paul Davies**, professor at Arizona State University, likewise confesses his discomfort: *"I belong to the group of scientists who do not subscribe to a conventional religion but nevertheless deny that the universe is a purposeless accident [...] the physical universe is put together with an ingenuity so astonishing that I cannot accept it merely as a brute fact. There must, it seems to me, be a deeper level of explanation. Whether one wishes to call that deeper level 'God' is a matter of taste and definition."*[185]

- **Paul Dirac** makes a similar argument at a conference in 1971: *"If physical laws are such that to start off life involves an excessively small chance, so that it will not be reasonable to suppose that life would have started just by blind chance, then there must be a god."*[186]

- The celebrated American astronomer **Allan Sandage**, winner of the prestigious Crafoord Prize in Astronomy, enters the fray with the following consideration. In the final analysis, he tells us, the mysterious tuning of the Universe leaves us no choice: *"I find it quite improbable that such order came out of chaos. There has to be some organizing principle. God to me is a mystery, but is the explanation for the miracle of existence, why there is something instead of nothing."*[187]

---

183. Lee Smolin, *The Life of the Cosmos* (Oxford: Oxford University Press, 1997), 44–45.

184. Richard Feynman, *QED: The Strange Theory of Light and Matter* (Princeton, NJ: Princeton University Press, 1985), 152.

185. Paul Davies, *The Mind of God* (New York: Simon and Schuster, 1992), 15.

186. Helge S. Kragh, *Dirac: A Scientific Biography* (Cambridge: Cambridge University Press, 1990), 257.

187. J.N. Wilford, "Sizing Up the Cosmos: An Astronomer's Quest," *New York Times* (March 12, 1991). https://www.nytimes.com/1991/03/12/science/sizing-up-the-cosmos-an-astronomer-s-quest.html.

- At the risk of upsetting his materialist colleagues, **Steven Weinberg** affirms: *"Life as we know it would be impossible if any one of several physical quantities had slightly different values."*[188]
- **Arno Penzias**, winner of the 1978 Nobel Prize in Physics, writes: *"Astronomy leads us to a unique event, a universe which was created out of nothing, one with the very delicate balance needed to provide exactly the conditions required to permit life, and one which has an underlying (one might say 'supernatural') plan."*[189]
- **Fred Hoyle**, English astronomer who coined the terme "Big Bang" and who was a contemporary of Robert Dicke, had already concluded at that time that: *"A common sense interpretation of the facts suggests that a superintellect has monkeyed with physics, as well as with chemistry and biology, and that there are no blind forces worth speaking about in nature. The numbers that we calculate from the facts seem to me overwhelming enough to make this conclusion almost indubitable."*[190] **He also summed up his position in a blurb from the magazine *Nature*:** *"The chance that higher life forms might have emerged in this way is comparable with the chance that a tornado sweeping through a junk-yard might assemble a Boeing 747 from the materials therein."*[191]
- **John Lennox**, Professor of Mathematics at Oxford, explains: *"The fundamental constants that govern what the Universe is like have to be within very, very finely defined limits in order to have a Universe like we have, in which carbon-based life is possible, and we call that the fine-tuning of the Universe. For example, the ratio of the electromagnetic force to gravity in the early*

---

188. Steven Weinberg, "Life in the Universe," *Scientific American*, October 1994, http://vv.arts.ucla.edu/teaching/software/lifeintheuniverse/early_proto/weinberg.html.

189. Quoted in, *Cosmos, Bios, Theos*, ed. Henry Margenau and Roy Abraham Varghese (La Salle, IL: Open Court, 1992), 83.

190. Fred Hoyle, "The Universe: Past and Present Reflections," *Annual Review of Astronomy and Astrophysics* 20:1–35 (1982): 16. https://www.annualreviews.org/content/journals/10.1146/annurev.aa.20.090182.000245.

191 From "Hoyle on Evolution," *Nature* 294 (12 November 1981): 105.

> *Universe has to be accurate to about $1/10^{40}$ in order that we can have the chemistry of the Universe as we now see it. Now $1/10^{40}$: to get some idea of that, let's imagine that we covered, say, the whole of Russia with small coins, and we built the piles of coins over the whole of Russia to the height of the Moon, and then we took a billion systems like that, and we painted one of the coins red and we asked you to blindfold a friend and go and find it. They'd have got about $1/10^{40}$ of finding it, so it's a very small probability."*[192]

- Similarly, **Max Planck**, the legendary founder of the science of the infinitely small that is quantum mechanics, affirmed: *"All matter originates and exists only by virtue of a force which brings the particles of an atom to vibration and holds this most minute solar system of the atom together…We must assume behind this force the existence of a conscious and intelligent mind."*[193]
- **Freeman Dyson**, physics professor at Princeton, admits: *"The more I examine the universe and the details of its architecture, the more evidence I find that the universe in some sense must have known we were coming."*[194]

Cosmic fine tuning is thus a centerpiece of our panorama of proofs for the existence of a creator God. There exists no reasonable materialist response to the extreme improbability of the Universe's structure and functioning.

We will first review the history that led to this discovery before describing in detail the astonishing improbabilities that mark the various stages of the development of our Universe.

---

192. John Lennox, "Design of the Universe," interview in Glad You Asked, small group evangelism resource (2008). https://www.youtube.com/watch?v=48BqosNvS5U.
193. Max Planck, "Das Wesen der Materie" ["The Nature of Matter"], speech delivered in Florence, Italy, 1944, in *Archiv zur Geschichte der Max-Planck-Gesellschaft* (Boltzmannstraße 14, D-14195 Berlin-Dahlem).
194. Freeman Dyson, *Disturbing the Universe* (New York: Harper & Row, 1979), 250.

I. The story of the discovery of the anthropic principle

John D. Barrow, Professor of Mathematics at Cambridge and expert on the anthropic principle, wrote the standard reference work on the subject in 1988, together with Frank Tipler. Entitled *The Anthropic Cosmological Principle*, this extraordinary, six-hundred-page tome details two hundred examples of fine-tuning.[195] In the book's introduction, Barrow and Tipler insist that the last few decades have witnessed the most impressive discoveries about the Universe ever made. Although the concept of the "anthropic principle" was not officially proposed until the 1970s (by Australian physicist Brandon Carter), the emergence of this new proof for God's existence took almost a century.

Robert Dicke (1916–1997)

It all began at the legendary Princeton University toward the end of the 1930s. There young physicist Robert Dicke first conceived the idea of incredible "settings" in our Universe that led inevitably to the birth and evolution of life. At a series of talks in 1937, Dicke heard the great Paul Dirac (1931 Nobel Laureate) speak about certain "*strange coincidences*"[196] between the great physical constants upon which the Universe is based: for example, the ratios of the cosmic ray to the proton ray and that of the force of gravity to the electromagnetic force are close to 10 to the power of 40. In that same year, Dicke had the opportunity to meet Einstein, who had been a professor at Princeton since 1933. On this question, the theorist of Relativity already held a clear position: "*God does not play dice.*" Chance does not rule the world.

---

195. John D. Barrow and Frank J. Tipler, *The Anthropic Cosmological Principle* (Oxford: Oxford University Press, 1996).

196. "*You could say that this is a strange coincidence. But it is quite difficult to believe it. One thinks that there must be a link between these two great numbers, a link that we cannot currently explain that we will be able to explain in the future.*" "A New Basis for Cosmology," *Proceedings of the Royal Society of London*, Series A 165, no. 921 (April 5, 1938): 199.

As the years went by and these speculations continued to swirl in his mind, Dicke kept coming back to the same question: *What role does chance play in the Universe?* This and similar thoughts would irresistibly lead Dicke to discover one of the most astounding principles in physics (even if he did not himself name it): the anthropic principle.

In the 1960s, Dicke's reputation at Princeton was on the rise. He sounded the depths of the Universe with radio-astronomical tools of his own invention. He had carefully read George Gamow's works on the hot Big Bang and the traces this primordial explosion had left in space. With the help of his student James Peebles and his colleagues Roll and Wilkinson, he made the first precise predictions regarding the existence of fossil radiation, which was ultimately detected in 1964 by Robert Wilson and Arno Penzias only a few miles from Princeton. Peebles would go on to win the 2019 Nobel Prize for revisiting and refining Gamow's calculations of this fantastic discovery, which Stephen Hawking considered to be "*the most important of all time.*"

While still in the thick of his research into the very first instants of the Universe, Dicke spent his spare time cataloguing a host of bizarre observations. In the first place, he was intrigued by the fact that the Universe's age—about fourteen billion years—is not a matter of chance. Why? Because the fundamental physical laws observed from the Earth link the lifespan of the Sun to the evolution of intelligent life. Dicke and several of his more audacious colleagues became convinced that the evolution of intelligent life had required the Universe to be at least as old as it is today.

One year later, Dicke took another step. This time he spoke out more vocally and to a wider audience, claiming that Fred Hoyle's witty remark that the Universe is a big setup had more to it than met the eye. He threw himself into his calculations. After a series of complicated equations, he concluded that the initial conditions at the moment of the Big Bang were tailored with dizzying precision and that our existence is something of a miracle. Naturally, his colleagues were somewhat unsettled by his conclusions. Many turned their backs on him. But the facts were there in the equations: even the tiniest variation in a single one of the numerous cosmological parameters at the basis of the Universe—a two instead of a three in the nth decimal place of a number—and neither space-time nor life as we know it would ever have emerged. For Dicke, the case was closed, and he never changed his mind: the Universe is not a result of chance.

Subsequently, a number of researchers and scientists would observe the same fact: the fine-tuning of the initial conditions, laws, forces, constants, and structures that make up our Universe are very finely adjusted, sometimes with breathtaking precision, and if this were not the case, we would not be here to speak of it.

II. What are these mysterious fine-tunings?

It is time to look into the details of the recent scientific discoveries that reveal this incredible adjustment of the initial data, constants, structures, and laws of our world. We will present a dozen examples, following the same stages of the development of our Universe that we did in Chapter 5 on the Big Bang.

1. In the beginning, the coupling constants that determined the relationships between the four fundamental interactions were very precisely fixed

Most scientists believe that at at the beginning of time,[197] that is, at the moment of the Big Bang, the four fundamental forces that shape our reality were united as one. Two of them—the strong and weak forces—act

---

197. After 10^{-43} seconds, which is, given our understanding of physics, the smallest conceptually useful unit of time, (0.0000000000000000000000000000000000000000001 seconds), our Universe had, according to the standard model of the Big Bang, a radius of 10^{-33} centimeters—in other words, billions of billions of times smaller than a hydrogen atom. In that instant, its temperature was unimaginably hot: 10^{32} degrees, or 100,000 billion billion billion degrees. Its energy was just as precise and immense: 10^{19} giga-electronvolts (GeV). But the most extraordinary thing is that in that primordial epoch where neither matter nor any structure or element existed yet, everything was already fantastically *ordered* and *calibrated*.

on the extremely small, at the scale of elementary particles. The other two—the electromagnetic and gravitational forces—produce significant large-scale forces (like those between bodies as great as stars).

These four forces sustain all life's processes at the most profound level. The strong force, for example, keeps the atoms of your body from scattering into a cloud of elementary particles. At the other extreme, the electromagnetic force is constantly at work in the depths of your brain, producing your thoughts and transmitting them from one neuron to another. The same phenomenon takes place in your muscles (in your heart, for example, whose pace is regulated by electric pulses). In sum, you can think and walk thanks to a force that came into being at the instant of the Big Bang, 13.8 billion years ago.

Now here is where things get interesting. These four forces that underpin everything you see around you—and also the entirety of the Universe—are themselves based on... numbers. More precisely, these are pure and dimensionless numbers (physicists call them "coupling constants"). These numbers are called "pure" because as the relative ratio of the four different interactions, they hold true independently of any unit of measure, and their value can only be discovered through measuring devices.

What exactly are our four fundamental forces and the numbers attached to them? Where do their respective values come from? It's a mystery. Their incredible fine-tuning still seems to be the result of a miracle.

If the strong force has a value of one, the electromagnetic force is 137 times weaker. Then comes the weak force (0.000001), a million times weaker than its cousin, the strong force. Finally, gravity plunges into an abyss: one thousand billion billion billion billion times weaker than the strong nuclear force. How can we explain this incredible but highly precise fall of exactly thirty-nine orders of magnitude, rather than forty-three or thirty-five? Why have these constants maintained these same values since the birth of the Universe, rather than some other value? Where do they come from? No matter how hard we look, it is impossible to find answers—so much so that in the final analysis, as the Nobel Laureate Richard Feynman remarked regarding the value of the pure number that is the basis for the electromagnetic force: "*It's one of the greatest damn mysteries of physics: a magic number that comes to us with no understanding by man. You might say the 'hand of God' wrote that number, and 'we don't know how He pushed his pencil.'*"[198]

At any rate, the extent and precision of this calibration is dizzying. The same holds for the other three pure numbers that correspond to the other three great forces of the Universe. Seized by the same emotion, the great theoretical physicist Dirac went beyond his militant atheism when he said in a 1963 issue of *Scientific American* something that soon became well known internationally: "*One could perhaps describe the situation by saying that God is a mathematician of a very high order, and He used very advanced mathematics in constructing the universe.*"[199]

Paul Dirac (1902–1984)

His observation is essentially very simple: if these four pure numbers that regulate the destiny of the four elemental forces of the Universe did not fall within a very narrow range, nothing we know would exist. Not the book you hold in your hands, not the flowers in your garden, not your dog or cat, not even our world. And the Universe as a whole would have generated only the lightest elements, like helium and hydrogen. None of the heavy elements capable of supporting life or complex chemistry could ever have been formed.

---

198. Richard Feynman (1918–1988), professor of Physics at Caltech, Nobel Laureate in Physics, and pioneer of quantum mechanics, in *QED: The Strange Theory of Light and Matter* (Princeton, NJ: Princeton University Press, 2006), 129.

199. Paul Dirac, "The Evolution of the Physicist's Picture of Nature," *Scientific American* (May 1963): https://blogs.scientificamerican.com/guest-blog/the-evolution-of-the-physicists-picture-of-nature/.

To conclude, notice that all the constants—all the cosmological parameters mentioned above—are said to flow from the breaking of the primordial unity between the four forces at the instant of the Big Bang.

---

2. In the first instant after the Big Bang, the ratio between the amount of energy in the Universe and its rate of expansion had to be fixed with phenomenal precision

At the instant of the Big Bang, the Universe already had a fixed quantity of energy and a precise, unchanging rate of expansion. Now, the ratio between the two is extremely important because if the Universe's expansion had been even slightly weaker at that moment, the newborn cosmos would have collapsed in upon itself under the force of gravitation well before it could reach its current size. Inversely, if the initial expansion had been even a bit more rapid, the stars would not have had time to form, and the Universe would now be composed entirely of gases.

"Bob Dicke told us that if you thought about the universe at one second after the beginning, the expansion rate really had to be just right to 15 decimal places, or else the universe would either fly apart too fast for any structure to form or re-collapse too fast for any structure to form," explains the famous cosmologist Alan Guth, Professor of Physics at MIT and one of the fathers of Inflation Theory.[200]

In addition, in order for the Universe to permit the generation of atoms, the stars, and complex life, first, according to the classical model, the initial density of the Universe had to be set within a very narrow range, tuned with a remarkable precision that we are now able to calculate.[201]

---

200. "10 Questions for Alan Guth, Pioneer of the Inflationary Model of the Universe," interview by Christina Couch, July 1, 2016. See also Alan Guth, "Inflation and the New Era of High-Precision Cosmology," *MIT Physics Annual* (2002): 2–39, https://physics.mit.edu/wp-content/uploads/2021/01/physicsatmit_02_cosmology.pdf.

201. If the real average density of the Universe were equal to a certain critical density, it would

One nanosecond (a billionth of a second) after the Big Bang, this relationship between the average density and the critical density of the Universe must have only differed from 1 to the twenty-fourth decimal place, that is, 1.000 000 000 000 000 000 000 001, as reported by Geraint Lewis and Luke Barnes.[202]

In the Planck era, 10^{-43} seconds after the Big Bang, this value must have been even closer to 1 and is written as 1. 000 000 000 000 000 000 000 000 000 000 000 000 000 000 000 000 000 000 001. This negligible deviation from 1 only appears at the sixtieth decimal. Shocked by such an incredible result, George Smoot could not refrain from bringing this to the attention of his most skeptical colleagues: *"That is so close to 1 that reasonable people think it is not simply a matter of chance—something requires that Omega is 1 to all those decimal places."*[203]

The Buddhist cosmologist Trinh Xuan Thuan also demonstrated that in order to explain the flatness or absence of curvature of the Universe, the mass density had to be fixed with a precision of the order of $1/10^{60}$! We might compare the improbability of this to *"the accuracy that would be required for an archer to be able to hit a 1 square centimetre target 15 billion*

Trinh Xuan Thuan, born in 1948

---

be flat (Euclidian); if the density were greater, it would present a closed curve (spherical); if it were smaller, it would form an open curve (hyperbolic). But today, we can use the WMAP and Planck astronomical satellites to determine that the Universe is "almost flat." If it were completely flat, that would mean that the relationship between its real and critical densities would be precisely equal to 1. Now, however, the average calculated density (on the order of 1.0002) is very slightly greater than this critical density, and this means that the space-time in which we live must have a closed form, that of a sphere with three dimensions. Consequently, one can imagine that, if we could charge straight ahead at a speed much greater than the speed of light, we would end up coming back to where we started, from behind.

202. Geraint F. Lewis and Luke A. Barnes, *A Fortunate Universe* (Cambridge: Cambridge University Press, 2016), 167.

203. George Smoot and Keay Davidson, *Wrinkles in Time* (New York: Harper Perennial, 1993), 161–162.

light years away." In other words, the probability is near zero.[204]

Alan Guth conceived of the moment of cosmic inflation (see point 6 in Chapter 6) as a solution to this enigma of the calibration of the flatness of the Universe, as well as other important questions faced by physicists (the absence of magnetic monopoles, the homogeneity of the Universe affirmed after 380,000 years, the order of greatness of distances between the star clusters of galaxies). Today, this solution is favored by the majority of cosmologists, though it has not yet been confirmed through experimental research. If confirmed, it would still require the extremely precise calibration of the sequence of events that from these first moments allowed for the emergence of a Universe favorable to life.

Let's be clear: this fact alone would justify calling the emergence of the Universe a miracle, rather than mere chance...but there is more.

3. The fine-tuning of the weak force allows antimatter to disappear

At the very beginning of the Universe, matter and antimatter were created in equal parts. What differentiates matter from antimatter is their respective charge (an electron has a charge of -e, an antielectron a charge of +e). These entities canceled each other out and produced energy in the form of radiation. If the initial ratio of matter and antimatter had been conserved, there would have been total destruction: colossal explosions of energy leaving nothing in the Universe but radiation. The Universe would have been sterile and devoid of matter.

Why isn't this what happened? This question went unanswered for a long time since matter and antimatter seem to behave symmetrically during interactions. However, an initial clue appeared one day in 1964 amidst research on the disintegrations of particles called kaons: the

---

204. Trinh Xuan Thuan, "Interview: Trinh Xuan Thuan Talks to Neda El Khazen," *The UNESCO Courier* 47, no. 5 (n.d.): 4–7, 50, illus.

discovery of a violation of this symmetry (called "CP violation"). But this effect was too small to explain by itself the quasi-disappearance of antimatter.

Recently a new avenue of research has opened up, resulting in an article published in *Nature* (May 15, 2020) devoted to this question.[205] The article focuses on the oscillations of neutrinos. Neutrinos, or antineutrinos, are emitted during the weak interaction, as is the case for the radioactive beta. It had long been believed that these particles had no mass. But then it was observed that they didn't have zero mass, but rather a minute mass that allowed for a phenomenon called oscillation. This discovery earned the Nobel Prize 2015 for Takaaki Kajita and Arthur B. McDonald. This phenomenon provided the conditions for the asymmetry between matter and antimatter.[206]

Admittedly, these very recent experiments still have yet to be refined and confirmed. However, we note that these avenues of research have to do with extremely fine phenomena (CP violation of quarks, the miniscule mass of neutrinos, the oscillations of flavor). In theory, all of particle physics would be able to function without them. Yet, they appear to contribute to the asymmetry of matter and antimatter. There could, however, be another explanation. If true, this explanation would entail phenomena so infinitesimal that they have not yet been discovered, even though particle physics has been capable of detecting incredibly precise phenomena (to the order of 10^{-11}) for 60 years. It is stupefying that such

---

205. "*Violation of this CP symmetry was first observed in 1964, and CP violation in the weak interactions of quarks was soon established. Sakharov proposed that CP violation is necessary to explain the observed imbalance of matter and antimatter abundance in the Universe. It has been shown that CP violation in leptons could generate the matter-antimatter disparity through a process called leptogenesis.*" "Constraint on the Matter–Antimatter Symmetry-Violating Phase in Neutrino Oscillations," Nature 580 (2020): 339–44.

206. "*Using these data, the T2K collaboration measured the probability that a neutrino would oscillate between different physical properties that physicists call 'flavours' during its journey. The team then ran the same experiment with antineutrinos, and compared the numbers. If matter and antimatter are perfectly symmetrical, the probabilities should be the same. The results, however, suggest they are not.*" *Nature* 580 (2020): 305. As the previously cited article states, "*Our results indicate CP violation in leptons and our method enables sensitive searches for matter antimatter asymmetry in neutrino oscillations.*"

an essential phenomenon to our existence as the disappearance of antimatter would find its source in causes so obscure that they have evaded the grasp of particle physicists for 60 years. This is another example of hyper-fine-tuning, one which turns out to be absolutely essential.

4. The masses of the fundamental structures of the Universe (electrons, protons, neutrons) are also adjusted to perfection

From 10^{-6} to 10^{-4} seconds, quarks, which had just formed a few moments earlier along with neutrinos and electrons, combined to form protons and neutrons. What caused these structures to appear? In short, the laws of the Universe caused their emergence, but those laws had to be minutely adjusted to do so. In fact, in 2015 a dizzying discovery showed that the mass of a proton (938.27 MeV) is nearly identical to the mass of a neutron (939.57 MeV): a mere 0.14% lighter. If this miniscule gap had been even slightly different, no life or even complex chemistry would exist, and therefore we would not be here.[207]

Fortunately, the neutron is slightly more massive than the proton, and this allows a free neutron to decay rapidly into a proton while the proton itself remains stable.

If the reverse were true, protons would decay into stable neutrons, fusion reactions would be based on neutrons, and the only matter in the Universe would be neutronium. The chemical elements would never form, and no life would be possible.[208]

These and numerous other instances of fine-tuning[209] prompted the

---

207. The nuclei of all hydrogen atoms would have been immediately transformed into neutrons and neutrinos, and so the structure of matter could not have been organized: instead of stars, the Universe would contain only fragments of atoms swirling in the darkness. The absence of hydrogen would render the synthesis of atomic nuclei impossible. Neither carbon nor oxygen nor any of the heavy elements indispensable to the existence of matter and life could have emerged. Such a Universe would have been populated here and there by scattered stars that went out quickly, like little candles, without generating a single planet.

208. Craig J. Hogan, "Quarks, Electrons and Atoms in Closely Related Universes," in *Universe or Multiverse?*, ed. Bernard Carr (Cambridge: Cambridge University Press, 2009), 221-230.

209. For example, the stability of the proton could have come about in another way: the reaction

famous English materialist Stephen Hawking to observe: *"The laws of science, as we know them at present, contain many fundamental numbers, like the size of the electric charge of the electron and the ratio of the masses of the proton and the electron... The remarkable fact is that the values of these numbers seem to have been very finely adjusted to make possible the development of life."*[210]

Already at this early stage of the Universe's development, we are surrounded by remarkable fine-tunings. Each one is stupefying enough in itself...but that's not all.

5. Keeping the highly unstable neutron going: another essential fine-tuning

Between 10^{-6} and 10^{-4} seconds, quarks combined to form protons and neutrons. Here again we encounter numbers that defy comprehension: a proton's lifespan could be 10^{34} years—10 million billion billion billion years—while a free neutron sitting right beside it dies after only a brief fifteen minutes. Therefore, after the Big Bang, the neutron had only fifteen minutes—exactly 878 seconds—to group together with the proton in the first atomic nuclei during the very brief period of first nucleo-

---

between a proton and an electron could have produced a neutron and a neutrino. What makes this reaction impossible is that the mass difference between the neutron and the proton is greater than the mass of the electron, which is 0.511 MeV. While a free neutron decays rapidly, the neutrons inside most atomic nuclei do not decay (as long as they are not submitted to beta radioactivity), ensuring the stability of these atomic nuclei. For this stability to be possible, the mass of a neutron has to be less than the sum of the mass of a proton, an electron, and the nuclear binding energy. This gives us another, higher limit for the mass difference between a proton and neutron, to the order of 10 MeV. In conclusion, the mass difference between a proton and a neutron must be absolutely restricted to a number between 0.511 and 10 MeV: this is precisely the case because its real value is 1.29 MeV. Researchers have calculated that the proton is exactly 938 million times more massive than the neutrino. While the difference is colossal, it is still extremely precise. In other words, if this huge mass difference was even one hundredth of a millionth larger or smaller, the Universe would have remained a formless mist of elementary particles that could never have evolved.

Further, we now know that the formation of deuterium depends on the crucial mass difference between neutrons and protons. Deuterium is necessary for the formation of helium, which is indispensable for the formation of carbon, without which, once again, we would not be here to talk about any of this.

210. Stephen Hawking, *A Brief History of Time* (New York: Bantam Books, 1998), 125.

synthesis. At the end of this short "Hadron epoch," further creation of protons and neutrons became impossible.

So what keeps these unstable neutrons in existence? Again, the answer is the velocity of the expansion of the Universe. Since the proton-neutron structure cannot survive above a certain temperature, the Universe has to expand at a rate rapid enough to diminish its average energy density, thereby quickly lowering the temperature.[211] Otherwise, all neutrons would disappear. Scientists estimate that the fine-tuning allowed hydrogen to predominate. With this regulation of the expansion, only one neutron in seven is preserved,[212] allowing the Universe to be open to life. If the expansion were a little slower, all neutrons would have disappeared, preventing the formation of heavier atoms; and if it had been a little faster, all protons would have bonded with neutrons—which would have been produced in equal numbers—to create helium nuclei, and there would have been no hydrogen, needed for the stars to exist.

In sum, without this unbelievably well-calibrated adjustment, we would not be here speaking about it.

6. The cosmological constant: one of the finest of fine-tunings

The "cosmological constant" that Einstein added to his Relativity equations in order to maintain the stability of the Universe but later recognized as the "*greatest mistake of his life*" after observing the expansion of the Universe (see Chapter 5), seems to exist after all, but it is very tiny. And this is where things get truly amazing. In calculating the value of this notorious constant, the positive contributions esti-

---

211. Bharat Ratra and Michael S. Vogeley, "The Beginning and Evolution of the Universe," *Publications of the Astronomical Society of the Pacific*, 120, no. 865 (March 2008): 235-265.

212. Alpher and Herman, then Fermi and Turkervitch, calculated the relative abundance of these primordial elements, and the theory fits like a glove. For helium 4, the theoretical prediction was 25%, and it measures 25%. For deuterium, the prediction was 0.000026%, and it measures 0.000026%. For helium 3, the prediction was 0.000011%, and it measures 0.000011%. For lithium 7, the prediction was 0.00000000047%, and it was recently measured at 0.00000000015%, thanks to the James Webb satellite. And cosmologists have clues to follow in accounting for even this tiny discrepancy.

.00000000000000000000000000000000000
00000000000000000000000000000000000
00000000000000000000000000000000000
000000000000000000000000000000000138

The cosmological constant (in Planck mass units)

mated from the standard model of particle physics and the negative contribution of the so-called "naked" cosmological constant cancel each other out to the 122nd decimal place, matching the value that it's possible to physically measure. In other words, after the decimal point come 122 zeros, and the first non-zero appears in the 123rd place. In a 2012 TED talk, Brian Greene projected an image of the number: 0.00000000000000000000000000000000000000000000000000000000000000000000000000000000000000000000000000000000000000000000000001 38. He commented: *"This number is small. Expressed in the relevant units, it is spectacularly small. And the mystery is to explain this peculiar number. We want this number to emerge from the laws of physics, but so far no one has found a way to do that."*[213]

To put it another way, the constant contributing to our Universe's expansion had an extraordinarily slim chance—one in a billion, billion, billion, billion, billion, billion, billion, billion, billion, billion, billion, billion, billion—of achieving precisely the right value.

So we are faced once again with an extremely measured and precise calibration. Ultra low in value, the cosmological constant corresponds to a very small "vacuum energy" that acts as a repulsive force against gravity and produces the acceleration of the expansion of the Universe observed today. If the value had been slightly larger, the Universe would have expanded too rapidly for stars and galaxies to form, ma-

---

213. Brian Greene, "Why Is Our Universe Fine-Tuned for Life?" filmed March 2012 in Long Beach, CA, *TED* video, 21:48, https://www.youtube.com/watch?v=bf7BXwVeyWw.

king the emergence of any form of life impossible. Conversely, if the value had been slightly smaller, the cosmos would have collapsed in on itself long ago.

Leonard Susskind, Professor of Physics at Stanford and highly reputed materialist cosmologist, expresses his astonishment: "*Most fine-tunings are 1% sorts of things—in other words, if the thing is 1% different, everything is bad, and the physicist could say, 'Maybe those are just luck.' On the other hand, this cosmological constant is tuned to one part in ten to the one hundred twenty power ($1/10^{120}$). Nobody thinks that's accidental… That's the most extreme example of fine-tuning.*"[214]

Confronted with such precise and delicate tuning, even prominent atheist cosmologists have started to refer to it as a "miracle."

- Atheist **Stephen Hawking** used the word explicitly: "*The most impressive fine-tuning coincidence involves the so-called cosmological constant […] Physicists have created arguments explaining how it might arise due to quantum mechanical effects, but the value they calculate is about 120 orders of magnitude (a 1 followed by 120 zeroes) stronger than the actual value, obtained through the supernova observations. That means that either the reasoning employed in the calculation was wrong or else some other effect exists that miraculously cancels all but an unimaginably tiny fraction of the number calculated.*"[215]

- **Larry Abbott**, professor at Brandeis and another atheist cosmologist, declared: "*There must in fact be a miraculous conspiracy occurring among both the known and the unknown parameters governing particle physics […] the small value of the cosmological constant is telling us that a remarkably precise and totally*

---

214. Leonard Susskind in Sir Martin Rees, "What We Still Don't Know: 'Are We Real?'" *television episode*, 48:28, aired 2004 on Channel 4 (UK), https://youtu.be/oyH2D4-tzfM, beginning at 17:17.

215. Stephen Hawking and Leonard Mlodinow, *The Grand Design* (New York: Bantam Books, 2010), 161–162.

unexpected relationship exists among all the parameters of the standard model, the bare cosmological constant and unknown physics."[216]

- **Robert Laughlin**, Professor of Physics at Stanford, 1998 Nobel Laureate, and an atheist, claimed: *"The fact that it is so small tells us that gravity and the relativistic matter pervading the universe are fundamentally related in some mysterious way that is not yet understood, since the alternative would require a stupendous miracle."*[217]

- **Alexander Vilenkin**, the openly materialist director of the Tufts Institute of Cosmology, is equally fascinated: *"The cosmological constant problem is one of the most intriguing mysteries that we are now facing in theoretical physics."*[218]

The precision of the tuning behind the cosmological constant is jaw-dropping. But there is still more.

7. The tuning of the strong and weak nuclear forces is equally astonishing and indispensable

Between one second and fifteen minutes after the Big Bang, the nucleosynthesis of all the light elements took place once and for all: hydrogen (92% of the matter in today's Universe), helium (8%) and deuterium (which could only be created in the extreme conditions present during the first minutes of the Big Bang), and a small quantity of lithium, beryllium, and boron.

The synthesis of the first atomic nuclei was possible thanks to the strong

---

216. Larry Abbott, "The Mystery of the Cosmological Constant," *Scientific American* 258, no. 5 (May 1988): 112, https://www.jstor.org/stable/24989092.

217. Robert Laughlin, *A Different Universe: Reinventing Physics from the Bottom Down* (New York: Basic Books, 2005), 123.

218. Alexander Vilenkin, "Anthropic Approach to the Cosmological Constant Problems," *International Journal of Theoretical Physics* 42, no. 6 (June 2003): 1193.

nuclear force, a short-range interaction one thousand billion, billion, billion, billion times greater than the force of gravity. The weak force also had to be precisely set to allow life. Stephen Hawking explains: *"For example, if the other nuclear force, the weak force, were much weaker, in the early universe all the hydrogen in the cosmos would have turned to helium, and hence there would be no normal stars; if it were much stronger, exploding supernovas would not eject their outer envelopes, and hence would fail to seed interstellar space with the heavy elements planets require to foster life."*[219]

Incredible, isn't it? If you thought that there couldn't be any more surprises in store, you'd be wrong. There is yet another reason to be amazed at the Universe's mysterious tuning at the moment of its emergence, a tuning some scientists don't hesitate to call "supernatural."

8. The tremendous conservation of beryllium, which enables the generation of the carbon essential for life

The conditions behind the generation of carbon at the heart of stars are even more impressive. To understand these conditions, let's look at the conclusion drawn by Nobel Prize-winning physicist Steven Weinberg—an avowed agnostic—in his 1992 bestseller, *Dreams of a Final Theory*. His book came like an electric shock, igniting a furious debate among the scientific community. One of the phenomena he presented was simply beyond our power of comprehension. All the fuss was over the strange behaviour of beryllium 8. The most common isotope of carbon, carbon-12, forms within stars in a two-step process: first, two helium nuclei collide and combine to form the highly unstable isotope beryllium-8, which has a half-life of 0.0000000000000001 seconds. If beryllium-8 does

---

[219] Steven Hawking and Leonard Mlodinow, *The Grand Design* (New York: Bantam Books, 2010), 160. The reason is that in this case, hydrogen would disappear in a few minutes and no elements lighter than iron would form. But if the strong force were weaker, no element heavier than hydrogen or helium could form. Finally, in neither hypothetical situation could any molecules form, and so the evolution of our Universe would have been stopped in its tracks. The strong nuclear force deserves, therefore, to be counted among the parameters without which the Universe could not have formed and which are controlled by extremely fine mechanisms.

not absorb another helium nucleus within this incredibly short timeframe, it decays and becomes unavailable for carbon formation. Fred Hoyle was the first to understand this, but by the 1950s he had also seen that this interaction is only possible in the context of an excited state of carbon at a very precise energy level (7.65 MeV), at that time unknown. So Hoyle, in 1954, predicted that this excited state of carbon must exist, since without it life would not exist.[220] Later, he managed to convince an initially skeptical Willy Fowler, nuclear physicist at Caltech, to carry out the necessary testing, which confirmed that the excited state of carbon with the precise and highly improbable energy level and resonance predicted did in fact exist. Militant atheist Hoyle was so impressed by this that he ended his life a deist. As early as 1957, he spoke out about his discovery of this providential phenomenon, at the risk of falling out with his colleagues at the very progressive University of Cambridge: "*I do not believe that any scientist who examined the evidence would fail to draw the inference that the laws of nuclear physics have been deliberately designed with regard to the consequences they produce inside the stars.*"[221] An affirmation that, once again, leaves no place for chance.[222] As Steven Weinberg points out along the same lines: "*If the energy of this state of the carbon nucleus were too large or too small, then little carbon or heavier elements would be formed in stars, and with only*

---

220. And so Hoyle predicted the existence of this excited state on the basis of the argument that, if it did not exist, life as we know it could not have arisen in the Universe. Thus, for the first time, what would later be called the "anthropic principle" was used as a predictive argument for a scientific discovery (some authors refer to this as "anthropic prediction").

221. On this subject, see E. Margaret Burbidge, Geoffrey R. Burbidge, William A. Fowler, and Fred Hoyle, "Synthesis of the Elements in Stars," *Reviews of Modern Physics* 29 (1957): 547–650. In 1982, Fred Hoyle reiterated his belief that this precise resonance was proof of the existence of a "*a superintellect.*"

222. We could also quote Stephen Hawking: "*Today, we can create the computer models that tell us how the rate of the 'triple alpha' reaction [of carbon synthesis] depends on the magnitude of nature's fundamental forces. Such calculations show that changes as small as 0.5% for the strong nuclear force, or 4% for the electromagnetic force, would result in the disappearance of almost all the carbon and all the oxygen in every star, and thus the possibility of life as we know it.*" Stephen Hawking and Leonard Mlodinow, *The Grand Design* (New York: Bantam Books, 2010). And there's even more: the reaction that synthesizes oxygen-16 from the fusion of carbon-12 and helium-4 involves the oxygen-16 nucleus in an excited state. If this excited state had a slightly higher energy, the reaction would be accelerated, eventually converting almost all the carbon into oxygen, without leaving enough for the oxygen-16 nucleus to be converted into oxygen.

hydrogen and helium there would be no way that life could arise."[223]

And without it, the heaviest elements in the Periodic Table, i.e., elements which formed between 3 and 5 billion years later in the explosion of the first-generation stars that became supernovas, would not exist either.

9. The "magical" adjustment of the electromagnetic force has also astounded the greatest scholars

During the first 380,000 years that followed the initial fifteen minutes of its existence, the Universe continued its rapid expansion, dominated by the agitation of highly energized photons coupled to ionized matter. These photons were like "prisoners," incapable of striking out on their own because of the high temperature. But at the end of this period, electrons began to associate with existing nuclei. As a result, matter became electrically neutral, and radiation decoupled from matter, allowing the emission and propagation of the first light over long distances and the emergence of the visible Universe.[224] This association of electrons to atomic nuclei was made possible by the electromagnetic force, which is 137 times smaller than the strong force.

The electromagnetic force is determined and governed by the "*fine-structure constant*," introduced in 1916 by the German physicist Arnold Sommerfeld (close friend of Einstein and the academic mentor to Nobel Laureates Wolfgang Pauli and Werner Heisenberg). Why does the fine-structure constant have the exact value of **0.0072973525376**? Nobody knows. What we do know is that if the constant changed by a few percent, life could not exist in the universe.[225]

---

223. Steven Weinberg, *Dreams of a Final Theory* (New York: Knopf Doubleday, 2011), 221.

224. Robert Lea, "*The 1st light to flood the universe can help unravel the history of the cosmos. Here's how.*" Space.com, 10 July, 2023. https://www.space.com/universe-first-light-cosmic-microwave-background-history-cosmos.

225. If a magical hand tampered with the electromagnetic force, the whole Universe would cease to exist. If the force were slightly stronger, electrons would repel atoms; if it were weaker, electrons couldn't remain within their atoms. In both cases, molecules would have no way of forming, and the Universe would be sterile. Once more, the evidence of incredibly precise fine-tuning is impossible for the modern physicist to avoid.

As already mentioned, this phenomenon haunted the German physicist **Max Born**, the 1954 Nobel Laureate in Physics and quantum mechanics superstar: *"If alpha [the fine-structure constant] were bigger than it really is, we should not be able to distinguish matter from ether [the vacuum, nothingness], and our task to disentangle the natural laws would be hopelessly difficult. The fact, however, that alpha has just its value 1/137 is certainly no chance but itself a law of nature. It is clear that the explanation of this number must be the central problem of natural philosophy."*[226]

We saw earlier how the Nobel Laureate **Richard Feynman** was fascinated by this mysterious constant that seems to arise from nowhere: *"[The constant] has been a mystery ever since it was discovered more than fifty years ago, and all good theoretical physicists put this number up on their wall and worry about it."*[227]

10. Cosmic microwave background anisotropy is also fine-tuned

380,000 years after the Big Bang, the liberation of the first visible light produced what we now call cosmic microwave background radiation (CMBR), in near-perfect thermal equilibrium at 3,000 kelvin. That's about the temperature of the surface of our Sun. But CMBR is not completely uniform. In 1992, tiny variations called *"anisotropy"* were detected in it by George Smoot and John Mather. The term anisotropy refers to the fact that the echo of this primordial radiation of the Universe is not perfectly homogeneous. Although the magnitude may seem insignificant—only some ten thousandths of a degree![228]—these

irregularities were indispensable for the future evolution of the Uni-

---

226. Arthur I. Miller, *Deciphering the Cosmic Number: The Strange Friendship of Wolfgang Pauli and Carl Jung* (New York: Norton, 2009), 253.

227. Richard Feynman, *QED: The Strange Theory of Light and Matter* (Princeton: Princeton University Press, 1988), 129.

228. The most recent measurements made by the Planck satellite give us a temperature of 2.72538 K with a deviation (Delta T/T) of 0.000066, which leads to a range between 2.72529 and 2.72547.

The cosmic microwave background radiation shows a Universe that is initially very homogeneous, with very small temperature differentials, between 2.72538 and 2.72556 K. A breathtakingly fine tuning, essential to the existence of galaxies and thus of life

verse into stars and galaxies.

Computer simulations show that if the temperature variation 380,000 years after the Big Bang had been slightly greater, our Universe could have transformed into a giant field of black holes. On the other hand, if the anisotropy had been slightly less, then instead of shaping the Earth, planets, and stars, the Universe would have become a formless cloud of gas.

When the astronomical satellite WMAP was launched in 2001 to succeed the device sent up by Smoot and Mather, David Wilkinson (Dicke's former collaborator at Princeton) confidently exclaimed: "*I am confident that our probe will soon confirm that nothing in the universe has been left to chance.*"[229]

Brilliant mathematician **Steve Carlip** from the University of California has arrived at the same conviction and expresses his views very openly. After analyzing the temperature differences between the hot points (in red in the "baby picture" of the Universe taken by

---

229. Cf. Igor and Grichka Bogdanov, *Le Visage de Dieu* (Paris: J'ai Lu, 2011), 178.

satellites) and the cold regions (in blue), the mathematician drew this strong conclusion: *"When the Universe was very young and very hot, the ordinary matter was almost entirely a plasma of ionized hydrogen. The small perturbations in the density of this plasma propagated as what were essentially sound waves, with speeds that can be predicted accurately from ordinary laboratory physics. This resulted in correlations of the density at predictable distances."*[230] Professor Carlip continues with enthusiasm: *"In particular, the angular correlation power spectrum— the measure of the relationship between the temperature fluctuations at different angles— is a rather complicated curve, which nevertheless agrees exquisitely with theory [...] the plasma was denser at crests of sound waves, and less dense at troughs— and these show up in the measured CMB temperature"*[231]

Even more striking, **Roger Penrose**, winner of the 2020 Nobel Prize in Physics, has found that the observed entropy of the Universe is extremely improbable compared to all the values it could have had. He found 10 to the 10th power of 101 mass-energy configurations that correspond to an ordered Universe like ours. But he also showed that there are even more configurations, 10 to the 10th power of 123, which generate universes dominated by black holes. Since 10 to the 10th power of 101 is only a tiny fraction of 10 to the 10th power of 123, Penrose concluded that the conditions which could give rise to a universe favorable to life are extremely rare in comparison with the total number of possible configurations that might have existed at the beginning of the Universe. Dividing 10 to the 10th power of 101 by 10 to the 10th power of 123 always yields 10 to the 10th power of 123. And so the hyperexponential number

---

230. Steve Carlip and Tony Pagano, "Is the Earth the fixed center of the Universe? (The motion of the earth around the sun causes a doppler shift in the light of the Cosmic Microwave Background (CMB).)" The TalkOrigins Archive Post of the Month: July 2010. From a conversation thread originally posted 13 July 2010: The CMB is isotropic and its energy distrubution [sic] is that of perfect blackbody (at thermal equilibrium). https://www.talkorigins.org/origins/post-month/2010_07.html.

231. Ibid.

calculated by Penrose, 1 out of 10 to the 10th power of 123, provides a quantitative measure of the fine-tuning of the initial conditions of the Universe. As **Paul Davies** observed, *"The really amazing thing is not that life on earth is balanced on a knife-edge, but that the entire universe is balanced on a knife-edge, and would be total chaos if any of the natural 'constants' were off even slightly."*[232] What an understatement!

After receiving the 2006 Nobel Prize for his striking pictures of the first light in the Universe, **George Smoot** devoted his entire lecture to the theme, saying: *"If you're religious, it's like looking at God. The order is so beautiful and the symmetry so beautiful that you think there is some design behind it. I saw the Universe at its very beginning. I saw this anisotropy which allowed the Universe to exist."* Smoot and his team are reported as saying that they had *"pinpointed the oldest and largest cosmic structures yet known, the first anomalies in a seamless universal soup that signalled the beginning of galaxies."*[233] Smoot went on to explain that the Big Bang was not a disordered, cataclysmic event at all, as we have mentioned, but rather a very carefully organized process consisting of a particular number of sequential events that unfold over the course of time: the evolution of the Universe *"was written in its beginnings, in its cosmic DNA, if you will."*[234]

But there's more...

11. The Planck Constant—which universally regulates the energy

---

232. Paul Davies, "The Anthropic Principle," *Horizon* series, season 23, episode 17, aired May 18, 1987, on BBC, quoted in Stephen C. Meyer, "The Return of the God Hypothesis," *Journal of Interdisciplinary Studies* 11, no. 1–2 (1999): 141.

233. Rae Corelli, Marci McDonald, and Hilary Mackenzie, "'Looking at God,'" *Maclean's* (May 4, 1992): 38–39, available at https://web.archive.org/web/20221229200538/ – https://archive.macleans.ca/article/1992/5/4/looking-at-god.

234. George F. Smoot, "Cosmic Microwave Background Radiation Anisotropies: Their Discovery and Utilization," The Nobel Prizes 2006, ed. Karl Grandin (Stockholm: The Nobel Foundation, 2007), https://www.nobelprize.org/uploads/2018/06/smoot_lecture.pdf.

levels of all atoms—deserves the nickname "theological constant" because, without it, all chemistry would be impossible

380,000 years after the Big Bang, conditions became suitable for the formation of atoms. Miraculously, they arose with predetermined energy levels, exactly identical always and everywhere. Werner Heisenberg, one of the founders of quantum mechanics, was astonished as early as the 1920s by the stability of properties of bodies unexplainable through Newtonian mechanics. *"Only quite different natural laws can help us to explain why atoms should invariably rearrange themselves and move in such a way as to produce the same substances with the same stable properties."*[235] Had this not been the case, as many atoms would have arisen as there were initial states, depending on particular local conditions. In such a world, chemical reactions simply would not be able to take place. For a reaction to occur, elements have to be homogenous and energetically compatible. Now, thanks to the Planck constant, all the atoms of each element have exactly the same energy level; this is the basis of quantum mechanics. If the laws of the Universe had not been programmed this way, the world around us would not exist. Seeing that the Planck constant yields a Universe whose structure permits chemistry, some have called it *"the theological constant,"* since it seems to be such an integral part of an intelligently designed plan engineered by God.

Conclusion

These are only a handful of the most incredible fine-tunings that serve to illustrate the "anthropic principle." The sum total of all these physical improbabilities is mathematical confirmation that the Universe is not the result of chance. That a creator God exists is the only obvious conclusion. This proof is, in our opininon, as strong as the evidence predicting the heat death of the Universe and of cosmology as we know it. Let us not forget that these different proofs are perfectly

---

235. W. Heisenberg, *Physics and Beyond: Encounters and Conversations*, trans. Arnold J. Pomerans (New York: Harper & Row, 1972), 21.

independent of each other.

All this strong evidence accords with the findings of many other researchers and scientists. It is directly based on the most cutting-edge discoveries in physics and leads to very clear and simple conclusions. But we can go just a little bit further with the help of American astronomer Robert Wilson, who won the 1978 Nobel Prize for his 1964 discovery of the first light in the Universe: *"Certainly if you are religious, I can't think of a better theory of the origin of the universe to match with Genesis."*[236]

---

[236]. Robert Wilson, interviewed by Fred Heeren in *Show Me God: What the Message from Space is Telling Us About God* (Miamitown: Day Star Publications, 1997), 157.

10.

The Multiverse: Theory or Loophole?

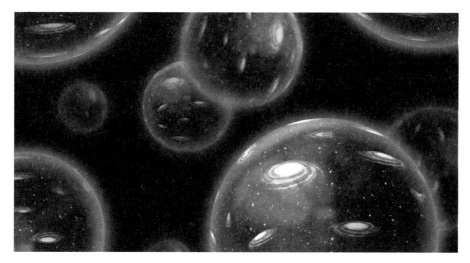

The anthropic principle puts us up against a wall. We cannot escape the metaphysical questions raised by these extraordinarily precise fine-tunings. But where do they come from? Could someone be behind it all?

There are really only two possible explanations: a creator God or pure chance. The probability that our Universe arose as a result of chance alone has been estimated at least at 1 in 10^{60}. For such an unlikely event to occur, one would have to assume the existence of approximately 10^{60} independent universes,[237] each governed by different physical laws. This implies an extraordinary multitude of universes existing "elsewhere," along with a universe-creating mechanism that we know

---

237. We can get some idea how big that number is by contemplating the fact that there are only 10^{23} grains of sand in the Sahara. If we follow the calculations of string theory, we get at an even greater figure: 10^{500} additional universes.

nothing about—one with the inherent ability to vary the fundamental constants of these universes, selecting them from an appropriate range.[238]

As foolish and unverifiable as this hypothesis may seem, it has been surprisingly well studied. Once we exclude the possibility of a creator God, there remains no other rational response to the problem posed by the anthropic principle. Nevertheless, a number of alternative theories have appeared, whose main lines we will trace in this chapter.

Hugh Everett and the "many-worlds interpretation"

In 1954, Hugh Everett was discussing the paradoxes of quantum physics with a classmate at Princeton. In the course of the conversation, he threw out the hypothesis of multiple universes as a kind of joke, quipping that perhaps every time the wave function collapsed a new world was born. He did serious work on this original solution to the problem of the superposition of quantum states, which became the subject of his thesis. In 1957 he published an article explaining the Universe like a many-branched tree that creates new universes constantly like a branch gives off new shoots. Although this unorthodox theory became a major source of inspiration for science fiction writers, it had no real impact on the scientific community. Eventually Everett left academia and was recruited by the Pentagon to do military research.

Alan Guth and "Cosmic Inflation Theory"

Alan Guth laid out his inflation theory in 1979 to explain why the Universe is so flat and homogenous. He postulated that the Universe grew very rapidly at first, by a factor of 10^{26} or more in the tiny span of 10^{-35} seconds, a rate that permitted it to smooth out naturally as it stretched, ensuring the extraordinary homogeneity observable today.

---

238. See, for example the video "The Multiverse itself is fine-tuned" at: https://youtu.be/b2pD-QY9zmlo.

THE MULTIVERSE: THEORY OR LOOPHOLE?

Arvind Borde (1940–), Alan Guth (1947–), and Alexander Vilenkin (1949–)

This part of the theory was well received by the scientific community, but its further developments were much more speculative. Like the Russian researcher Andrei Linde, Alan Guth found it difficult to imagine that inflation was limited to the first moments of the life of the Universe and that we can plausibly suppose that an infinity of universes are continuously generated in a fractal pattern. The result has been dubbed a "bubble universe" (chaotic inflation theory).

Guth imagined our Universe as one among innumerable universes, each following its own set of physical laws, universes that never stop emerging, like "champagne bubbles," out of "quantum fluctuations." According to Guth and Linde's multiverse theory, the fact that our particular Universe has conditions favorable to the emergence of life is only a matter of chance.

From the 2000s, we moved from "the theory of inflation" to an inflation of theories relating to the multiverse

More than fifty theories have been elaborated on the basis of Guth's hypothesis alone, not counting other versions. Some are entirely far-fetched, but many others continue to be studied today, like the highly speculative loop quantum gravity theory, or string theory, despite its increasing complexity and the growing doubts it raises among physicists.

These theories fall into two types:

- Those that conceive of multiverses in parallel dimensions that are entirely and forever inaccessible to us. Within this type we find:

- Theories that conceive of the Universe as evolving in dimensions parallel to our space-time.
- Theories that suppose that these parallel universes exist entirely "beyond" our space-time and have no connection to it.

- Those that imagine successive multiverses on the same timeline as our Universe, difficult to access, but that we may one day be able to consider scientifically. Here too, there are two sub-categories:
  - Theories that conceive of "bubble universes" continuously emerging in an expanding space. They imagine our oberservable Universe as just one bubble among a multitude of others, inaccessible to us and out of the reach of our Universe, as they move away from us much faster than the speed of light.
  - Theories postulating cyclic models that exist within our Universe.

No matter how many universes we imagine, an "initial singularity" (the Big Bang) remains necessary in all cases

In the late twentieth century, Hawking and Penrose developed several singularity theorems. These prove that universes that meet certain minimum energy conditions must start from an initial spatial singularity (that is, a point where the volume of the universe approaches zero).

Early mathematical models based on simplified scenarios—such as a universe with a uniform density throughout—suggested such a beginning, but many cosmologists expected that more factually accurate models would lead to different conclusions, allowing for a past and eternal Universe to exist. As it turned out, Hawking and Penrose demonstrated that the standard cosmological model, even with its more realistic parameters, indicates that the Universe began with a spatial singularity.

However, Hawking and Penrose did not study inflationary models that generate bubble universes.[239] Could these get us away from the necessity

---

239. Paul Steinhardt, one of the creators of inflationary cosmology, now believes that inflationary cosmologies are not correct. See Anna Ijjas, Paul J. Steinhardt, and Abraham Loeb, "Pop Goes

of a beginning? It turns out that the answer is no. In 2003, Alan Guth, Arvind Borde and Alexandre Vilenkin proved the absolute necessity of an "initial singularity" in time (a beginning of the Universe) for all realistic cosmological models, including the inflationary model. Vilenkin said: *"It is said that an argument is what convinces reasonable men and a proof is what it takes to convince even an unreasonable man. With the proof now in place, cosmologists can no longer hide behind the possibility of a past-eternal universe. There is no escape: they have to face the problem of a cosmic beginning."*[240] In other words, multiverse theories do not solve the problem of how our Universe began in the first place. No matter how many universes these theories put forward, the absolute beginning of the multiverse still demands an explanation.[241]

In other words, the multiverse theory might offer an explanation for the fine-tuning found in our world, but even the dizzying proliferation of universes does not resolve the problem of how our Universe began in the first place.

The multiverse as an escape mechanism

As with the previous chapter, this list of highly original explanations is impressive not only for its length, but also for how much work it takes to create and maintain these theories.

---

the Universe," *Scientific American* 316, no. 2 (January 17, 2017): 32-39.

240. Alexander Vilenkin, *Many Worlds in One* (New York: Hill and Wang, 2006), 176. See also his 2013 conference for the Copernicus Center for Interdisciplinary Studies, "Many Worlds in One": https://www.youtube.com/watch?v=8CChnwOsg9I.

241. Quantum mechanics (through the discovery of quanta) has shown that infinity does not exist in the infinitely small either. As seen in Chapter 7, this was already strongly affirmed by David Hilbert, one of the greatest of mathematicians, in June 1925, during a conference organized by the Mathematical Society of Westphalia: *"Our principal result is that the infinite is nowhere to be found in reality. It neither exists in nature nor provides a legitimate basis for rational thought."* P. Benacerraf and H. Putnam, *Philosophy of Mathematics: Selected Readings* (Cambridge: Cambridge University Press, 1984; 1983), 201. Science chases infinities. Every time it sees equations that diverge and point away from infinity, it says, "No, this is not Physics. The models must be revised."

The fine tuning and the origin of the Universe present significant challenges for materialist intellectuals. If they could discover reasonable scientific solutions to these issues, they would likely abandon these speculative theories, many of which are inherently unverifiable and sometimes resemble science fiction more than actual science. The remaining theories continuously seek hypothetical experimental evidence to support them. In the end, the various multiverse theories function as escape mechanisms that distract us from the real metaphysical questions at stake.[242]

"Theories of everything in general, and string theory in particular, puzzled me more and more, because they are 'tamper-proof:' no experience can prove them wrong. I realized that people accept string theory for ideological reasons. It was a terrible shock to me, because I thought that scientists refused any form of ideology. That's far from the case,"[243] explains Robert Laughlin, Professor of Physics at Stanford and 1998 Nobel laureate.

---

242. Stanford physicist Leonard Susskind, one of the founders of string theory and the multiverses associated with it, describes the reason behind the construction of this immense theoretical structure like this: "*If, for some unforeseen reason the [string] landscape turns out to be inconsistent—maybe for mathematical reasons, or because it disagrees with observation, [then] as things stand now we will be in a very awkward position. Without an explanation of nature's fine tunings we will be hard pressed to answer the ID [intelligent design] critics.*" Cited in Stephen Meyer, *The Return of the God Hypothesis* (New York: Harper, 2020), 345.

243. Translated from Robert Laughlin, *La Recherche* (February 2007). See also Robert Laughlin, *A Different Universe: Reinventing Physics from the Bottom Down* (New York: Basic Books, 2005).

11.

Preliminary Conclusions: One Small Chapter for Our Book, One Giant Leap for Our Argument

The time has come to put to work the scientific method we explained in Chapter 2, "What is Evidence?" To prove a thesis, the scientific method begins with a hypothesis, the implications of which are then confirmed or disproved by what we observe in the real world.

The thesis "there is no creator God; the world is purely material" generates the three logical consequences listed below (cf. Chapter 3):

> **If the Universe is exclusively material, then:**
>
> 1. The Universe cannot have a beginning.
>
> 2. The Universe cannot have an end like thermal death, because such an end implies a beginning.
>
> 3. Deterministic laws are purely the result of chance, and therefore it is extremely unlikely that they would be favorable to life.

People have known about these implications for as long as the thesis has been around. But they held little scientific interest for thinkers and researchers since it seemed like they would always remain far beyond human understanding. No one could ever have imagined that one day we might be able to test them.

Against all expectations, these three implications have now come within the reach of science. Starting at the end of the twentieth century, progress in physics, mathematics, technology, and more recently computing began to allow cosmologists to draw conclusions about these subjects. Today we know that the three propositions listed above are very difficult to sustain. This means that the thesis they flow from—"there is no creator God; the Universe is exclusively material"—is equally problematic. A reasonable person must, therefore, adopt the opposite thesis: "A creator God exists."

A single valid proof is enough to disprove the hypothesis of a purely material Universe

But cosmology allows us to establish two separate proofs:

- **The Universe had a beginning.** This we know, most notably from thermodynamics and the Borde-Guth-Vilenkin theorem, which is based on Hawking and Penrose's work on initial singularity.
- **The laws of the Universe are very favorable to human life,** and the complex, minute fine-tuning of these physical laws is extremely improbable, as demonstrated by the anthropic principle.

These undeniable truths contradict the implications that follow from the theory of a purely material Universe. According to the rules of logic, a single one of these two proofs is enough to invalidate the theory.

The two proofs are even stronger because they are totally independent of one another

Firstly and fundamentally, the fact that the Universe had a beginning and that its structure and laws are improbable are two distinct facts with no relation between them.

Furthermore, their results are not correlated because they were determined by independent methods.

This double independence reinforces their value as proof, because the falsity of one has no impact on the truth or falsity of the other. This significantly lowers the probability that the two are simultaneously false.

The Big Bang greatly reinforces these two proofs

The discovery of the Big Bang is perfectly coherent with the reality that the Universe had a beginning. It is the final, long-awaited missing puzzle piece in the theory. It fits perfectly, exactly where we'd expect it.

Short on arguments, some detractors question it on grounds that are not even scientific. They say it is too similar to the Bible's creation account. Some major figures in science have pointed out this unfair move:

- British physicist George Thomson, who shared the 1937 Nobel Prize: *"Probably every physicist would believe in a creation if the Bible had not unfortunately said something about it many years ago and made it seem old fashioned."*[244]

- American physicist Robert Wilson, 1978 Nobel co-laureate in physics, as already mentioned in a preceding chapter: *"Certainly if you are religious, I can't think of a better theory of the origin of the universe to match with Genesis."*[245]

- Arno Penzias, 1978 co-laureate in Physics: *"The best data we have are exactly what I would have predicted, had I nothing to go on but the five books of Moses, the Psalms, the Bible as a whole."*[246]

Proof through slander and harassment

As we have seen, the attacks and harassment suffered by the supporters of a beginning to the Universe and of its fine-tuning—two proofs that disprove the theory of an exclusively material Universe—paradoxically give them even more legitimacy.

---

244. Sir George Paget Thomson, "Continuous Creation and the Edge of Space," *New Republic* 124 (1951): 21-2.

245. Robert Wilson, interviewed by Fred Heeren in *Show Me God: What the Message from Space is Telling Us About God* (Miamitown: Day Star Publications, 1997), 157.

246. A. Penzias, "Clues to Universe Origin Expected," in a *New York Times* article by Malcolm W. Browne, 12 March 1978. https://www.nytimes.com/1978/03/12/archives/clues-to-universe-origin-expected-the-making-of-the-universe.html.

If the two proofs were false, uncertain, or debatable, or if their consequences were not evident, then Gamow, Bronstein, Landau, Frederiks, Kozyrev, and many others would not have been attacked and persecuted, exiled, thrown into the Gulag, or executed.

Materialists would not have advocated for every conceivable alternative theory or invested excessive time in theories they label as "speculative," which some of us, with less regard for their sentiments, refer to as "science fiction."[247]

Remember, if Galileo was put under house arrest, it was because he was right. Violence begins when people run out of arguments.

This violence against scientists serves as indirect evidence, particularly because it was perpetrated by intellectuals, politicians, and fellow scientists who were well-equipped to evaluate both the validity of these new discoveries and the genuine threat they represented to their interests.

These cosmological proofs are so recent that they aren't yet common knowledge

Keep in mind that this revolution in cosmology and its metaphysical implications is extremely recent. These proofs are not yet part of everyone's mental landscape, though they are slowly making headway.

- The confirmation of thermodynamics dates to the early twentieth century.
- The Big Bang was confirmed in 1964.
- The anthropic principle was first formulated in 1973, and the confirmations are still rolling in.
- The Big Crunch was only disproved in 1998.
- In 2003, the Borde-Guth-Vilenkin theorem showed that the Universe can't have an infinite past: there must have been an initial singularity.

---

247. According to Karl Popper, whatever is irrefutable is unscientific, but this overly restrictive definition does not fit with the usual use of the word "science."

The thesis of a creator God on the edge of Ockham's razor

In the fourteenth century, the erudite monk William of Ockham laid down the principle that when we're faced with two explanations, the simpler one is more likely to be true. This principle has become known as Ockham's razor, and it is an important tool for scientists today. For many centuries, it has been used to argue that, because the hypothesis of God is not necessary, it should not be believed.

But today the same principle can be turned against materialists.

The multiverse: the materialist's final trump card

Materialist thinkers today cling to the multiverse as their last hope. This theoretical construct certainly stretches the imagination, but it does not offer a solution to the problem of improbability posed by the fine-tuning of the Universe. All the multiverse theories in fact require processes of universe creation which must themselves be extremely fine-tuned in order to generate multiple universes, some of which would be favorable to life.[248]

In other words, the challenge posed by improbability is simply transferred to the mechanisms that generate new universes, without providing any ultimate explanation of their fine-tuning.

Nor do these models provide any solution to the question of the beginning of the Universe: the problem is simply shifted back to a "mother" universe about which we know nothing at all. But it is impossible to go back infinitely in time in this way, and so there must be an absolute beginning to all this, as the previously cited Borde-Guth-Vilenkin theorem also demonstrates.

In short, the materialist multiverse hypothesis turns out to be purely speculative and completely powerless to counter the thesis of a creator God.[249]

---

248. Cf. the aforementioned video, "The Multiverse Itself Is Fine-Tuned": https://youtu.be/b2p-DQY9zmlo.

249. Stephen Meyer, *Return of the God Hypothesis: Three Scientific Discoveries That Reveal the*

Quotations on the subject:

In his book *The Secret Melody*, Trinh Xuan Thuan remarks how *"For myself, I am prepared to bet on the existence of a supreme being. The hypothesis of a multiplicity of imaginary, unverifiable universes violates my sense of simplicity and economy."*[250]

"The multiverse hypothesis is alleged to be the last resort for the desperate atheist."[251] (Neil Manson, Professor of Philosophy at King's College, Cambridge).

"This 'multiverse' view represents the failure of our grand agenda and seems to me contrary to the prescribed simplicity of Ockham's Razor, solving our lack of understanding by multiplying unseen entities into infinity."[252] (Gregory Benford, Professor of Physics and Astronomy at the University of California Irvine, specialist in space exploration).

"Even if the probability of finding a universe with a weak cosmological constant is very low, if we have an infinity of universes, it will happen somewhere [. . .] But is this a solution or a desperate explanation?"[253] (George Efstathiou, Professor of Cosmology, Director of the Cambridge Institute of Astronomy).

Mind Behind the Universe (New York: HarperOne, 2021), 326-347.

250. Trinh Xuan Thuan, *The Secret Melody: And Man Created the Universe*, trans. Storm Dunlop. (Oxford: Oxford University Press, 1995), 249.

251. "Introduction" to Neil A. Manson, *God and Design: The Teleological Argument and Modern Science* (New York: Routledge, 2003), section entitled "The Much-Maligned Multiverse," 18. Without taking sides, philosopher of science Neil Manson first used this expression to summarize the main criticism against the hypothesis of multiple universes. His neutrality does not, however, prevent Manson, in collaboration with Michael Thrush, from denouncing certain weaknesses of the multiverse hypothesis: "As Leslie suggests: '[For a cosmological hypothesis], the capacity to reduce astonishment is a fairly good criterion of accuracy.' But the theory of multiple universes does not seem to satisfy this criterion. [. . .] Suppose as many universes as you desire, this does not make it more plausible that ours has the characteristics allowing life, or the fact that we are here. Thus, our chance of existing in a universe allowing life does not give us reason to suppose that there are a multitude of universes." Michael Thrush and Neil A. Manson, "Fine-Tuning, Multiple Universes, and the 'This Universe' Objection," *Pacific Philosophical Quarterly* 84, no. 1 (2003).

252. Gregory Benford, *What We Believe but Cannot Prove*, ed. John Brockman, (New York: Harper Perennial, 2006), 226.

253. George Efstathiou, Director of the Cambridge Institute of Astronomy and member of the scientific team of the European Space Agency's Planck satellite, during a talk at McGill University,

"Here is the cosmological proof of the existence of God—the design argument of Paley—updated and refurbished. The fine tuning of the universe provides prima facie evidence of deistic design. Take your choice: blind chance that requires multitudes of universes or design that requires only one [...] Many scientists, when they admit their views, incline toward the teleological or design argument."[254] (Edward Harrison, Professor of Astrophysics at the University of Amherst).

A creator God is a much more simple and well-founded solution than the thesis of multiverses. A rational mind should favor it. It is simpler because if we have to choose between one creative being and a dizzying inflation of universes, the more rational choice is the one that limits unnecessary hypotheses. Let's remember that the multiverses belong only to the final group in our classification of evidence, group 6, theories with no known or observable implications. On the other hand, the thesis of a creator God, and its opposite are part of group 5 and generate clear, logical, and numerous implications that we can compare with the real world.

Is that the last word?

We have now reached the end of the first part of our survey. Cosmology alone has provided us with sufficient evidence for the existence of a creator God. We could stop writing, and you could stop reading here. But it would be a shame to limit ourselves to a single angle of analysis.

"A Cosmic Coincidence. Why is the Universe Conducive to Life?" (Montréal: McGill University, January 15, 2007).

254. Edward Harrison, *Masks of the Universe: Changing Ideas on the Nature of the Cosmos* (Cambridge: Cambridge University Press, 1985), 201 and 263.

12.

Biology: The Incredible Leap from Inert to Living Matter

Cosmological discoveries have allowed us to discredit the thesis of an entirely material Universe. And recent advances in several other fields further confirm this conclusion. One such discovery, this time in the field of Biology, is the incredible leap from inert to living matter.

I. A gap that ultimately turned out to be an abyss

Less than a century ago, what was thought to be a mere gap between inert and living matter was revealed to be a deep abyss. Modern biology has shed light on the incredible complexity of even the smallest living cell, which in fact bears a resemblance to an ultra-sophisticated factory. We now know that the transition to life was an extremely unlikely and complex event that took place over a relatively short span of time. Consequently, the thesis that life emerged by pure chance in a Universe that was not specifically designed to be hospitable to it is no longer tenable.

A biological fine-tuning is added to the cosmological fine-tuning

The only possible rational explanation for the natural appearance of life on our planet is that life resulted from laws of the Universe that remain unknown to us, or at least from as-yet unrecognized effects of finely-tuned laws that prevailed at the time this leap occurred. That is to say, in addition to the cosmological fine-tuning of the Universe, there is a second anthropic principle that applies to the emergence of life.

Some established facts

Though no one knows exactly how life appeared on our planet, there are three facts about which we are absolutely certain:

1. The transition from inert matter to life as we know it must have happened, because we are here now.

2. This occurred at least 3.8 billion years ago, less than 700 million years after the formation of the Earth.

3. This enormous leap required the coordination of a great number of extremely improbable factors. The likelihood of this has been calculated by scientists, and we will see later that the improbability is dizzying.

Chance alone cannot explain the appearance of life

The goal of this chapter is to show that only the "*invisible hand*" of precisely regulated laws, or effects of the laws of nature that we still do not understand, could have guided this astonishing process. First matter organized itself in an ever more elaborate way into quarks, atoms, molecules, and polymers, leading to the complex languages of RNA, DNA, and proteins which would allow the replicative, reproductive, and metabolic functions necessary for the emergence of life. All these functions came together to form the first unicellular living organisms. These, in the course of evolution, led to the earliest common ancestor of all terrestrial living beings, which scientists call LUCA (Last Universal Common Ancestor): a cellular system that was itself remarkably complex, characterized by a universal genetic code linking DNA/RNA to proteins, with the ribosome playing the role of "translator."

The fact that the laws of matter and energy are fine tuned for life poses a major obstacle for the materialist, as the claims of materialism mean denying both that the Universe was particularly favorable to the appearance of life and that this constitutes another strong argument for the existence of a creator God.

How has our thinking about the appearance of life evolved over time?

Since people in past centuries had little understanding of the complexity of living organisms, philosophers and scientists did not devote much study to the transition from inert matter to life.

Plato took it for granted that the Universe and every living being were composed of the four elements: fire, air, water, and earth.[255]

Aristotle opted for continuity, thinking, for example, that eels came from river silt.[256] He also wrote: "*Nature passes so gradually from inanimate to animate things, that from their continuity their boundary and the mean between them is indistinct. The race of plants succeeds immediately that of inanimate objects.*"[257]

Lucretius described the emergence of life as a simple operation: "*Thus then the new Earth first of all put forth grasses and shrubs, and afterward begat the mortal generations, there upsprung—in modes innumerable—after diverging fashions.*"[258]

In the eighteenth century, Diderot and other thinkers supposed that matter itself was alive: "*Every animal is more or less a human; every mineral more or less a plant, every plant more or less an animal.*"[259]

Leibniz thought the same: "*in the smallest particle of matter there is a world of creatures, living beings, animals, entelechies,*[260] *souls. Each*

---

255. Plato, *Timaeus*, 56b.

256. See Aristotle, *History of Animals*, 3.2; *Aristotle's History of Animals*, trans. Richard Cresswell (London: George Bell and Sons, 1887).

257. Aristotle, *Aristotle's History of Animals*, trans. Richard Cresswell (London: George Bell and Sons, 1887), 8.1.

258. Lucretius, *De rerum natura* 5.5, in On *the Nature of Things*, trans. William Ellery Leonard (New York: E. P. Dutton, 1916).

259. Denis Diderot, *Rameau's Nephew and D'Alembert's Dream*, trans. and introduction by Leonard Tancock (New York: Penguin Books, 1966), 181.

260. According to the *Encyclopaedia Britannica*, an entelechy is "*that which realizes or makes actual what is otherwise merely potential.*" The entry goes on to note that Leibniz "*called his monads (the ultimate reality of material beings) entelechies in virtue of their inner self-determined activity.*"

portion of matter may be conceived as like a garden full of plants and like a pond full of fishes. [. . .] Thus there is nothing fallow, nothing sterile, nothing dead in the universe, no chaos, no confusion save in appearance."[261]

From antiquity onwards, people spoke of the "*spontaneous generation*" of life from inert matter. In the mid-nineteenth century, biologist Félix-Archimède Pouchet imagined: "*It must be specified that by spontaneous generation we are only claiming that, under the influence of still unexplained forces, there occurs, either in animals themselves or elsewhere, a synthetic event that tends to group molecules, to impose on them a special mode of vitality that eventually gives rise to a new being that is related to the event, from which its elements were originally drawn.*"[262]

In 1861, Louis Pasteur's discovery of microorganisms put an end to belief in spontaneous generation. But the transition from inert to living matter was still viewed as something that could one day be explained or even replicated.

When Darwin approached the question in 1871, in a letter to his colleague Joseph Hooker, he wrote: "*But if (and oh what a big if) we could conceive in some warm little pond with all sorts of ammonia and phosphoric salts,—light, heat, electricity etc. present, that a protein compound was chemically formed, ready to undergo still more complex changes, at the present day such matter would be instantly devoured, or absorbed, which would not have been the case before living creatures were formed.*"[263]

Between 1950 and 1970, scientists attempted to replicate the original conditions for the emergence of life on Earth. They tried to recreate the

---

261. Gottfried Wilhelm Leibniz, *Monadology*, trans. Robert Latta (Oxford: Clarendon Press, 1865), 66-69.

262. Translated from *Hétérogénie ou Traité de la génération spontanée* (Paris: J.-B. Baillière et Fils, 1859), 7–8.

263. See *The Correspondence of Charles Darwin*, vol. 19, letter no. DCP-LETT-7471.

soup of chemicals spoken of by Darwin and others,[264] to see what might arise when this "primordial soup" was heated, dried, ground up, and subjected to electric currents.

As we will see later, these attempts did succeed in creating some of the building blocks of life. Despite this, these preliminary results represent only a very small step towards understanding the genesis of life.

In fact, since those pioneering days, a growing number of scientists from a wide range of fields including physics, chemistry, biochemistry, biology, geology, and earth sciences are now working in astrobiology,[265] a field that aims to understand the history of life in the Universe. However, despite the amazing progress made over the last fifty years, the gap between inert and living matter seems to be widening endlessly, beyond anything we could have imagined.

A preliminary analogy: a car and a spare part

In order to appreciate the enormity of this leap, the following analogy[266] will help us visualize the difference between the most complex inert matter of the theoretical primordial soup (some building blocks of proteins) and the simplest known life form (a single-celled organism, such as a bacterial cell). The relation between the two will give us some idea of the distance separating the two banks of the gap, and thus of the magnitude of the leap required to cross it.[267]

---

264. David W. Deamer and Gail R. Fleischaker, *Origins of Life: The Central Concepts* (Boston: Jones & Bartlett, 1994).

265. Shawn D. Domagal-Goldman, Katherine E. Wright, et al., "The Astrobiology Primer v2.0," *Astrobiology* 16 (August 2016): 561-653. See https://www.liebertpub.com/doi/abs/10.1089/ast.2015.1460.

266. This analogy is of limited scope because, from a scientific point of view, physical size is a very imperfect measure of complexity.

267. George Ellis, "Why Reductionism Does Not Work?" in *Wider den Reduktionismus: Ausgewählte Beiträge zum Kurt Gödel Preis 2019*, ed. O. Passon and Ch. Benzmüller (Wiesbaden: Springer Spektrum, 2019). https://doi.org/10.1007/978-3-662-63187-4.

If we take this artificially constructed, complex, inert matter, a protein about 3×10^{-9} meters (10×10^{-9} feet) in diameter, and scale it up 10 million times, we end up with a small mechanical part measuring three centimeters (a little over one inch) in every direction, with a volume of $3 \times 3 \times 3 = 27$ cm³ (about 1.5 cubic inches). If we then enlarge the simplest unit of living matter, say a single-celled bacterium measuring 2×10^{-7} meters, on the same scale as the protein, it would measure two meters (six and a half feet) in every direction. In other words, it would be the size of a car with a volume of 8 m³ (or 10 cubic yards), which is 300,000 times bigger than the 27 cm³ (1.5 cubic inch) spare part.

The leap in size is equivalent to going from a mechanical part of 27 cm³ to a mechanism of eight million cm³. In other words, the largest inert matter produced to date represents $1/300,000^{th}$ of what we'd have to synthesize to get the smallest possible single-celled organism. That is, the equivalent of one word out of all the words in two books like the one you are holding in your hands. To continue to assess the immense gap to be bridged, we have to look more closely at the most complex piece of inert matter that has been obtained by experimental means and compare it with the smallest conceivable living being.

Between inert and living matter, the same ratio as between a spare part and a car...

II. The most complex inert matter obtained by experimental means

Between 1950 and 1970, experiments were run on "prebiotic soups," that mimicked the conditions of the emergence of life on Earth. The result? A few building blocks of proteins.

> The imaginative non-specialist reader might picture scientists busy around bubbling cauldrons, stirring the scum of a strange liquid—the prebiotic soup from which bits of living matter are supposed to emerge. The actual experiments undertaken by leading chemists, biochemists, and biologists were much less picturesque but still quite exciting.
>
> - The best-known of these experiments was conducted by **Stanley Miller** in 1953. It approximated what was thought at the time to be the atmosphere of the early earth, bringing methane, ammonia and hydrogen to a boil in water and subjecting the mixture to a high voltage electrical current imitating lightning. This atmosphere must, moreover, necessarily be deprived of any oxygen, which would destroy the amino acids. The experiment yielded thirteen of the twenty-two stable amino acids that are the building blocks of living things. Amino-acids are molecules made up of fifteen to forty chemical elements—mainly carbon, hydrogen, oxygen, and nitrogen—that when linked together in long chains of several hundred to several thousand elements constitute the complex macromolecules that are proteins, essential to all living organisms.
>
> - In 1958, at Florida State University, **Sidney W. Fox** and his student Kaoru Harada succeeded in creating amino acid polymers through the thermal condensation of glutamate and aspartate, two amino acids previously identified in Miller's experiment. Although resembling proteins, these polymers, called "proteinoids," were not, however, made up like normal proteins. Dissolved in water, the proteinoids spontaneously formed microdroplets that suggested possible ancestral forms of protocells, but, not surprisingly, these microstructures were nonfunctional.

- Next, in 1960, **Dr. Joan Oró** of the University of Houston mixed cyanhydric acid and ammonia in water. By heating the solution for twenty-four hours at 90°C, he was able to synthesize adenine (A)—another building block of life—which is one of the four nucleotide bases that combine to form RNA and DNA.

- In 1961 **Melvin Calvin**, a professor at the University of California at Berkeley, used a cyclotron to run a stream of accelerated electrons at high speed through a mixture of primitive gases. He managed to synthesize amino acids, along with sugar, urea, and fatty acids.

- In 1963, **Dr. Cyril Ponnamperuma** used Berkeley's cyclotron to irradiate a mixture of ammonia, methane, and water, simulating what was thought at that time to be the conditions of the early atmosphere. He obtained a relatively good yield of the adenine nucleobase (A). In collaboration with Carl Sagan of Stanford University, he synthesized adenosine,[268] through ultraviolet irradiation of a mixture of ribose and adenine, and then the corresponding nucleobase triphosphate in the presence of phosphoric acid and polyphosphate ester under similar conditions.

- **Sidney W. Fox**, already cited for his work on proteinoids in 1958, was later able to synthesize amino acids by bringing a mixture of methane, ammonia, and water vapor to a temperature of 1,000°C.

These experiments initiated a conceptual revolution by showing that the "primordial soup" by itself is not sufficient to produce life

Since then, and especially over the last twenty years, further progress has been made.[269] For example, organic molecules involved in the composition of living matter have been identified in meteorites, comets, and interstellar clouds.

---

268. Adenosine is the nucleoside that results from the combination of a ribose with the nucleobase adenine. Nucleoside triphosphates, such as adenosine triphosphate, are the nucleotides used as building blocks by RNA or DNA polymerases to synthesize RNA and DNA molecules.

269. Shawn D. Domagal-Goldman, Katherine E. Wright, et al., "The Astrobiology Primer v2.0," *Astrobiology* 16 (August 2016): 561–653, https://www.liebertpub.com/doi/abs/10.1089/ast.2015.1460.

These important discoveries show that the chemistry that flows from the laws of the Universe is conducive to the emergence of the basic molecules of life. In other words, the building blocks of life appear to be an inevitable result of the Universe.

However, it is more difficult to understand how these molecules were selected and built up on Earth. Other environmental factors must have been essential to promoting the chemical selection of the molecules that characterize life, and scientists imagine that this could have happened not only in "small pools" but also perhaps in space, in the Earth's atmosphere, on mineral surfaces, or in hydrothermal vents.

We have not proposed a single scenario for the origin of life, but we can suggest that none of these hypotheses excludes the others. The very complex chemistry leading to life was probably built from numerous and very different laws of the Universe.

These laws lead to the appearance of the first building blocks of life, which are energy wells, that is to say, naturally occurring stable combinations. Under certain conditions, as we have seen, these bricks form spontaneously and then remain stable and solid. Nevertheless, if prebiotic chemical processes can plausibly lead to small informational peptides and RNAs, it is also clear that none of these molecules are generated by the simple brewing of a "primordial soup." There is still a long way to go to understand the history of life in the Universe.

Hubert P. Yockey, Professor of Theoretical Physics at the University of California Berkeley and specialist in information theory as it applies to biology and the origin of life, came to this conclusion: *"Virtually all biochemists agree that life on earth arose spontaneously from nonliving matter[. . .]This belief is simply a matter of faith in strict reductionism and is based entirely on ideology."*[270]

Philippe Labrot of the Centre de Biophysique Moleculaire (Molecular Biology Center) at the CNRS of Orleans made the same observation:

---

270. Information Theory and Molecular Biology (1992), 284, available digitally at https://archive.org/details/informationtheoro000yock/page/284/mode/2up?q=%22simply+a+matter+of+faith%22.

"The construction of one living cell from extremely simple molecules [such as those synthesized by Stanley Miller] is a feat of incredible complexity. The advances made by chemists since the historic experiments of Stanley Miller in 1953 seem risibly inadequate to the task."[271]

III. The simplest life-form

The development of the electron microscope has allowed us to explore reality on a very small scale. Scientists have confirmed that all living beings are composed of at least one cell, and that even the smallest of these single-celled organisms (0.2 microns for the smallest imaginable bacteria, compared to the 20 microns or 2 hundredths of a millimeter of a human cell) demonstrates organization at a remarkable level of complexity.

The single-celled organism is the smallest form of life, but even its structure is highly complex

The smallest living thing that we can currently imagine, even a little smaller than the smallest living thing known to date, would have a minimum size of 0.2 micrometers (2×10^{-7} meters). This organism could only live, develop, and reproduce thanks to a very dense and organized internal structure. Inside its already complex cellular membrane, it would need, at minimum, a DNA genome of at least 250 genes, around 150,000 pairs of nucleotide bases assembled in a very precise order. It would also need an RNA system and a ribosome to interpret the RNA. Finally, this complex system would have to be able to produce at least 180 types of proteins, dozens of enzymes, and locomotory organelles. All of this equipment just for a single cell, the tiniest one possible!

After twenty years of research, John Craig Venter, an expert in microbiology, genetics, and "synthetic biology," made the following remark

---

[271]. Translated from Labrot's article "Chimie prébiotique." Labrot was awarded the Stanley Miller prize by the International Society for the Study of the Origin of Life (IS-SOL) for his remarkable contributions to the study of the origins of life: https://www.nirgal.net/ori_life2.html.

about this incredible complexity: *"We're showing how complex life is, even in the simplest of organisms."*²⁷²

To date, no definite intermediate steps have been identified between the most complex inert matter and the least complicated life forms currently known

Viruses are even smaller than the smallest single-celled organisms, but they are not strictly speaking living beings. They are parasites that can't live or replicate unless they are attached to a living being.

Nevertheless, it remains a possibility that at some intermediate stage there existed other beings capable of self-replication that disappeared after the first life forms arose. Just because they are nowhere to be found does not mean they never existed.²⁷³

Meanwhile, the smallest quasi-living being known by science to date is the *bacterium Nasuia deltocephalinicola*, discovered in 2013, which has 225 genes for a genome of 112,000 pairs of nucleotides.²⁷⁴ In reality, it is not complex enough an organism to survive on its own; it has to acquire certain nutrients from a living host. So it is a parasite, but one that possesses all the machinery of a living being capable of reproducing itself: DNA, RNA, and a ribosome to build its own proteins.

---

272. "Design and synthesis of a minimal bacterial genome," *Science* 25, Vol. 351 (Mar 2016).

273. In fact, since RNA is the only type of natural molecule that acts both as a template and as an enzyme, it has been proposed that an RNA polymerase could, at the origin of life, have initiated its own replication as well as the production of other functional RNA. This led to the famous "RNA world hypothesis" for the origin of life on Earth (see Gerald F. Joyce and Jack W. Szostak, "Protocells and RNA Self-Replication," *Cold Spring Harbor Perspectives in Biology* 10:a034801. See https://cshperspectives.cshlp.org/content/10/9/a034801.full.pdf and https://www.ncbi.nlm.nih.gov/pmc/articles/PMC6120706/. It is likely, however, that this "RNA world" was an already complex system that took advantage of proteins and other organic molecules, perhaps lipids. The Holy Grail in this field of research would be uncovering the first organism capable of self-replication and discovering how the laws of the Universe could have made it emerge.

274. Gordon M. Bennett and Nancy A. Moran, "Small, Smaller, Smallest: The Origins and Evolution of Ancient Dual Symbioses in a Phloem-Feeding Insect," *Genome Biology and Evolution* 5, no. 9 (2013): 1675–1688, https://doi.org/10.1093/gbe/evt118.

But DNA, RNA, and ribosomes cannot exist independently in nature. Lacking the protective and reproductive mechanisms proper to a living being, they rapidly decay outside the protective membrane of the cell. This means that these three fundamental structures of life can only exist within a cell and can only be produced when the cell duplicates itself. The cell cannot exist without DNA/RNA, which themselves cannot exist independently outside the cell. Science does not yet have a clear explanation for this.

So we return to the original question: what natural mechanism could have led to the emergence of the first cell?

In order to visualize this tiny single-celled organism, let's go back to our car analogy

Let's take the smallest conceivable life form, our bacterium of 0.2 microns, one hundred times smaller than a human cell.[275] If we enlarge it ten million times, it will be the size of a two-meter car. On the same relative scale, the hundreds of thousands of proteins that make it up will each be the size of one- to ten-centimeter car parts. What would we see then?[276]

- Hundreds of openings not unlike a boat's portholes appear on the cell's surface, opening and closing to permit a continual circulation of materials into and out of the cell.

- As we enter the structure through one of these orifices, a world of immense complexity comes into view. A network of corridors and pipes branches out in every direction from the cell's perimeter, leading to information processing units and assembly lines.

---

275. To get a better idea of the internal complexity of cells, consult the following websites. For the simplest free bacterial cell, a Mycoplasma, see: https://ccsb.scripps.edu/gallery/mycoplasma_model/; for eukaryotic cells, see the video "The Inner Life of the Cell" at https://xvivo.com/examples/the-inner-life-of-the-cell/.

276. By increasing the size of a cell in this way, the water molecules that constitute up to 70% of the total content of the cell would each be the size of a small 3 mm polystyrene ball. The cell and all its other internal constituents would actually be obscured in a thick fog. But for the following description we assume that the water molecules are transparent.

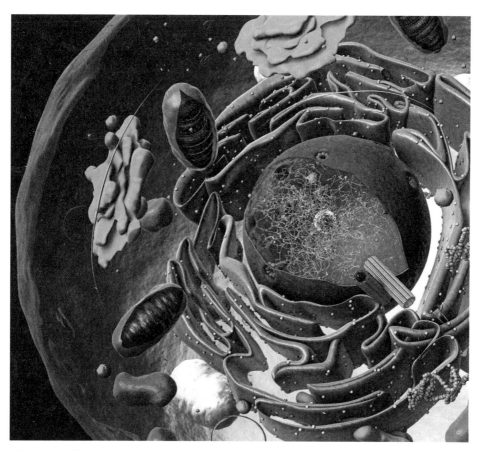
The extraordinary complexity of a simple cell

- Cells lacking a nucleus (prokaryotes) are much larger and more complex than those with a nucleus (eukaryotes), which have the equivalent of a central memory bank. The memory bank is comparable to a spherical chamber, like a geodesic dome, storing dozens of meters of twisted chains of very fine DNA molecules well stacked in orderly rows.

- We marvel at the highly controlled movement of so many objects passing down so many corridors.

- Machines zoom about like robots in every direction. We notice that even the simplest functional parts of the cell, the protein molecules, are astonishingly complex pieces of molecular machinery, each composed of thousands of atoms in a highly organized configuration.

The life of the cell depends on the coordinated activity of numerous RNA molecules and at least 180 different proteins.

- We notice that the cell's workings share many features with our own advanced machines: an artificial language and a decoding system, a data bank for storing and extracting information, control systems that direct the automated assembly of parts and components, automatic quality control devices, assembly procedures based on prefabrication and modular construction.[277]

We're watching a whirring machine working at full steam. It is infinitely more complex than a car—it is a completely automated factory in perpetual movement,[278] with the extraordinary capacity to duplicate itself entirely within a matter of minutes or hours. What the industrial executive can only dream of is achieved by the tiniest of our cells.

We are still far from understanding the natural mechanisms that led to the synthesis of this smallest-conceivable form of life, whose untold complexity continues to astonish scientists. According to microbiologist Werner Arber, 1978 Nobel Prize laureate: *"The most primitive cells may require at least several hundred different specific biological macromolecules. How such already quite complex structures may have come together, remains a mystery to me."*[279]

But if the cell as a whole is one of the greatest scientific enigmas, its principal components—DNA, RNA, proteins, ribosomes, and enzymes—have their own share of mystery.

---

277. Of all the cell's robots, the ribosome is certainly the most remarkable, because its complex engine is responsible for producing the components of all the other robots, made of proteins. As such, many copies of it fill the cell, up to a third of the cell's total mass.

278. One of the fastest bacterial cells is Ovobacter propellens. It has a speed of 200 cell body lengths per second, or 1000 μm/s. Transposed to a macroscopic scale, our blown-up 2-meter-long cell would move at a speed of 400 meters per second, about 1440 km per hour (or ~1,000 miles per hour). To appreciate the speed of some of the basic cellular operations, see http://book.bionumbers.org; Ron Milo et al., *Nucleic Acids Research* 38, suppl. 1, 2010, pp. D750-D753.

279. Werner Arber, "The Existence of a Creator Represents a Satisfactory Solution," in H. Margenau and R. A. Varghese (ed.), *Cosmos, Bios, Theos: Scientists Reflect on Science, God, and the Origins of the Universe, Life,* and *Homo sapiens,* part II, chap. 2 (Chicago: Open Court, 1992), 141.

1. DNA: Double helix, double mystery

The discovery of the double-helix structure of DNA in 1953, attributed to James Watson and Francis Crick, marked a turning point in research. It revealed the existence of a single encoding language common to all life forms.

DNA, or deoxyribonucleic acid, is a long chain composed of four molecules formed from nucleobases: adenine (A), thymin (T), cytosine (C), and guanine (G). Its beautiful double-helix structure opens up to allow the genetic information to be copied.

From bacteria all the way up the hierarchy of life from plants to animals to humans, every form of life on Earth makes use of this unique "language of life."

The 'genetic message' carried by DNA is coded information that, along with epigenetic sources of biological information, programs every stage of your development and physical traits, from basic cell function to the color of your eyes. Scientists call it a "genetic message" because it really does carry a message: an intelligible, carefully composed text that gives sequences of instructions called "genes."

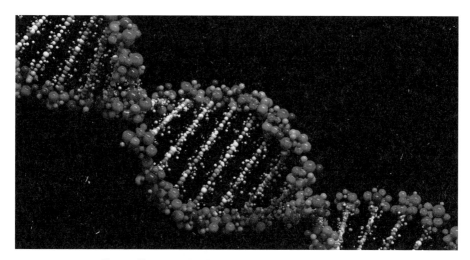

DNA: an astounding collection of information

DNA represents the most sophisticated body of information that we know of in the Universe

The way DNA stores information is a true technological marvel.

If we wrote out all the information stored in the DNA of a single nucleus measuring six millionths of a millimeter, it would take us one million pages, or more than thirty times the length of the *Encyclopedia Britannica*. If all the books ever written by humans (roughly 30,000,000 times the *Encyclopedia Britannica*) were stored using this format, they would fit in a teaspoon.

We are nowhere close to inventing such an efficient data-storage device today, as the density of information in DNA is 40,000 billion times greater than humans' best efforts now in the 21st century.

Hubert P. Yockey, of Berkeley, writes on this topic: "*It is highly relevant to the origin of life that the genetic code is constructed to confront and solve the problems of communication and recording by the same principles found both in the genetic information system and in modern computer and communication codes. There is nothing in the physico-chemical world that remotely resembles reactions being determined by a sequence and codes between sequences.*"[280]

Pierre Sonigo, expert on molecular genetics and Research Director at Inserm, the French equivalent of the USA's NIH, adds: "*DNA seems like a historical cause, the beginning of the story, but no one knows where it comes from, or how it produces life.*"[281] Coming from an atheist scientist, this confession of ignorance should make us think.

---

280. Hubert P. Yockey, "Origin of Life on Earth and Shannon's Theory of Communication," *Computers & Chemistry* 24, no. 1 (January 2000): 105, 200, https://doi.org/10.1016/S0097-8485(00)80010-8.

281. Translated from Pierre Sonigo, "Dieu, la science et la religion," *La Recherche* 14 (January-March 2004).

In 2003, the revelation of the human genome confirmed the astonishing complexity of this "genetic message"[282]

Daniel Cohen, former Professor of Genetics at the Université d'Évry and founder and scientific director of the Genset Corporation (genetic engineering), was one of the very first to map the human genome. He admitted to the French weekly *Le Point*: "*The genome is a program written in a very sophisticated language. Could such a complex language arise by accident? We are free to suppose that, but we can't demonstrate it. Personally, I went from being an atheist to agnostic in the space of one year because if this language is not the result of chance, then I suspect that one day we will be able to demonstrate it. Can you imagine what a sea-change that would be?*"[283]

Christian de Duve, 1974 Nobel laureate in physiology, a biochemist who specializes in cellular biology, claims that the likelihood DNA arose from "pure chance" is one of those "*improbabilities so incommensurably high that they can only be called miracles, phenomena that fall outside the scope of scientific inquiry.*"[284]

Francis Crick (1916–2004) and James Watson (1928–)

---

282. For a long time, it was thought that DNA genes only encoded proteins. However, recent advances in biology have shown that most of the genetic information encoded in the human genome codes for the production of RNA as end products. There is probably no such thing as useless, junk DNA. See: P. P. Amaral, M. E. Dinger, T. R. Mercer, and J. S. Mattick, "The Eukaryotic Genome as an RNA Machine," *Science* 319 (2008): 787-1789.

283. Translated from *Le Point*, October 21, 1995.

284. Cited in Robert Shapiro, "A Simpler Origin for Life," *Scientific American*, February 12, 2007, https://www.scientificamerican.com/article/a-simpler-origin. See also Christian de Duve, *Blueprint for a Cell: The Nature and Origin of Life* (Burlington, NC: N. Patterson, 1991).

Further, Sir Francis Crick, a molecular biologist who won the 1962 Nobel Prize in chemistry for discovering DNA, and is not a believer, admitted that DNA was a mystery: *"A structure like DNA could not have appeared by chance."*[285]

Scientists are unanimous on this point: the origin and development of this "genetic code" constitute a significant double enigma.

2. Proteins or the 3-D letters of the alphabet of life

Proteins are highly complex macromolecules built out of 22 amino acids, which are like the letters used to compose genetic phrases.

A cell cannot exist without proteins because, to take but one example, the proteins at the heart of the cell, along with RNA, execute many essential tasks. These proteins are composed of hundreds or even millions of smaller chemical entities called "amino acids," which are particularly special because they are lateralized: of the 2,000 amino acids found in nature, about half are oriented to the right and half to the left,[286] but only 22 of them, all left-oriented, are useful for life. These combine to form long protein chains. Biologists have compared them to the twenty-six letters of the alphabet, because these "letters" can form a huge number of sequential combinations that function like "words" and "phrases" in the genetic code. If the letters are arranged correctly, they are a comprehensible and useful text. Otherwise, it is gibberish. There are more than 80,000 distinct proteins,[287] and each one rests on a different specific combination of the twenty-two amino acids.

---

285. Quotation translated from the French text. See *Life Itself: Its Origin and Nature* (New York: Simon & Schuster, 1981), 88.

286. In technical language, dextrorotatory and levorotatory, depending on whether they deviate the plane of the polarized light clockwise or counterclockwise, respectively.

287. According to the 2022 update of the Protein Data Bank (https://www.rcsb.org).

The proteins must also be arranged in a particular three-dimensional (3-D) configuration

Proteins are effective in performing cellular tasks like recognizing other molecules and catalyzing a wide range of chemical reactions in part due to their complex and irregular three-dimensional structure. It is the specific arrangement of amino acids within a protein sequence that specifies a particular 3-D configuration. In theory, a long chain of amino acids should be able to take on an infinite number of 3-D configurations, but it has recently been shown that the laws of nature exert physico-chemical constraints that reduce 3-D layouts to less than a thousand possible shapes.[288]

3. The ribosome: a mysterious translator

The ribosome is the incredibly complex piece of machinery responsible for the synthesis of proteins encoded from messenger RNAs, which are temporary copies of DNA genes. While scientists are working to understand how this mechanism functions, its origin continues to elude all understanding. Atheist George Church, Professor of Genetics at Harvard and Director of the Centre for Computational Genetics at MIT, expresses his astonishment on the subject of this "translator" composed essentially of RNA and proteins: *"The ribosome, both looking at the past and at the future, is a very significant structure—it's the most complicated thing that is present in all organisms. [. . .] If I were to be an intelligent design defender, that's what I would focus on; how did the ribosome come to be? If we take all the life forms we have so far, isn't the minimum for the ribosome about 53 proteins and 3 polynucleotides? And hasn't that kind of already reached a plateau where adding more genomes doesn't reduce that number of proteins? That's really marvelous. We need to understand that better. Nobody has constructed a ribosome*

---

288. M. J. Denton, C. J. Marshall, M. Legge, "The Protein Folds as Platonic Forms: New Support for the Pre-Darwinian Conception of Evolution by Natural Law," *Journal of Theoretical Biology* 219, no. 3, 7 (December 2002): 325-342.

that works well without proteins."[289] These proteins themselves are synthesized thanks to a ribosome made up of about two-thirds RNA.

4. Enzymes: incredible time-savers

Enzymes are a crucial kind of protein. Without them, the many chemical reactions that take place in cells every second would take a prohibitively long time.

Richard Wolfenden, Professor of Biochemistry at the University of North Carolina, expert in enzymes and biochemical kinetics and member of the American Academy of Sciences, notes that without the proper enzyme at work in "*the decarboxylation of orotidine 5'-monophosphate decarboxylase (OMPDC),*" a reaction that is "*absolutely essential*" for the synthesis of DNA and RNA, it would take 78 million years to process just half the substrate.[290]

Enzymes get the job done in 18 milliseconds, 10^{17} times faster.[291] And like most enzymes, OMPDC is a protein whose synthesis depends on DNA and RNA.

In the same way, it would normally take 2.3 billion years to process half the substrate needed for another reaction crucial for the synthesis of chlorophyll and hemoglobin,[292] but thanks to their shape-recognition abilities, enzymes make the reaction take place in a few milliseconds. How can we explain the existence of enzymes and their presence at the right place at the right time? No one yet knows.[293]

---

289. *Life: What a Concept!*, ed. J. Brockman (Edge Found), 76-78. Available online: https://www.edge.org/event/life-what-a-concept.

290. A. Radzicka and R. Wolfenden, "A Proficient Enzyme," *Science* 267 (1995): 90-93.

291. "The Rate of Hydrolysis of Phosphomonoester Dianions and the Exceptional Catalytic Proficiencies of Protein and Inositol," *Proceedings of the National Academy of Sciences* 100, no. 10 (2003): 5607–5610, https://doi.org/10.1073/pnas.0631607100.

292. Charles A. Lewis Jr. and Richard Wolfenden, "Uroporphyrinogen Decarboxylation as a Benchmark for the Catalytic Proficiency of Enzymes," *Proceedings of the National Academy of Sciences* 105, no. 45 (November 11, 2008): 17328–17333.

293. It should also be noted that most enzymes are proteins whose synthesis depends on information

5. Two languages and a translator

Since the year 2000, we have known that DNA, proteins, and ribosomes form part of **a unique, sophisticated, and coordinated coding system**.

The 2009 Nobel Prize in Chemistry was awarded to three researchers, Venki Ramakrishnan (Cambridge), Thomas A. Steitz (Yale) and Ada E. Yonath (Israel), who worked on this specific subject: "How does the translator function between the two languages, that of DNA [4 letters] and that of proteins [22 letters]?"[294]

A 2009 press release notes that the prize was awarded in recognition of research into "*...one of life's core processes: the ribosome's translation of DNA information into life.*" The press release continues: "*Ribosomes produce proteins, which in turn control the chemistry in all living organisms. [...] But the DNA molecule is passive. If there was nothing else, there would be no life.*"

In fact all life requires these two languages (DNA and proteins) to be in place, each with its own alphabet (4 bases for DNA, 22 amino acids for proteins), vocabulary, and grammar. The ribosome's translation system is the crucial intermediary between these two languages.[295]

The analogy of language is well suited to explaining how DNA works. The four bases group together in threes to form codons with $4 \times 4 \times 4 = 64$ possible combinations of words, but we notice that only 61 are used to give, through genes, instructions that program our development. What about the 3 unused codons? They function as punctuation ("codon stops") to signify the beginning and end of "phrases." The rules of language are nothing new.

---

stored in the DNA. However, protein synthesis also depends on one of the ribosomal RNAs, which is actually responsible for catalyzing the formation of the peptide bond that links the amino acids together. In fact, protein enzymes are made by an RNA of the ribosome.

294. "The Ribosome Translates the DNA Code into Life," *NobelPrize.org*, Nobel Media AB, April 6, 2021, https://www.nobelprize.org/prizes/chemistry/2009/press-release/.

295. It principally relies on ribosomal RNA molecules that are directly involved in the formation of peptide bonds but also in the process of decoding messenger RNAs with adaptor RNAs, called "transfer RNA"!

This extremely complex system of two languages and their translator must have existed with LUCA (Last Universal Common Ancestor), the common ancestor of all living beings, and has changed very little over the past 3.8 billion years.

> Jacques Monod, a militant atheist, wrote: *"The big problem is the origin of the genetic code and the mechanism of translation. Actually, it is more of an enigma than a problem. The code has no meaning unless it is translated. The translation machine of the modern cell possesses at least fifty macromolecular parts that are also coded in DNA. That means the code can only be translated by products that are the result of a translation. It's the modern version of the chicken and the egg paradox. When and how did the loop close? That is an exceedingly difficult question to think about."*[296]

> Hubert P. Yockey observed: *"The genetic information system is segregated, linear and digital. It is astonishing that the technology of information theory and coding theory has been in place in biology for at least 3.85 billion years."*[297]

How could nature alone have put in place more than 3.85 billion years ago the entire ultra-complex system that characterizes life on Earth? How could this code, the only key to life on Earth as we know it, have created itself and become operational all at once, without any subsequent evolution? All these questions remain unanswered.

6. Metabolism: the whirlpool of life

A living being cannot be reduced to the sum of its parts, each more improbable than the next. The movement underlying the kinetic and stable chemical system that we call "life" itself has to be set in motion.

Metabolism is a series of coordinated movements and highly complex

---

296. Translated from *Le Hasard et la Nécessité* (Coll. Points, Le Seuil, 1970), 181–182.

297. "Origin of Life on Earth and Shannon's Theory of Communication," *Computers & Chemistry* 24, no. 1 (2000): 105–123.

biochemical operations that have to take place without ceasing for an organism to stay alive. If we gather together all the cell's parts at the same place and time, and in the right order, we still have only a dead cell: the equivalent of a factory filled with machines that aren't switched on. A living structure must be capable of autonomous movement and precise activities—constantly assimilating, eliminating, and renewing the material elements that compose it—while also performing a wide range of complex operations to adapt, regenerate if necessary, reproduce, and evolve.

Metabolism involves setting in action a number of catalytic nanomachines crucial to carrying out numerous interactions and coordinated chemical processes every second, and this coordination so essential to life is still very difficult to explain.

In the last fifty years, we have discovered that the complexity of life surpasses anything we could have imagined. Today's leading figures in science have been humbled by this fact.

- Sir Francis Crick, a militant atheist and winner of the 1962 Nobel Prize in Chemistry for the discovery of DNA, explains the situation: *"An honest man, armed with all the knowledge available to us now, could only state that in some sense, the origin of life appears at the moment to be almost a miracle, so many are the conditions which would have had to have been satisfied to get it going."*[298]

- Harold Urey, 1934 Nobel prize winner in Chemistry and atheist, admits: *"All of us who study the origin of life find that the more we look into it, the more we feel it is too complex to have evolved anywhere. We all believe, as an article of faith, that life evolved from dead matter on this planet. It is just that its complexity is so great, it is hard for us to imagine that it did."*[299]

---

298. *Life Itself: Its Origin and Nature* (New York, Simon & Schuster, 1981), 88.

299. Harold C. Urey, quoted in *Christian Science Monitor* (January 4, 1962): 4.

- Franklin Harold, Professor of Microbiology at the University of Washington, ended his work *The Way of the Cell: Molecules, Organisms, and the Order of Life* with these words: *"The origin of life appears to me as incomprehensible as ever, a matter for wonder but not for explication."*[300]

- Lynn Margulis, microbiologist at the University of Massachusetts, member of the American Academy of Sciences and a convinced atheist, underlines the importance of the qualitative leap from inert to living matter: *"To go from a bacterium to people is less of a step than to go from a mixture of amino acids to that bacterium."*[301]

- Michael Denton, former Director of the Genetics Center in Sydney, Professor of Biochemistry at the University of Otago and expert on evolution writes: *"Is it really credible that random processes could have constructed a reality, the smallest element of which—a functional protein or gene—is complex beyond our own creative capacities, a reality which is the very antithesis of chance, which excels in every sense anything produced by the intelligence of man?"*[302]

To get some idea about what nature supposedly accomplished all by itself, let's calculate the probability of a protein emerging by pure chance

There are three reasons why we should calculate this probability:

- It is mathematically approximable.
- Many specialists have already done so, and we present the probabilities they've calculated.
- Proteins belong to the realm of inert matter, so the probability of a protein emerging is not affected by the laws of evolution—the

---

300. Franklin M. Harold, *The Way of the Cell: Molecules, Organisms, and the Order of Life* ((Oxford: Oxford University Press, 2001), 251.

301. As quoted by J. Horgan in *End of Science* (Boston: Addison Wesley, 1995), ch. 5.

302. Michael Denton, *Evolution, A Theory in Crisis* (Chevy Chase: Adler and Adler, 1986), 342.

calculation of the probability of their appearance is not affected by natural selection.

Protein chains consist of between 150 and several million amino acids. What is the minimum probability that a chain composed of 1,000 of the 22 types of amino acids could come together in the right order? Our calculation won't take into account the question of proper 3-D arrangement (extremely improbable) or the success rate of the connection of amino acids through peptide bonds (which only happens successfully half the time), and we'll also ignore the fact that to be used in a living system, an amino acid must have a left enantiomer conformation (not the equally likely right confirmation).[303] Even when we make the task this much simpler, the probability comes out to be one chance in 22^{1000}, in other words about one chance in 10^{1500}.

The number is so large it is impossible to comprehend. To put it in perspective, the likelihood is less than winning the lottery (by drawing the right six numbers between 1 and 49 in order—at one chance in ten billion) every week in a row for 150 weeks[304]—more than two years![305] It is radically impossible for this to happen by chance. Practically speaking, it would take longer than the lifespan of the Universe for a single protein to form by accident out of available amino acids. And since the simplest life form has at least two hundred proteins, we are dealing with improbabilities of staggering proportions, especially since a living

---

303. As we have seen, life on earth is made of exclusively left-handed amino acids. But the chemical reactions that produce these molecules can also make them right-handed (deflecting polarized light to the right). Scientists talk about about the homochirality of life. It is a curious fact that is not well understood.

304. A more rigorous calculation would need to account for multiple sequences of amino acids corresponding to the same functional protein. Such analyses have been performed on different proteins, and many proteins correspond to functional sequences that are so rare that they could never form by chance. See Brian Miller, "A Percolation Theory Analysis of Continuous Functional Paths in Protein Sequence Space Affirms Previous Insights on the Optimization of Proteins for Adaptability," *PLOS ONE* 19, no. 10 (December 2024): e0281789.

305. Even if it were possible in theory, in practice you would never get away with it. If you won three times in a row, you would be arrested for fraud. Then if you protested that your luck was simply a matter of chance, the police would not be convinced and would require another explanation.

cell needs not only two hundred proteins but also DNA, RNA, genes, enzymes, and a good deal of other factors like metabolism.

Professor Frank Salisbury, former Director of Plant Science at Utah State University and member of the American Association for the Advancement of Science (AAAS), has estimated the probability of the formation of a molecule of DNA through natural chemical reactions on 10^{20} *"hospitable"* planets over the course of four billion years at one chance in 10^{415}.[306] Simply put, in practical reality there is no chance at all.

As we shall see later, there are others who have calculated the probability that the two thousand enzymes useful for life could self-assemble by chance at one in $10^{40,000}$, a number that makes virtually no sense in the real world. The possibility that a hundred proteins could self-assemble in parallel, by chance, has been calculated at the ridiculously low chance of one in $10^{2,000}$.

Finally, other scientists estimate at one chance in $10^{340,000}$ the probability that the simplest living cell could arise by chance. The number is dizzying.[307]

The discovery of this near-absolute impossibility is reflected in the facts: there has been no large-scale international effort to recreate the conditions that might have led to the emergence of life

Since the 1970s, scientists have ceased their attempts to obtain life from inert matter. Their resignation provokes a realization: yes, we are today faced with an enigma that surpasses our understanding.

With the help of state governments, scientists often manage to find the time and money to undertake important research that advances our understanding in matters of crucial scientific importance.

---

306. Frank B. Salisbury, "Natural Selection and the Complexity of the Gene," *Nature* 224 (1969): 342–343, https://doi.org/10.1038/224342a0.

307. See below.

- The **Apollo** program launched in the 1960s to get a man on the moon required four hundred thousand people working for ten years, with a total budget of 25 billion dollars (169 billion in today's money). As is well-known, this effort was successful.

- To decrypt the **human genome**, twenty-five thousand researchers worked for twelve years with a total budget of 3 billion euros. They also succeeded.

- **Iter**, an international nuclear fusion research and engineering project whose goal is to use nuclear fusion to produce unlimited free energy for the world, has an operating budget of more than 10 billion euros. The project is still in progress.

However, no international program is trying to recreate and experiment with the conditions that would allow us to observe a natural start of life. Why is this? Probably because so far no one has proposed a single credible scenario for how to cross the chasm that separates living and inert matter. Discovering the origins of life would be a huge leap for human knowledge, but, in the absence of even a minimal roadmap, not even the most princely budget can work miracles.

Eugene Koonin, director of an evolutionary genetics library at the National Center for Biotechnology Information, assesses the current state of scientific thought on the origins of life: *"The origin of life field is a failure—we still do not have even a plausible coherent model, let alone a validated scenario, for the emergence of life on Earth. Certainly, this is due not to a lack of experimental and theoretical effort, but to the extraordinary intrinsic difficulty and complexity of the problem. A succession of exceedingly unlikely steps is essential for the origin of life, from the synthesis and accumulation of nucleotides to the origin of translation; through the multiplication of probabilities, these make the final outcome seem almost like a miracle."*[308]

---

308. *The Logic of Chance: The Nature and Origin of Biological Evolution* (Upper Saddle River, NJ: FT Press, 2012), 391.

However, some programs actively work on "synthetic biology"

There is a sharp distinction between the stalled state of research into the spontaneous emergence of life and the flourishing science of synthetic biology.

Synthetic biology is a robust scientific field that aims to recreate the various components necessary for life out of inert matter. This is a question of copying existing life using the chemical elements present in nature and knowing in advance the desired result, using all our intelligence, all our knowledge, and the extraordinary tools of modern technology.

Although it does not concern itself with the spontaneous emergence of life that takes place in nature, without the intervention of human understanding or modern technological intervention, synthetic biology is nevertheless a fascinating and rapidly developing discipline.

- In September 2017 researchers at seventeen labs in the Netherlands formed the group Building a Synthetic Cell (BaSyC) with the goal of constructing a synthetic cell that can grow and divide, within a ten-year projected time frame. The project is funded by a government grant of 18.8 million euros.[309]

- In September 2018, the US National Science Foundation (NSF) announced the launch of its first synthetic cell program, financed to the tune of ten million dollars.

- Several European researchers have also proposed to construct a synthetic cell within the framework of the European Commission's flagship programs in Future and Emerging Technologies, whith a one billion euro budget.

Even if we can hope to succeed one day in synthesizing a living organism, understanding the conditions under which life emerged naturally remains entirely speculative and outside our comprehension.

---

309. Find a presentation of the group's work here: http://www.basyc.nl/.

According to leading scientists, it's no fluke

- **Anthony Flew**, one of the greatest philosophers of the twentieth century, became an atheist when he was fifteen years old and spent almost his whole life fighting for the atheist cause, most famously writing *Theology and Falsification*. After fifty-four years of militant atheism, he made a profound reversal at 81 years old. He explains: *"I think the DNA material has [. . .] shown, by the almost unbelievable complexity of the arrangements which are needed to produce life, that intelligence must have been involved in getting these extraordinarily diverse elements to work together."*[310]

- **Chandra Wickramasinghe**, professor of applied mathematics and astronomy at University College Cardiff and former collaborator with Fred Hoyle, concludes: *"The likelihood of the formation of life from inanimate matter is 1 to a number with 40,000 noughts after it ($10^{40,000}$)... It is big enough to bury Darwin and the whole theory of evolution. There was no primeval soup, neither on this planet nor any other, and if the beginnings of life were not random, they must therefore have been the product of purposeful intelligence."*[311]

- **Christian de Duve** thinks that chance has to be excluded as a hypothesis: *"I have opted for a Universe that is meaningful and not empty of meaning. Not because I want it to be so, but because this is how I interpret the scientific data we have."*[312]

- **Hubert P. Yockey** is also struck by the improbability of coding the emergence of DNA, pointing out that *"natural selection would have to explore 1.40×10^{70} different genetic codes to discover the universal genetic code found in nature."*[313]

---

310. Antony Flew and Roy Abraham Varghese, "How Did Life Go Live?," in *There Is a God*, Part II, chap. 7 (New York: HarperCollins, 2007), 256.

311. F. Hoyle and C. Wickramasinghe, in *The Biochemist*, 21 (6) (1999): 11-18. And F. Hoyle and C. Wickramasinghe, *Evolution from Space* (New York: Simon & Schuster, 1984), 14.

312. "Does the Universe Have a Purpose? No." *Templeton.org*, John Templeton Foundation, accessed November 18, 2013, https://www.templeton.org.

313. Hubert P. Yockey, *Information Theory and Molecular Biology* (Cambridge: Cambridge Uni-

- **Biologists** have calculated the probability that the thousand enzymes necessary for life could come together in an ordered manner to form a living cell over the course of an evolution of several billion years at less than one chance in $10^{1,000}$. The famous Fred Hoyle (MIT) went even further, declaring: *"Life cannot have had a random beginning... The trouble is that there are about 2,000 enzymes, and the chance of obtaining them all in a random trial is only one part in $10^{40,000}$, an outrageously small probability that could not be faced even if the whole universe consisted of organic soup."*[314]

- Biochemist **Michael Denton** also tried his hands at calculating the probabilities: *"To get a cell by chance would require at least one hundred functional proteins to appear simultaneously in one place. That is one hundred simultaneous events each of an independent probability which could hardly be more than 10^{-20}, giving a maximum combined probability of 10^{-2000}."*[315]

- The molecule of cytochrome c, present among almost all living things and essential for cellular respiration, is composed of one hundred amino acids. **Hubert P. Yockey** has calculated that the probability of this molecule coming into existence through the random linking up of amino acids is one in 10^{65}. Even if there were enough amino acids to fill all the world's oceans (around 10^{42} molecules), while taking account of the constraints due to chirality (the exclusion of right-handed amino acids), the probability plummets to chance in 10^{94}, as he explained at the Tacoma Conference (New York, June 1988).[316]

- **Robert Shapiro**, professor of biochemistry at New York University, focuses his research on bacteria. He has estimated at one chance in

---

versity Press, 1992), 180-183.

314. Fred Hoyle and Chandra Wickramasinghe, *Evolution from Space* (London: J.M. Dent, 1984).

315. Michael Denton, *Evolution: A Theory in Crisis and Nature's Destiny* (Bethesda, MD: Adler & Adler, 1986), 323.

316. Hubert P. Yockey, "Information Theory, Evolution, and the Origin of Life," *Information Sciences* 141, no. 3–4 (April 2002): 219–225.

BIOLOGY: THE INCREDIBLE LEAP FROM INERT TO LIVING MATTER

$10^{40,000}$ the probability that the two-thousand types of proteins present in a simple bacterium could form as a result of chance alone.[317]

- Finally, biophysicist **Harold Morowitz** of Yale University, an expert in the applications of computer science and thermodynamics in biology, calculated the probability that the simplest living cell arose as a result of chance alone at one chance in $10^{340,000}$, which for all practical purposes is evidently equal to zero.[318]

The numbers don't lie, and they all point one way: pure chance is not a reasonable hypothesis for explaining the transition from inert matter to the first living being

Consequently, the various laws and physico-chemical constraints that govern the appearance of life point necessarily to the existence of an intelligent designer.

The preceding chapters on cosmology have already shown that the fine-tuning of the Universe had a one in 10^{120} chance of turning out as it did. As if this number weren't crazy enough already!

But biology has yielded even more outlandishly long odds. The probability that life emerged by chance is mind-blowingly small: $10^{340,000}$, but nothing in the observable Universe can exceed 10^{120}!

The fine-tuning of the laws of the Universe is a genuine proof of the existence of a designer. The argument is convincing *"beyond all reasonable doubt."*

Fred Hoyle, himself a convert from atheism to deism, realized that only a psychological or ideological block can explain why some people refuse

---

317. R. Shapiro, *Origins: A Skeptic's Guide to the Creation of Life on Earth* (New York: Summit Books, 1986), 127.

318. Cited in Mark Eastman and Chuck Missler, *The Creator Beyond Time and Space* (Costa Mesa, CA: TWFT, 1996), 61. See the online article: https://www.jashow.org/articles/the-evolution-of-life-probability-considerations-and-common-sense-part-3/#cite_ref-12. Later Morowitz spoke of one chance in 10 to the power of 100 billion. See H. J. Morowitz, *Energy Flow in Biology* (New York: Academic Press, 1968), 67.

to admit the possibility of a creative intelligence: *"Once we see, however, that the probability of life originating at random is so utterly miniscule as to make it absurd, it becomes sensible to think that the favorable properties of physics on which life depends are in every respect deliberate [. . .] It is therefore almost inevitable that our own measure of intelligence must reflect [. . .] higher intelligences [. . .] even to the limit of God [. . .] such a theory is so obvious that one wonders why it is not widely accepted as being self-evident. The reasons are psychological rather than scientific."*[319]

The fact that we can quantify the incredible leap from the inert to the living is strong evidence for the existence of a creator God

A committed materialist can explain the existence of life in one of two ways:

- By stubbornly persisting in the assumption that this leap is a result of chance alone, not due to a fine-tuning of the laws of the Universe. In that case, he prefers to accept a hypothesis whose probability is so infinitesimal that for all intents and purposes it is impossible, to accepting the existence of a creator God, which by all accounts is an enormously more probable hypothesis.

  One could consider this choice entirely irrational, based more on a categorical rejection of the existence of a creator God than as the result of mature reflection.

- Or, he could view this passage as the outcome of a Universe governed by laws that, among countless possible multiverses, happens to be uniquely conducive to the emergence of life. This is the final trump card we referred to earlier, and it is really just another way of refusing to look the scientific facts in the face. It is to opt for science fiction.

---

319. Fred Hoyle and Chandra Wickramasinghe, *Evolution from Space* (New York: Simon and Schuster, 1984), 14.

On the other hand, accepting the hypothesis of a creator God makes it possible to choose from two explanatory models:[320]

- We can postulate that the leap from inert to living matter took place according to natural laws set in place by a creator God in the very beginning. According to this model, life is the result of precise programming on the part of God, who carefully designed cosmological development to permit life as we know it to develop.

- Or if it's impossible to believe that any number of natural laws could have brought about such a giant leap, we must hold that this particular step was the result of a special intervention by the same creator God.

One piece of evidence that has surprised many scientists is that the passage from inert to living matter seems to have taken place only once in the history of our Earth, and many believe that all living beings are descended from one common ancestor, the celebrated single-celled organism known as LUCA.[321] Nobel laureate Pierre-Gilles de Gennes has expressed his astonishment: *"What intrigues me is the fact that the genetic code is always the same, while life has adapted itself in many ways. I have a hard time believing that darwinian selection could result in only one type of code."*[322]

---

[320]. By analogy, it is possible to compose music directly as the author of the score or by setting up a program that can itself produce creations, as David Cope did with his EMI (Experiments in Musical Intelligence).

[321]. Today, many in the scientific community think that LUCA was an extremely complex cellular system whose genetic code was almost identical to ours. However, LUCA was probably not the first living organism on Earth: it probably appeared after the emergence of the first living cells, through an evolutionary process that took place over a hundred million years. However, the origin of life and the way in which life developed before the appearance of LUCA have long seemed to be shrouded in mystery.

[322]. Translated from an interview given to the Belgian journal *Le Soir*, "Un Nobel décoiffant," October 1996, https://www.lesoir.be/art/les-grands-temoins-un-nobel-decoiffant-pierre-gilles-de_t-19961017-ZoCRND.html. See also his Nobel lecture: https://www.nobelprize.org/prizes/physics/1991/gennes/lecture.

The conclusions drawn from biology join those we have made from cosmology as as additional evidence for the necessity of the existence of a creator God.

At this point, the reader might want to know what scientists themselves think about all these discoveries. The next chapter will answer this question. One hundred citations from leading figures in science will reveal the difficulties and the questions that cutting-edge advances have raised within the scientific community. To our knowledge, this is the first time that these citations, combined with precise references, have been published together in one place.

13.

One Hundred Essential Citations From Leading Scientists

The following one hundred quotes from leading scholars in various scientific fields highlight the surprising and thought-provoking nature of these discoveries. They are essential because they uncover the metaphysical significance of the insights these discoveries provide about reality and existence. Furthermore, they illustrate how weak and even discredited the old materialist frameworks of thought are in comparison.

Nietzsche's claim that God is dead, held by the scientists of the nineteenth and twentieth centuries, has itself suffered a mortal wound. In the science today, God is more alive than ever.

I. Cosmology – Physics – Chemistry

An asterisk indicates that the citation has been previously referenced.

1. *__Alexander Vilenkin__ (1949–), Professor of Theoretical Physics at Tufts University and director of the Tufts Institute of Cosmology, on his theorem demonstrating the necessity of an absolute beginning to the universe: *"It is said that an argument is what convinces reasonable men and a proof is what it takes to convince even an unreasonable man. With the proof now in place, cosmologists can no longer hide behind the possibility of a past-eternal universe. There is no escape: they have to face the problem of a cosmic beginning."*[323]

---

323. Alexander Vilenkin, *Many Worlds in One* (New York: Hill and Wang, 2006), 176.

2. ***Roger Penrose** (1931–), Professor of Mathematics at Oxford and winner of the 2020 Nobel Prize in Physics: "*But in order to start off the universe in a state of low entropy—so that there will indeed be a second law of thermodynamics—the Creator must aim for a much tinier volume of the phase space.*" [*Several pages of argumentative calculations follow.*] "*This now tells us how precise the Creator's aim must have been: namely to an accuracy of one part in 10 to the power of 10^{123}. This is an extraordinary figure. [. . .] Even if we were to write a 'o' on each separate proton and on each separate neutron in the entire universe—and we could throw in all the other particles as well for good measure—we should fall far short of writing down the figure needed.*"[324]

3. ***Richard Feynman** (1918–1988) physicist, pioneer of quantum mechanics and 1965 Nobel laureate in Physics: "*There is a most profound and beautiful question associated with the observed coupling constant, e—the amplitude for a real electron to emit or absorb a real photon. It is a simple number that has been experimentally determined to be close to 0.08542455. [. . .] It has been a mystery ever since it was discovered more than fifty years ago, and all good theoretical physicists put this number up on their wall and worry about it. [. . .] It's one of the greatest damn mysteries of physics: a magic number that comes to us with no understanding by man. You might say the 'hand of God' wrote that number, and 'we don't know how He pushed his pencil'.*"[325]

4. **Christian Anfinsen** (1916–1995), Professor of Chemistry at Harvard and winner of the 1972 Nobel Prize in Chemistry: "*I think only an idiot can be an atheist. We must admit that there exists an incomprehensible power or force with limitless foresight and knowledge that started the whole universe going in the first place.*"[326]

---

324. Sir Roger Penrose, *The Emperor's New Mind: Concerning Computers, Minds and The Laws of Physics* (Oxford: Oxford University Press, 1989), 340 and 344.

325. Quoted in Richard P. Feynman, *QED: The Strange Theory of Light and Matter* (Princeton, NJ: Princeton University Press, 1986), 129.

326. Anfinsen, cited in *Cosmos, Bios, Theos*, ed. Henry Margenau and Roy A. Varghese (Chicago: Open Court, 1997), 139.

5. **Alfred Kastler** (1902–1984), winner of the 1966 Nobel Prize in Physics and the inventor of the laser: *"The idea that the world, the material universe, created itself, seems to me patently absurd. I do not conceive of the world without a creator, which is to say without a god. For a physicist, a single atom is so complicated, so pregnant with intelligence, that the materialist universe simply makes no sense."*[327] **Again:** *"There is no chance of explaining the emergence of life and its evolution by the interaction of chance forces. Other forces are at work."*[328]

6. **Werner Heisenberg** (1901–1976), who invented quantum mechanics and won the 1932 Nobel Prize in Physics: *"Atomic physics has turned science away from the materialistic trend it had during the nineteenth century."*[329]

7. *****Robert Laughlin** (1950–), Professor of Physics at Stanford and winner of the 1998 Nobel Prize in Physics: *"The theories of 'everything' in general, and string theory in particular, left me more and more perplexed all the time, because they are 'unfalsifiable:' no experiential data can prove that they are wrong. I realized that people accept string theory for ideological reasons. It was a terrible shock to me, because I thought that scientists shunned any form of ideology. That's far from the case."*[330]

8. *****Robert W. Wilson** (1936–), the radio astronomer who discovered cosmic microwave background radiation and won the 1978 Nobel Prize in Physics: *"Certainly there was something that set it all off. Certainly, if you are religious, I can't think of a better theory of the origin of the universe to match with Genesis."*[331]

---

327. Translated from J. Duquesne's interview with Alfred Kastler in a special edition of the French weekly *L'Express*, "Dieu et les Français," August 12, 1968.

328. Translated from Alfred Kastler, *Cette étrange matière* (Paris: Stock, 1976).

329. Werner Heisenberg, *Physics and Philosophy: The Revolution in Modern Science* (London: George Allen & Unwin, 1959), chap. 4, 59.

330. Translated from Robert Laughlin, "Les lois physiques ressemblent à un tableau impressionniste," interview with Alexander Hellemans, *La Recherche II*, February 2007; cf. Robert Laughlin, *A Different Universe: Reinventing Physics from the Bottom Down* (New York: Basic Books, 2005).

331. Interview of Wilson cited in F. Heeren, *Show Me God* (Bonner Springs, KS: Day Star, 1995).

9. **William D. Phillips** (1948–), specialist in the laser cooling of atoms and 1997 Nobel Laureate in Physics: "*Why is the universe so finely tuned for the existence of life? More to the point, why is the universe so finely tuned for the existence of us? [. . .] Does this constitute legitimate scientific evidence for an intelligent creator? It may. But it is not universally compelling.*"[332]

10. **Robert Millikan** (1868–1953), the physicist who calculated the charge of the electron and the Planck constant, winner of the 1923 Nobel Prize in Physics: "*A lifetime of scientific research has convinced [me] that there is a divinity who is shaping the destiny of man.*"[333]

11. ***George Thomson** (1892–1975), British physicist, Nobel co-laureate 1937: "*Probably every physicist would believe in a creation [of the Universe] if the Bible had not unfortunately said something about it many years ago and made it seem old-fashioned.*[334]

12. **Arthur Schawlow** (1921–1999), professor at Stanford, co-inventor of the laser, and 1981 Nobel laureate in Physics: "*The world is just so wonderful that I can't imagine it just having come by pure chance.*"[335]

13. **Robert Jastrow** (1925–2008), astrophysicist, professor at Columbia, and director of NASA: "*For the scientist who has lived by his faith in the power of reason, the story ends like a bad dream. He has scaled the mountains of ignorance, he is about to conquer the highest peak; as he pulls himself over the final rock, he is greeted by a band of theologians who have been sitting there for centuries.*"[336]

---

332. *Science and the Search for Meaning, Perspectives from International Scientists*, ed. Jean Staune (West Conshohocken, PA: Templeton Press, 2006), 198.

333. *The Autobiography of Robert A. Millikan* (New York: Arno Press, 1980). As cited in the *Observance of Rural Life Sunday by 4-H Clubs, 1952: Theme, Serving as Loyal Citizens Through 4-H* (United States Department of Agriculture Extension Service, 1952), 10.

334. Sir George Paget Thomson, "Continuous Creation and the Edge of Space," *New Republic* 124 (1951): 21–22.

335. Arthur Schawlow, *Optics and Laser Spectroscopy, Bell Telephone Laboratories, 1951-1961, and Stanford University Since 1961*, Regional Oral History Office, The Bancroft Library (Berkeley, CA: University of California, 1998), 19.

336. Robert Jastrow, *God and the Astronomers* (New York: W. W. Norton, 1978), 116, https://www.nytimes.

ONE HUNDRED ESSENTIAL CITATIONS FROM LEADING SCIENTISTS   267

14. **Paul Davies** (1946–), cosmologist and astrobiologist, Professor of Theoretical Physics at the University of Adelaide (and formerly at Cambridge): *"The Multiverse Theory seems to have become the main explanation for scientists to justify the remarkable ability of our Universe to welcome life. But this theory is problematic for me."*[337] *"The death of scientism, of its determinism, of its dream of a transparent science that can access the secret of the Universe, was a kind of torture for the Nobel Prize Laureates who lived the quantum adventure."*[338]

15. **Antony Hewish** (1924–2021), Professor of Astronomy at Cambridge and winner of the 1974 Nobel Prize in Physics for the discovery of pulsars: *"I believe in God. It makes no sense to me to assume that the Universe and our existence is just a cosmic accident, that life emerged due to random physical processes in an environment which simply happened to have the right properties. [. . .] God certainly seems to be a rational Creator. That the entire terrestrial world is made from electrons, protons and neutrons and that a vacuum is filled with virtual particles demands incredible rationality."*[339]

16. ***Arno A. Penzias** (1933–2024), Nobel Prize in Physics 1978: *"In order to achieve consistency with our observations we must assume not only creation of matter and energy out of nothing, but creation of space and time as well. The best data we have are exactly what I would have predicted, had I nothing to go on but the five Books of Moses, the Psalms, the Bible as a whole. The Big Bang is a moment of discrete creation from nothing!"*[340]

17. **Richard Smalley** (1943–2005), Professor of Chemistry at Rice University and 1996 Nobel laureate in Chemistry: *"Although I suspect I will*

---

com/1978/06/25/archives/have-astronomers-found-god-theologians-are-delighted-that-the.html.

337. Paul Davies, *Superforce: The Search for a Grand Unified Theory of Nature* (New York: Simon & Schuster, 1985).

338. Paul Davies, "Symmetry and Beauty," in *Superforce* (New York: Simon & Schuster, 1984), 51.

339. Antony Hewish, letters to T. Dimitrov, May 27 and June 14, 2002.

340. Lecture at Northern Illinois University, 1992, quoted in Jerry Bergman, "A. Penzias: Astrophysicist, Nobel Laureate," (1992).

never fully understand, I now think the answer is very simple: it's true. God did create the universe about 13.7 billion years ago, and of necessity has involved Himself with His creation ever since. The purpose of this universe is something that only God knows for sure, but it is increasingly clear to modern science that the universe was exquisitely fine-tuned to enable human life."[341]

18. **Charles Townes** (1915–2015), physicist, professor at Berkeley, winner of the 1964 Nobel Prize in Physics, and former director of NASA: *"I strongly believe in the existence of God, based on intuition, observations, logic, and also scientific knowledge."*[342] **Again:** *"Many [cosmologists] have a feeling that somehow intelligence must have been involved in the laws of the universe."*[343] **Also:** *"Determinism no longer holds water... Biologists have not yet learned the limits of their knowledge, though it is staring them in the face."*[344]

19. **Marc Halévy** (1953–), physicist, pupil of Prigogine: *"Let's assume, along with Stephen Hawking and Steven Weinberg, that the development of our universe since the Big Bang is the result of pure chance alone. This universe, guided totally by chance, managed to synthesize a self-replicating RNA molecule. The probability that this synthesis took place according to the laws of chance can be calculated. It is thus also possible to calculate the time it would take for a chance-based universe to do it. It turns out that this time is millions of millions of times greater than the age of our present universe. [. . .] Thus, the chance hypothesis is refuted in its own language: the calculation of probabilities."*[345]

---

341. Letter read at the Alumni Weekend banquet, May 2005, Hope College, Holland, Michigan. His declining health prevented him from attending in person.

342. Charles Townes, letter to T. Dimitrov, May 24, 2002.

343. Cited in Sharon Begley, "Science Finds God," *Newsweek* 132, issue 3 (Jul 20, 1998): 49.

344. Translated from "Sommes-Nous Les Enfants Du Hasard?" 5ᵉ Colloque de l'Association des Scientifiques Chrétiens, 2003.

345. Cited in *Implications philosophiques et spirituelles des sciences de la complexité*, conference at the *Université interdisciplinaire de Paris* (UIP), March 2009.

20. **George Smoot** (1945–) Astrophysicist and cosmologist, professor at Berkeley, Nobel laureate 2006: *"The most cataclysmic event that we could imagine—the Big Bang—appears, when closely examined, to be precisely orchestrated."*[346] On the detailed images provided by the WMAP satellite, he said: *"If you're religious, it's like looking at God. The order is so beautiful and the symmetry so beautiful that you think there is some design behind it."*[347] Then, in 2006, when he received the Nobel Prize for his striking images of the first light of the Universe: *"It is like seeing God... I saw the Universe at its very beginning, I saw the anisotropy that allowed the Universe to exist."*[348]

21. **Donald Page** (1948–), Professor of Physics and Cosmology at the University of Alberta: *"In view of all the evidence, including both the elegance of the laws of physics, the existence of orderly sentient experiences, and the historical evidence, I do believe that God exists and think the world is actually simpler if it contains God than it would have been without God."*[349]

22. **Trinh Xuan Thuan** (1948–), Buddhist astronomer, Professor of Astrophysics at the University of Virginia and research associate at the Institut d'Astrophysique de Paris: *"One of the most surprising discoveries of modern cosmology is the realization that the initial conditions and physical constants of the universe had to be adjusted with exquisite precision if they are to allow the emergence of conscious observers. This realization is referred to as the 'anthropic principle.' Change the initial conditions and physical constants ever so slightly, and the universe would be empty and sterile; we would not be around to discuss it. The precision of this fine-tuning is nothing short of stunning: The*

---

346. George Smoot and Keay Davidson, *Wrinkles In Time: The Imprint of Creation* (London: Abacus, 1995), 135.

347. As reported by Rae Corelli, Marci McDonald, and Hilary Mackenzie, "'Looking at God,'" *Maclean's*, May 4, 1992, 38–39, available at https://web.archive.org/web/20221229200538/https://archive.macleans.ca/article/1992/5/4/looking-at-god.

348. George Smoot and Keay Davidson, *Wrinkles In Time: The Imprint of Creation* (London: Abacus, 1995), 135.

349. Donald Page, "Guest Post: Don Page on God and Cosmology," Sean Carroll (blog), March 20, 2015, https://www.preposterousuniverse.com/blog/2015/03/20/guest-post-don-page-on-god-and-cosmology/

initial rate of expansion of the universe, to take just one example, had to have been tweaked to a precision comparable to that of an archer trying to land an arrow in a 1-square-centimeter target located on the fringes of the universe, 15 billion light-years away!"[350] **Again:** *"I reject the hypothesis of chance because, aside from the meaninglessness and despair it would entail, I cannot conceive that the harmony, symmetry, unity, and beauty we perceive not only in the world, from the delicate pattern of a flower to the majestic architecture of galaxies, but also… in the laws of Nature, are the mere results of chance. If we accept the hypothesis of a single universe—our own—we must also postulate a Primary Cause that from the outset tweaked the laws of physics and initial conditions in order for the universe to become conscious of itself." "If we accept the hypothesis of a single universe—our own—we must also postulate a Primary Cause."*[351]

23. ***Edward Harrison** (1919–2007), Professor of Astrophysics at the University of Massachusetts at Amherst: *"Here is the cosmological proof of the existence of God—the design argument of Paley—updated and refurbished. The fine tuning of the universe provides prima facie evidence of deistic design. Take your choice: blind chance that requires multitudes of universes or design that requires only one." "Many scientists, when they admit their views, incline toward the teleological or design argument."*[352]

24. **Robert Dicke** (1916–1997) Professor of Physics at Princeton, discoverer of the fine-tuning of the Universe: *"The puzzle here is the following: how did the initial explosion become started with such precision, the outward radial motion become so finely adjusted as to enable the various parts of the Universe to fly apart while continuously slowing in the rate of expansion? There seems to be no*

---

350. Trinh Xuan Thuan, *Chaos and Harmony: Perspectives on Scientific Revolutions of the Twentieth Century*, trans. Axel Reisinger (Oxford: Oxford University Press, 2001), 235.

351. Ibid., 333.

352. Edward Harrison, *Masks of the Universe, Changing Ideas on the Nature of the Cosmos* (Cambridge: Cambridge University Press, 2003), 252, 263.

fundamental theoretical reason for such a fine balance. If the fireball had expanded only .1 percent faster, the present rate of expansion would have been 3×10^3 as great. Had the initial expansion rate been .1 per cent less and [sic] the Universe would have expanded to only 3×10^{-6} of its present radius before collapsing... No stars could have formed in such a Universe, for it would not have existed long enough to form stars."[353]

25. ***Lee Smolin** (1955–), materialist physicist: *"We must understand how it came to be that the parameters that govern the elementary particles and their interactions are tuned and balanced in such a way that a universe of such variety and complexity arises. [. . .] We should ask just how probable is it that a universe created by randomly choosing the parameters will contain stars. [. . .] The answer, in round numbers, comes to about one chance in 10^{229}."*[354] Again: *"The universe is improbable, and it is improbable in the sense that it has a structure which is much more complex than it would be if its laws and initial conditions were chosen more randomly."*[355]

26. ***Gregory Benford** (1941–), Professor of Physics and Astronomy at the University of California, Irvine: *"This 'multiverse' view represents the failure of our grand agenda and seems to me contrary to the prescribed simplicity of Ockham's Razor, solving our lack of understanding by multiplying unseen entities into infinity."*[356]

27. **Brian Greene** (1963–) Professor of Physics and Mathematics at Columbia, expert in string theory: *"The laws of physics are the closest thing in the world today that I would align with a theological perspective. I view the world as extraordinarily coherent and harmonious because of the*

---

353. Robert Dicke, *Gravitation and the Universe: Jayne Lectures for 1969*, vol. 78 (Philadelphia: American Philosophical Society, 1970), 62.

354. Lee Smolin, *The Life of the Cosmos* (Oxford: Oxford University Press, 1997), 44–45.

355. Lee Smolin and Ivan Briscoe, "Cosmological Evolution," *The UNESCO Courier* 54, no. 5 (May 2001), 34.

356. Gregory Benford, *What We Believe But Cannot Prove*, ed. John Brockman (New York: Harper Perennial, 2006), 226.

way the fundamental laws of physics are able, with just a few symbols etched on a piece of paper, to describe a wealth of phenomena, from subatomic particles to the edge of the universe. That resonates with me in a similar way as the order, organization and explanation for the universe that adherents of religion will ascribe to a deity. From my perspective, that order and organization come from fundamental physical laws; for others, that's the role that God plays."[357]

28. **Henry F. Schaefer** (1944–), Professor of Chemistry, Director of the Center for Computational Chemistry at the University of Georgia, and one of the most-quoted chemists in the world: *"Against powerful logic, some atheists continue to claim, irrespective of the anthropic constraints, that the universe and human life were created by chance."*[358]

29. **Werner Gitt** (1937–), Professor and Director of the Federal Institute of Physics and Technology, Brunswick: *"All experiences indicate that a thinking being voluntarily exercising his own free will, cognition, and creativity is required." "There is no known law of nature, no known process and no known sequence of events which can cause information to originate by itself in matter."*[359]

30. **Carlo Rubbia** (1934–) Professor of Physics at Harvard, Director of CERN, expert in particle physics, 1984 Nobel laureate: *"Speaking of the origin of the world brings us inevitably to think of creation, and, in considering nature, we find that it has an order too precise to have been determined by 'chance,' from confrontations between 'forces' that we—physicists—continue to maintain. However, I believe that the existence of a pre-established order of things is more evident among us*

---

357. Quoted in *The Evidence: God, the Universe & Everything* (Nashville, TN: The Apologetics Group, 2002), DVD. https://www.amazon.com/Evidence-God-Universe-Everything/dp/B000BS6XD4.

358. Henry F. Schaefer, "The Big Bang, Stephen Hawking, and God," the written version of the 2004 New College Lecture series' second public address, delivered on October 13, 2004, https://newcollege.unsw.edu.au/downloads/File/pdf/Lectures_Summaries/The Big Bang.pdf.

359. Werner Gitt, *In the Beginning Was Information*, trans. Jaap Kies, (Bielefeld, Germany: Christliche Literatur-Verbreitung, 1997), 65, 107, https://bruderhand.de/download/Werner_Gitt/Englisch-Am_Anfang_war_die_Info.pdf.

than elsewhere. We come to God by the path of reason, others follow the irrational path."[360]

31. **Derek Barton** (1918–1998), Professor of Chemistry at Imperial College and at Harvard, 1969 Nobel laureate in Chemistry: "*The observations and experiments of science are so wonderful that the truth that they establish can surely be accepted as another manifestation of God. God shows himself by allowing man to establish truth.*"[361]

32. **Jacques Demaret** (1943–1999), cosmologist specializing in the anthropic principle and professor at the University of Liège: "*The mere fact that we are present here in the universe tells us something about the value of the fundamental constants (e.g., the physical constants such as the Planck constant or the fine structure constant). Though scientists do not know why these constants have the specific values they do, what we are certain about is that if they varied even slightly, then life could never have emerged.*"[362]

33. **John Polkinghorne** (1930–2021), Professor of Mathematics and Physics at the University of Cambridge: "*The precise adjustment of the initial conditions of the Universe is an indispensable element of the existence of the world... Through their discoveries, scientists encounter the divine Logos.*"[363]

34. **Isidor Isaac Rabi** (1898–1988), winner of the 1944 Nobel Prize in Physics: "*Physics filled me with awe, put me in touch with a sense of original causes. Physics brought me closer to God. That feeling stayed with me throughout my years in science.*"[364]

---

360. "L'ADN le prouve: la vie sur terre n'a qu'un père," *Libéral*, December 23, 2011, https://www.uccronline.it/wp-content/uploads/2012/08/20111223rubbia.pdf.

361. Derek Barton, *Cosmos, Bios, Theos*, ed. Henry Margenau and Roy Abraham Varghese (Chicago: Open Court, 1997), 145.

362. Translated from a 1995 interview in *Libération*: www.liberation.fr/sciences/1995/02/15/question-a-jacques-demaret-l-homme-etait-il-obligatoire_123180

363. John Polkinghorne, "The Universe and Everything," quoted in *The Evidence: God, the Universe & Everything* (Nashville, TN: The Apologetics Group, 2002), DVD. https://www.amazon.com/Evidence-God-Universe-Everything/dp/B000BS6XD4.

364. Gerald Holton, "I. I. Rabi As Educator and Science Warrior," *Physics Today* 52 (Sept. 1999),

35. **Herbert Uhlig** (1907–1993), physical chemist and Professor of Metalurgy at MIT: *"The origin of the universe can be described scientifically as a miracle."*[365]

36. **Shoichi Yoshikawa** (1935–2010), Professor of Astrophysics at Princeton: *"I think that God originated the universe and life. Homo sapiens was created by God using a process which does not violate any of the physical laws of the Universe in any significant way."*[366]

37. **Antonino Zichichi** (1929–), physicist at CERN, President of the European Physical Society and of the World Federation of Scientists: *"Without science, we would have no retort for an atheistic culture that wishes we were nothing more than sons of chance."*[367]

38. ***Freeman Dyson** (1923–2020), physicist, astrophysicist, futurologist, and professor at Princeton: *"As we look out into the Universe and identify the many accidents of physics and astronomy that have worked together to our benefit, it almost seems as if the Universe must in some sense have known that we were coming."*[368]

39. ***Max Born** (1882–1970), physicist and Professor of Theoretical Physics at Göttingen, Nobel Prize in Physics, 1954: *"If alpha [the fine structure constant] were bigger than it really is, we should not be able to distinguish matter from ether [the vacuum, nothingness], and our task to disentangle the natural laws would be hopelessly difficult. The fact however that alpha has just its value 1/137 is certainly no chance but*

---

37. Also quoted in John S. Rigden, "Nearer to God," in *Rabi, Scientist and Citizen* (Cambridge, MA: Harvard University Press, 1987).

365. Herbert Uhlig, "The Origin of the Universe Can Be Described Scientifically as a Miracle," in *Cosmos, Bios, Theos*, ed. Henry Margenau and Roy Abraham Varghese (La Salle, IL: Open Court, 1992), 125.

366. Shoichi Yoshikawa, "The Hidden Variables of Quantum Mechanics Are Under God's Power," in *Cosmos, Bios, Theos*, ed. Henry Margenau and Roy Abraham Varghese (La Salle, IL: Open Court, 1992), 135.

367. Antonino Zichichi, "Scientific Culture and the Ten Statements of John Paul II," in *The Cultural Values of Science, Scripta Varia* 105 (Vatican City: Pontifical Academy of Sciences, 2003).

368. Freeman J. Dyson, *Scientific American* 224 (September 1971), 50.

itself a law of nature. It is clear that the explanation of this number must be the central problem of natural philosophy."[369]

40. **John Barrow** (1952–2020), Professor of Astronomy at Cambridge, speaking about theories of the Universe emerging *"from nothing:"* *"These theories necessarily assume the existence of far more than what we would usually consider 'nothing.' From the beginning, there must be laws of nature, energy, mass, geometry, and, implicitly, the world of mathematics and of logic. There must be a considerable underlying structure based on rationality [. . .] One must remain aware of the fact that many studies on cosmology are motivated by the desire to avoid an initial singularity."*[370]

41. **Arthur Compton** (1892–1962), Professor of Physics at Princeton and winner of the 1927 Nobel prize in Physics: *"If religion is to be acceptable to science it is important to examine the hypothesis of an Intelligence working in nature. The discussion of the evidences for an intelligent God is as old as philosophy itself. The argument on the basis of design, though trite, has never been adequately refuted. On the contrary, as we learn more about our world, the probability of its having resulted by chance processes becomes more and more remote, so that few indeed are the scientific men of today who will defend an atheistic attitude."*[371]

42. **George Efstathiou** (1955–) Professor of Cosmology, Director of the Institute of Astronomy at Cambridge, and member of the scientific team of the European Space Agency's Planck satellite: *"Even if the probability of developing a Universe is very low with a weak cosmological constant, if there were an infinity of universes it would have appeared somewhere [. . .] But is this really an explanation or a desperate answer?"*[372]

---

369. Max Born, quoted in Arthur I. Miller, *Deciphering the Cosmic Number: The Strange Friendship of Wolfgang Pauli and Carl Jung* (New York: W.W. Norton, 2009), 253.

370. John Barrow and Frank Tipler, *The Anthropic Cosmological Principle* (Oxford: Oxford University Press, 1986).

371. Arthur Compton, *The Freedom of Man* (New Haven, CT: Yale University Press, 1935), 73.

372. George Efstathiou, "An Anthropic Argument for a Cosmological Constant," *Monthly Notices of the Royal Astronomical Society* 274, no. 4 (1995): 73–76.

43. **Max Planck** (1858–1947), one of the founders of quantum mechanics, 1918 Nobel Laureate in Physics, and discoverer of the quantum structure of radiation: *"Metaphysical reality does not stand spatially behind what is given in experience, but lies fully within it."*[373] **Again:** *"Anybody who has been seriously engaged in scientific work of any kind realizes that over the entrance to the gates of the temple of science are written the words: You must have faith. It is a quality which the scientists cannot dispense with."*[374] **Also:** *"All matter originates and exists only by virtue of a force [. . .]. We must assume behind this force the existence of a conscious and intelligent spirit."*[375]

44. **Tony Rothman** (1953–), cosmologist and Professor of Physics at Wesleyan University: *"When confronted with the order and beauty of the universe and the strange coincidences of nature, it's very tempting to take the leap of faith from science into religion. I am sure many physicists want to. I only wish they would admit it."*[376]

45. **Max Tegmark** (1967–), astronomer at MIT: *"The [multiple] parallel universes allow us to ignore the fine tuning of the cosmos' initial conditions and of the fundamental constants. [. . .] At the end of the day, we are making our judgment based on nothing more than what we find most exquisitely to our taste."*[377]

46. **Dr. Allan Sandage** (1926–2010), one of the most celebrated astronomers of our day, who recognized his belief in God at the age of fifty: *"The world is too complicated in all parts and interconnections to be due to chance alone."*[378]

---

373. Max Planck, *Scientific Autobiography* (New York: Philosophical Library, 1949), 98.

374. Max Planck, *Where Is Science Going?* (London: Allen & Unwin, 1913), 214.

375. Max Planck, "Das Wesen der Materie" ["The Nature of Matter"], speech at Florence, Italy (1944), in *Archiv zur Geschichte der Max-Planck-Gesellschaft* (Boltzmannstraße 14, D-14195 Berlin-Dahlem).

376. Tony Rothman and George Sudarshan, *Doubt And Certainty: The Celebrated Academy Debates On Science, Mysticism, Reality* (Cambridge, MA: Helix Books, 1999).

377. Quoted in Pascual Jordan, "Creation and Development," in *Science and the Course of History* (New Haven, CT: Yale University Press, 1955), 108–119.

378. Cited in Allan Sandage, "A Scientist Reflects on Religious Belief," *Truth Journal* 1 (1985).

47. **Lothar Schäfer** (1939–2020), Professor of Quantum Chemistry at the University of Arkansas: *"At the root of all ordinary objects we find entities that possess the rudiments of consciousness. [. . .] In quantum leaps, quantum systems act spontaneously. A spirit is the only thing we know of that can act that way. One cannot use science as a basis for atheism—it's a lost cause."*[379]

48. *****Fred Hoyle** (1915–2001), cosmologist and astronomer: *"The chance that higher life forms might have emerged in this way is comparable with the chance that a tornado sweeping through a junk-yard might assemble a Boeing 747 from the materials therein."*[380] Hoyle and Wickramasinghe calculate the probability of the 2,000 enzyme proteins needed to accelerate the chemical reactions needed to cause life at 1 chance in $10^{40,000}$, calling this *"an outrageously small probability that could not be faced even if the whole universe consisted of organic soup."*[381] *"I have always thought it curious that, while most scientists claim to eschew religion, it actually dominates their thoughts more than it does the clergy."*[382] Again: *"I do not believe that any scientist who examined the evidence would fail to draw the inference that the laws of nuclear physics have been deliberately designed with regard to the consequences they produce inside the stars."*[383]

49. **Robert Kaita** (1952–), Professor of Physics and Astrophysics at Princeton: *"Science is impossible without accepting that we live in a 'caused' universe that ultimately should encourage one to acknowledge its 'first cause' or Creator."*[384]

---

379. Translated from Schäfer, "L'importance de la physique quantique pour la pensée de Teilhard de Chardin et pour une nouvelle vue de l'évolution biologique," (Rome: Colloque Teilhard, 2009), http://www.groupebena.org/IMG/pdf/L_Schafer_Physique_quantique.pdf.

380. Fred Hoyle, "Hoyle on Evolution," *Nature* 294 (November 12, 1981): 105.

381. Fred Hoyle and Chandra Wickramasinghe, *Evolution from Space: a Theory of Cosmic Creationism* (New York : Simon and Schuster, 1982), 24.

382. Cited by Paul Davies in *Mind of God: The Scientific Basis for a Rational World* (New York: Simon & Schuster, 1992), 223.

383. "Fred Hoyle," transcript of address by Hoyle in *Religion and the Scientists: Addresses Delivered in the University Church*, Cambidge, ed. Mervyn Stockwood (London: SCM Press Ltd, 1959), 64.

384. Robert Kaita, "What Does our Universe Reveal: A Designer, Creator, or Nothing at All?" in

50. **Wernher von Braun** (1912–1977), former director at NASA and inventor of the V-2 rocket (the first ballistic missile used during World War II): "*One cannot be exposed to the law and order of the universe without concluding that there must be design and purpose behind it all. [. . .] The better we understand the intricacies of the universe and all it harbors, the more reason we have found to marvel at the inherent design upon which it is based. [. . .] To be forced to believe only one conclusion—that everything in the universe happened by chance—would violate the very objectivity of science itself. Certainly there are those who argue that the universe evolved out of a random process, but what random process could produce the brain of a man or the system of the human eye?*"[385]

51. **Henry Lipson** (1910–1991), president of the Physics Department at the University of Manchester: "*I think we must admit that the only acceptable explanation is creation. I know that this is anathema to physicists, as indeed it is to me, but we must not reject a theory that we do not like if the experimental evidence supports it.*"[386]

52. **John O'Keefe** (1916–2000), NASA astronomer, planetary scientist, and a major leader in developing the American lunar science program during the early Project Apollo era: "*If the universe had not been made with the most exacting precision we could never have come into existence. It is my view that these circumstances indicate the universe was created for man to live in.*"[387]

53. **Vincent Fleury** (1963–), biophysicist, researcher at CNRS, and teacher at the École Polytechnique: "*What is consonant with physics as we currently understand it is either a God who fixed everything at the*

Science, Christian Perspectives For The New Millennium, ed. Paul Copan et al., (Addison, TX and Northcross, GA: ClM and RZIM, 2003), 137. See also the documentary series *The Evidence: God, the Universe and Everything*, directed by Robert Kaita, Fred Alan Wolf, and Elisabet Sahtouris (2001).

385. Quoted in Dennis R. Petersen, *Unlocking the Mysteries of Creation* (El Dorado, CA: Creation Resource Foundation, 1990), 63.

386. Henry Lipson, "A Physicist Looks at Evolution," *Physics Bulletin*, volume 31 (1980): 138.

387. John A. O'Keefe, "The Theological Impact of the New Cosmology," in Robert Jastrow, *God and the Astronomers* (New York: W. W. Norton, 1992), 118.

beginning, or one who intervenes in the reduction of a quantum wave packet. Everything else is physically impossible unless God does not respect the laws of physics."[388]

54. **Walter Kohn** (1923–2016), physics professor at the University of California, and winner of the 1998 Nobel Prize in Chemistry: *"There continue to be very deep epistemological questions about the significance of sharp scientific laws like the laws of quantum mechanics and the laws that govern the nature of chaos. Both of these fields have irreversibly shaken the 18th and 19th centuries' purely deterministic, mechanistic view of the world."*[389]

55. **Pierre Perrier** (1935–), mathematician, logician, and physicist: *"The argument that it was all simply a matter of time, and that there was a sufficient amount of it after the Earth had formed, flies in the face of numerical reality. The proponents of this or that metaphysics must take into account the combinatorial explosion that results from all possible configurations, while also respecting the chronologies and duration of formation of the real world that surrounds us, which forms the backdrop against which we may launch attempts to falsify the available theories."*[390]

56. **Bernard d'Espagnat** (1921–2015), Professor of Physics at the University of Paris: *"Quantum mechanics has delivered us from the lead cloak of deterministic materialism."* Again: *"Bohr has unmade what Copernicus had made. He has put man back at the center of his own representation of the universe."*[391]

57. **Stephen Hawking** (1942–2018), Professor of Mathematics at Cambridge (who, despite the following words, was an atheist all the same): *"If the*

---

388. Translated from Vincent Fleury, "Sur la toile," January 4, 2007.

389. Walter Kohn, "Dr. Walter Kohn: Science, Religion, and the Human Experience," interview by John F. Luca, *The Santa Barbara Independent*, July 26, 2001.

390. Translated from Pierre Perrier, in *La science, l'homme et le monde—Les nouveaux enjeux*, ed. Jean Staune (Paris: Presses de la Renaissance, 2008), 230.

391. Bernard d'Espagnat, *À la recherche du réel, le regard d'un physicien* (Paris: Dunod, 2015).

rate of expansion one second after the Big Bang had been smaller by even one part in a hundred thousand million million, the universe would have recollapsed before it ever reached its present size." *Again: *"The laws of science [. . .] contain many fundamental numbers. [. . .] The remarkable fact is that the values of these numbers seem to have been very finely adjusted to make possible the development of life."* Again: *"What is it that breathes fire into the equations and makes a universe for them to describe?"*[392] Finally: *"The odds against a universe like ours emerging out of something like the Big Bang are enormous... I think clearly there are religious implications whenever you start to discuss the origins of the universe. But I think that most scientists prefer to shy away from the religious side of it."*[393]

58. **Sir James Hopwood Jeans** (1877–1946), British physicist, astronomer, and mathematician: *"The stream of knowledge is heading towards a non-mechanical reality; the universe begins to look more like a great thought than like a great machine. Mind no longer appears as an accidental intruder into the realm of matter."*[394]

59. **Joe Rosen** (1937–), Professor of Physics at Tel Aviv University and the University of Central Arkansas: *"[The anthropic principle] is among the most important fundamental principles around, even, let me dare venture, the most basic principle we have."*[395]

60. **Nicolas Gisin** (1952–), Professor of Physics at the University of Geneva and specialist in quantum mechanics: *"Non-local correlations seem, in some way, to emerge from outside space-time. Who keeps track of who is entangled with whom, i.e. where is the information stored? Are there angels manipulating enormously huge Hilbert spaces? Let us emphasize that we take such questions very seriously."*[396]

---

392. Stephen Hawking, *A Brief History of Time* (New York: Bantam, 1988), 125.

393. Stephen Hawking, *A Brief History of Time* (New York: Bantam, 1988), 174; see also David Filkin, *Stephen Hawking's Universe: The Cosmos Explained* (New York: Basic Books, 1998), 109.

394. James Hopwood Jeans, *The Mysterious Universe* (New York: Macmillan, 1931), 137.

395. Joe Rosen, "The Anthropic Principle," *American Journal of Physics* 53 (1985): 335.

396. Nicolas Gisin and Florian Fröwis, "From Quantum Foundations to Applications and Back,"

61. **David Gross** (1941–), Professor of Theoretical Physics at the University of California, 2004 Nobel laureate: *"The danger of the anthropic principle is that it is impossible to disprove."*[397]

62. **Geoffrey Chew** (1924–2019), Professor of Theoretical Physics at Berkeley: *"Appeal to God may be needed to answer the 'origin' [of the Universe] question."*[398]

63. **Hubert Reeves** (1932–2023), popular astronomer and Professor of Astrophysics at the University of Paris and the University of Montreal: *"[The laws governing physical forces] possess remarkable properties. They seem to us to be 'fine-tuned' to promote complexity. Tiny variations of the numerical values that specify them would be enough to render the universe sterile. No life form, no complex structure would ever have arisen from it. [. . .] Not even an atom of carbon."*[399]

II Biology and Life Sciences

64. ***Sir Francis Crick** (1916–2004), codiscoverer of DNA in 1953, winner of the 1962 Nobel Prize in Medicine: *"At present, the gap from the primal 'soup' to the first RNA system capable of natural selection looks forbiddingly wide."*[400] *"An honest man, armed with all the knowledge available to us now, could only state that in some sense, the origin of life appears at the moment to be almost a miracle, so many are the conditions which would have had to have been satisfied to get

---

Philosophical Transactions of the Royal Society A 376, no. 2123 (July 13, 2018): 2, https://doi.org/10.1098/rsta.2017.0326.

397. "A Cosmic Coincidence: Why Is the Universe Just Right for Life?" lecture, McGill University, January 15, 2007.

398. Geoffrey Chew, "Appeal to God May Be Required to Answer the Origin Question," in H. Margenau and R.A. Varghese, eds. *Cosmos, Bios, Theos: Scientists Reflect on Science, God, and the Origins of the Universe, Life, and Homo Sapiens* (La Salle: Open Court, 1992), 36.

399. Translated from Hubert Reeves, *Dernières nouvelles du cosmos* (Points Science, 2002), 27.

400. Francis Crick, "Foreword," in *The RNA World*, ed. R.F. Gesteland and J.F. Atkins (Cold Spring Harbor, NY: Cold Spring Harbor Laboratory Press, 1993), 11–14.

*it going."*⁴⁰¹ **Again:** *"[We can't help but notice] the intrinsic beauty of the DNA double helix. [. . .] It is the molecule which has style."*⁴⁰²

65. **George Wald** (1906–1977), Professor of Sensory Physiology at Harvard, Nobel Prize in Medicine in 1967: *"And now for my main thesis. If any one of a considerable number of the physical properties of our universe were otherwise— some of them basic, others seemingly trivial, almost accidental—that life, which seems now to be so prevalent, would become impossible, here or anywhere."*⁴⁰³ **Again:** *"When it comes to the origin of life there are only two possibilities: creation or spontaneous generation. There is no third way. Spontaneous generation was disproved one hundred years ago, but that leads us to only one other conclusion, that of supernatural creation. We cannot accept that on philosophical grounds; therefore, we choose to believe the impossible: that life arose spontaneously by chance!"*⁴⁰⁴

66. **Richard C. Lewontin** (1929-2021), Harvard evolutionary biologist: *"We take the side of science in spite of the patent absurdity of some of its constructs, in spite of its failure to fulfill many of its extravagant promises of health and life, in spite of the tolerance of the scientific community for unsubstantiated just-so stories, because we have a prior commitment, a commitment to materialism. [...] Moreover, that materialism is absolute, for we cannot allow a Divine Foot in the door."*⁴⁰⁵

67. **Ilya Prigogine** (1917-2003), Nobel Prize in Chemistry in 1977, and Isabelle Stengers, philosopher, epistemologist (born in 1949): *"In biology, the opposition between anti-reductionists and reductionists has often taken the form of a rivalry between "fraternal enemies": proponents of internal finality versus those advocating for external finality.*

---

401. Francis Crick, *Life Itself: Its Origin and Nature* (New York: Simon and Schuster, 1981), 88.
402. Francis Crick, "The Double Helix: A Personal View," *Nature*, 26 (April 1974).
403. George Wald, address at the First World Congress for the *Synthesis of Science and Religion*, Bombay, India, 1986.
404. George Wald, "The Origin of Life," *Scientific American* 191, no. 2 (1954): 48.
405. Richard Lewontin, "Billions and Billions of Demons," *New York Review of Books*, January 9, 1997.

The idea of an immanent organizing intelligence is thus countered by a model of organization borrowed from the technology of the era (mechanical, thermal, or cybernetic machines), which immediately raises this objection: "Who" assembled the machine? Is this automaton governed by an external purpose?"[406]

68. ***Christian de Duve** (1917–2013), biochemist, winner of the 1974 Nobel Prize in Physiology: *"God plays dice because he is sure to win. [. . .] I opted for a meaningful universe, not a meaningless one. Not because I wanted it to be that way, but because that's how I interpret the scientific data we have. The Universe was 'pregnant with life,' and the biosphere with man."*[407]

69. **John Eccles** (1903–1997), neurologist, electrophysiologist, winner of the 1963 Nobel Prize in Medicine: *"I maintain that the human mystery is incredibly demeaned by scientific reductionism, with its claim in promissory materialism to account eventually for all of the spiritual world in terms of patterns of neuronal activity. This belief must be classed as a superstition."*[408] Again: *"I am constrained to attribute the uniqueness of the Self or Soul to a supernatural spiritual creation."*[409]

70. ***Werner Arber** (1929–), microbiologist and winner of the 1978 Nobel Prize in Medicine: *"Life only starts at the level of a functional cell. The most primitive cells may require at least several hundred different specific biological macromolecules. How such already quite complex structures may have come together remains a mystery to me. The possibility of the existence of a creator, of God, represents to me a satisfactory solution to this problem."*[410]

---

406. I. Prigogine and I. Stengers, *La Nouvelle Alliance*, 4. (cf. chrome-extension://efaidnbmnn-nibpcajpcglclefindmkaj/https://ens.hal.science/hal-02376126/document)

407. Christian de Duve, *Poussière de vie: une histoire du vivant* (Paris: Fayard, 1996).

408. Jojn C. Eccles, *Evolution of the Brain: Creation of the Self* (London: Routledge, 1991), 241.

409. Ibid., 322

410. Werner Arber, "The Existence of a Creator Represents a Satisfactory Solution," in *Cosmos, Bios, Theos,* ed. H. Margenau and R. Varghese (Chicago: Open Court, 1992), 141.

71. **Jacques Monod** (1910-1976), atheist, biologist and biochemist from the Pasteur Institute in Paris, Nobel Prize in Physiology or Médicine in 1965: *"The machinery by which the cell (at least the non-primitive cell, which is the only one we know) translates the code consists of at least fifty macromolecular components which are themselves coded in the DNA. Thus the code can not be translated except by using certain products of its translation. This constitutes a baffling circle; a really vicious circle, it seems, for any attempt to form a model or theory of the genesis of the genetic code."*[411]

72. **Ernst Chain** (1906–1979), professor at the Universities of Berlin, Cambridge, and Oxford, pioneer in the development of penicillin, and 1945 Nobel laureate in Physiology or Medicine: *"I would rather believe in fairies than in such wild speculation. [...] I have said for years that speculations about the origin of life lead to no useful purpose as even the simplest living system is far too complex to be understood in terms of the extremely primitive chemistry scientists have used in their attempts to explain the unexplainable that happened billions of years ago. God cannot be explained away by such naive thoughts."*[412]

73. **Simon Conway Morris** (1951–), Professor of Paleontology at Cambridge: *"The evolutionary routes are many, but the destinations are limited. [...] In essence, this approach posits the existence of something analogous to 'attractors,' by which evolutionary trajectories are channelled towards stable nodes of functionality. [...] the nodes of occupation are effectively pre-determined from the Big Bang."*[413] *"If the universe is actually the product of a rational Mind and evolution is simply the search engine that in leading to sentience and consciousness allows us to discover the fundamental architecture of the universe... then things not only start to make much better sense, but they are also much more interesting [...]*

---

411. In Don Batten, "Origin of life: An explanation of what is needed for abiogenesis (or biopoiesis)," http://www.esalq.usp.br/lepse/imgs/conteudo_thumb/Origin-of-life.pdf, page 11.

412. Quoted by Ronald W. Clark, *The Life of Ernst Chain: Penicillin and Beyond* (London: Weidenfeld & Nicolson, 1985), 147-148.

413. Simon Conway Morris, *Life's Solution* (Cambridge, 2004), 145, 309-310.

Farewell bleak nihilism; the cold assurances that all is meaningless. [. . .] We are indeed dealing with unfinished business. God's funeral? I don't think so."[414]

74. **Sarah Woodson** (1967–), Professor of Biophysics at Johns Hopkins University, speaking about the nanomachines (such as the flagellar motor) in various types of cells: "*The cell's macromolecular machines contain dozens or even hundreds of components. But unlike man-made machines, which are built on assembly lines, these cellular machines assemble spontaneously from their protein and nucleic-acid components. It is as though cars could be manufactured by merely tumbling their parts onto the factory floor.*"[415]

75. *****George Church** (1954–), atheist, Professor of Genetics at Harvard and MIT, and director of the Center for Computational Genetics: "*The ribosome [. . .] is the most complicated thing that is present in all organisms. [. . .] If I were to be an intelligent design defender, that's what I would focus on; how did the ribosome come to be? [. . .] But isn't it the case that, if we take all the life forms we have so far, isn't the minimum for the ribosome about 53 proteins and 3 polynucleotides? [. . .] That's really marvelous. [. . .] Nobody has constructed a ribosome that works well without proteins.*"[416]

76. **Wilder Penfield** (1891–1976), renowned neurologist specializing in the recall of memories through stimulation of the brain, professor at McGill University: "*After a professional lifetime spent in trying to discover how the brain accounts for mind, it comes as a surprise now to discover, during this final examination of the evidence, that the dualist hypothesis (body and soul) seems the more reasonable of the two possible explanations.*"[417]

---

414. Simon Conway Morris, "Darwin was Right. Up to a Point," *The Guardian*, February 12, 2009, https://www.theguardian.com/global/2009/feb/12/simon-conway-morris-darwin

415. Sarah Woodson, "Biophysics: Assembly Line Inspection," *Nature* 438 (2005): 566.

416. *Life: What a Concept!*, ed. J. Brockman (presentations given at the EDGE Foundation's 2007 summer event), 76-78: https://www.edge.org/conversation/george_church-george-church%E2%80%94life-what-a-concept.

417. Wilder Penfield, *Mystery of the Mind: A Critical Study of Consciousness and the Human Brain* (Princeton, NJ: Princeton University Press, 1975), 85.

77. **Ernst Mayr** (1904–2005), professor at Harvard and one of the most prominent and vocal defenders of Neo-Darwinism: *"The problem of the origin of life [. . .] poses a keen challenge. [. . .] The chances that this improbable phenomenon could have occurred several times is exceedingly small, no matter how many millions of planets in the universe."*[418]

78. *****Michael Denton** (1943–), former Director of Prince of Wales Hospital's genomics lab (Sydney, Australia), Professor of Biochemistry at the University of Otago (New Zealand): *"In short, the laws of physics are supremely fit for life and the cosmos gives every appearance of having been specifically and optimally tailored to that end." "There is simply no tolerance possible in the design of the celestial machine. For us to be here, it must be precisely as it is. The new picture that has emerged in twentieth-century astronomy presents a dramatic challenge to the presumption which has been prevalent within scientific circles during most of the past four centuries: that life is a peripheral and purely contingent phenomenon in the cosmic scheme. These advances in astronomy and physics have established what for Newton and generations of natural theologians was only an affirmation of belief: that there is indeed a deep and necessary connection between virtually every characteristic of the cosmic stage and the drama of life."*[419]

79. *****Daniel Cohen** (1951–), Professor of Genetics at the University of Évry and scientific director of Genset (a genetic engineering lab), helped map the human genome: *"The genome is a program written in a very sophisticated language. Could such a complex language arise by accident? We are free to suppose that, but we can't demonstrate it. Personally, I went from being an atheist to agnostic in the space of one year because if this language is not the result of chance, then I*

---

418. Ernst Mayr, *The Growth of Biological Thought: Diversity, Evolution and Inheritance* (Cambridge, MA: Belknap Press of Harvard University Press, 1982), 583–584.

419. Michael Denton, *Nature's Destiny: How the Laws of Biology Reveal Purpose in the Universe* (New York: The Free Press, 1998), 13-14.

suspect that one day we will be able to demonstrate it. Imagine what a sea-change that would be!"[420]

80. **Michael Behe** (1952–) Professor of Biochemistry at Lehigh University: "*The simplicity which one believed to be the foundation of life has been revealed to be a fantasy which has been replaced by systems of frightening complexity. The awareness that life was conceived by an intelligent being is a shock for we men of the twentieth century who were made to think that life was the result of simple, natural laws.*"[421]

81. **Dean H. Kenyon** (1939–), Professor Emeritus of Biology at San Francisco State University: "*It is absolutely incredible to observe on this microscopic scale such a finely-tuned mechanism, a device that bears the mark of intelligent design and manufacturing. We have the details of a highly complex molecular Universe which manages genetic information. And it is precisely in this new branch of molecular genetics that we see evidence of the most indisputable intelligent design on Earth.*"[422]

82. *****Pierre-Gilles de Gennes** (1932–2007), winner of the 1991 Nobel Prize in Physics: "*Another thing that intrigues me is the fact that the genetic code is always the same, whereas life has adopted many different solutions. I have difficulty believing that Darwinian evolution could give rise to only a single type of code.*"[423]

83. **Hubert P. Yockey** (1916–2016), Professor of Theoretical Physics at Berkeley and specialist in information theory applied to biology and the origins of life: "*The origin of life by chance in a primeval soup is impossible in probability in the same way that a perpetual motion*

---

420. Translated from *Le Point*, October 21, 1995.

421. Translated from Jean-Michel Olivereau in his epilogue to Michael Behe, *La Boîte noire de Darwin: L'Intelligent Design* (Paris: Presses de la Renaissance, 2009).

422. P. William Davis, Dean H. Kenyon, and Charles Thaxton, *Of Pandas and People: The Central Question of Biological Origins* (Dallas, TX: Haughton Pub. Co., 1993), 58.

423. Translated from F. Brochard-Wyart, D. Quéré, M. Veyssié, *L'Extraordinaire Pierre-Gilles de Gennes, prix Nobel de physique* (Paris: Odile Jacob Sciences, 2017).

machine is impossible in probability. The extremely small probabilities calculated in this chapter are not discouraging to true believers. [. . .] A practical person must conclude that life didn't happen by chance."[424]

84. **Georges Salet** (1907–2002), polytechnician, statistician, author of *Hasard et certitude*, which opposes the invincibility of chance as theorized by Jacques Monod: "*It is not those who think that living beings arose through the action of an Intelligent Being who make appeal to miracles, but those who deny it.*"[425]

85. **Roger Sperry** (1913–1994), neurologist and winner of the 1981 Nobel Prize in Medicine: "*I must take issue especially with the whole general materialistic-reductionist conception of human nature and mind that seems to emerge from the currently prevailing objective analytic approach in the brain-behavior sciences. [. . .] I suspect that we have been taken, that science has sold society and itself a somewhat questionable bill of goods.*"[426]

86. **Philippe Labrot** (1971–), researcher at the Center for Molecular Biophysics, CNRS (Orléans:) "*For some researchers, the fact that the living cell shows a terrifying level of complexity proves that it probably could not have come about in stages, but rather that it emerged fully formed out of nothing. The probability of such an event happening is comparable to the chances of a tornado blowing over a dump heap to create a functioning Airbus A320 out of scrap metal.*"[427]

87. **Yves Coppens** (1934–2022), paleontologist and paleoanthropologist: "*It is still astonishing that changes for the better happen just when*

---

424. Hubert Yockey, *Information Theory and Molecular Biology* (Cambridge: Cambridge University Press, 1992), 257.

425. Georges Salet, *Hasard et Certitude—Le transformisme devant la biologie actuelle* (Paris: Saint-Edme, 1972), 328. Quotation translated from Pierre Rabischong, "Prolégomènes sémantiques: la problématique de l'homme," in *Le programme Homme* (Paris: PUF, 2003).

426. Roger Sperry, *Science and Moral Priority* (New York: Columbia University Press, 1982), 28.

427. Translated from *Chimie prébiotique* (www.nirgal.net/ori_life2.html). The Stanley Miller prize of the International Society for the Study of the Origin of Life (ISSOL) was awarded to Philippe Labrot for his remarkable contribution to the study of the origins of life.

*they are needed! [. . .] In any case, chance does things better than can be believed."*428

88. **Ali Demirsoy** (1945–), Professor of Biological Sciences at Hacettepe University in Ankara: *"The probability of the formation of a Cytochrome-C sequence [an enzyme necessary for life] is as likely as zero. That is, if life requires a certain sequence, it can be said that this has a probability likely to be realized once in the whole universe [. . .] In fact, the probability of the formation of a protein and a nucleic acid (DNA-RNA) is a probability way beyond estimate. Furthermore, the chance of the emergence of a certain protein chain is so slight as to be called astronomic."*429

89. **Pierre-Paul Grassé** (1895–1985), Professor of Biology at the University of Paris, zoologist, and ethologist: *"The idea that man is the result of innumerable errors in the copying of DNA during molecular duplication [. . .] seems upon reflection absurd—not a damning fault in itself, but also contrary to reality, which condemns the idea."*430

90. **Johnjoe McFadden** (1956–), Professor of Molecular Genetics at the University of Surrey: *"But who's doing the tweaking? Another reason scientists are wary [of the anthropic principle] is that it seems to reverse the Copernican revolution and place humankind at the centre of the universe. Even worse, it could allow creationists to bring the 'G' word back into science."*431

91. **Stuart Kauffman** (1939–), Professor of Biophysics at the University of Vermont and specialist in complex systems: *"Consider all proteins of*

---

428. Translated from the report in *L'Express*, August 3, 1995, www.lexpress.fr/informations/paleontologie-yves-coppens- professeur-au-college-de-france_609043.html.

429. Ali Demirsoy, *Inheritance and Evolution* (Ankara, Turkey: Meteksan, 1984), 39, 64.

430. Translated from Pierre-Paul Grassé,"Evolution," in *Encyclopédie Universalis* (cf. http://academie-metaphysique.com/ paroles/epistemologie-1888/le-reductionnisme-11026.html) and his work: *L'évolution du vivant – Matériaux pour une nouvelle théorie transformiste* (Paris: Albin Michel, 1973).

431. Johnjoe McFadden, *Quantum Evolution: The New Science of Life* (New York: Harper Collins, 2000).

length 200 amino acids. There are 20 to the 200th power or 10 to the 260th power such proteins. Were the 10 to the 80th particles in the universe to do nothing but make proteins length 200 on the Planck time scale, it would require 10 to the 39th times the lifetime of the universe to make all these proteins just once."[432]

92. **Perry Reeves** (1945–), Professor of Chemistry at Abilene Christian University: *"When one examines the vast number of possible structures that could result from a simple random combination of amino acids in an evaporating primordial pond, it is mind-boggling to believe that life could have originated in this way. It is more plausible that a Great Builder with a master plan would be required for such a task."*[433]

93. **Francis Collins** (1950–), geneticist, Director of the American National Institute of Health, and specialist in the sequencing of the human genome: *"Belief in God can be an entirely rational choice, and the principles of faith are, in fact, complementary with the principles of science."*[434]

94. **Bruce Lipton** (1944–), professor at the University of Wisconsin: *"As a conventional biologist, I believed that I was merely a mechanism and that my life was nothing but the folds of molecules. And so I felt that I was only an accident, as Darwin had said. [. . .] my research in cellular biology showed me that I was far more than my physical reality. And that there was an energy, a Spirit or a God that controlled [the laws of] biology."*[435]

---

432. Foreword to Robert E. Ulanowicz, *A Third Window* (West Conshohocken, PA: Templeton Press, 2009), XII.

433. Cited in J. D. Thomas, *Evolution and Faith* (Abilene, TX: ACU Press, 1988), 81–82.

434. Francis Collins, "Introduction," in *Language of God* (New York: Simon & Schuster, 2006).

435. Bruce Lipton, *The Biology of Belief: Unleashing the Power of Consciousness, Matter & Miracles* (New York: Hay House, 2005).

III. Mathematics

95. **Kurt Gödel** (1906–1978), logician and Professor of Mathematics at Princeton: "*Mechanism in biology is a prejudice of our time which will be disproved. In this case, one disproof, in my opinion, will consist in a mathematical theorem to the effect that the formation within geological time of a human body by the laws of physics (or any other laws of similar nature), starting from a random distribution of the elementary particles and the field, is as unlikely as the separation by chance of the atmosphere into its components.*"[436] "*There is a scientific (exact) philosophy and theology (this is also more highly fruitful for science), which deals with the concepts of the highest abstractness. [. . .] Est Deus [God exists]...*"[437]

96. *****Paul Dirac** (1902–1984), one of the fathers of quantum mechanics and winner of the 1933 Nobel Prize in Physics: "*God is a mathematician of a very high order, and He used very advanced mathematics in constructing the universe.*"[438]

97. **Alexander Polyakov** (1945–), mathematician, professor at Princeton: "*We know that nature is described by the best of all possible mathematics because God created it.*"[439]

---

436. Kurt Gödel, letter to Hao Wang, 1972.

437. First printed in English in Hao Wang, *A Logical Journey: From Gödel to Philosophy* (Cambridge: MIT Press, 1996), an emended translation of Gödel's remarks is available in Eva-Maria Engelsen, "What Is the Link Between Aristotle's Philosophy of Mind, the Iterative Conception of Set, Gödel's Incompleteness Theorems and God? About the Pleasure and the Difficulties of Interpreting Kurt Gödel's Philosophical Remarks," in Gabriella Crocco and Eva-Marie Engelson, eds., *Kurt Gödel: Philosopher-Scientist* (Aix-en-Provence: Presses universitaires de Provence, 2016), http://books.openedition.org/pup/53605.

438. Paul Dirac, "The Evolution of the Physicist's Picture of Nature," *Scientific American* 208, no. 5 (May 1963).

439. Quoted in "People at the Frontiers of Science," *Fortune* 114, no. 8 (October 13, 1986): 57.

IV. Philosophy of Science

98. **Antony Flew** (1923–2010), Professor of Philosophy at the University of Reading and one of the greatest atheist philosophers of our time. He chose atheism at age fifteen and wrote against divine creation for more than fifty-four years (the title of his article "Theology and Falsification" is the most explicit). He publicly renounced this stance in 2004 and later said, with regret: *"As people have certainly been influenced by me, I want to try and correct the enormous damage I may have done."*[440] Again: *"The most impressive arguments for God's existence are those that are supported by recent scientific discoveries. [. . .] The argument to Intelligent Design is enormously stronger than it was when I first met it."*[441] Also: * *"What I think the DNA material has done is that it has shown, by the almost unbelievable complexity of the arrangements which are needed to produce [life], that intelligence must have been involved in getting these extraordinarily diverse elements to work together."*[442] *"When you examine RNA as a chemist, you simply experience such complete awe before a molecule that is so wondrous, as well as before its magnificent complexity, that you ask yourself the question: how is it possible that this structure appeared?"*[443]

99. **Edward Feser** (1968–), American atheist-turned-Catholic philosopher, Professor of Philosophy of Pasadena City College in Pasadena, California: *"I don't know exactly when everything clicked. There was no single event, but a gradual transformation. As I taught and thought about the arguments for God's existence, and in particular the cosmological argument, I went from thinking "These arguments are no good" to thinking "These arguments are a little better than they are given credit for" and then to "These arguments are actually kind of interesting."*

---

440. In "Has Science Discovered God?," a video from Antony Flew himself after his conversion.

441. Gary Habermas, "My Pilgrimage from Atheism to Theism: An Exclusive Interview with Former British Atheist Professor Antony Flew," *Philosophia Christi* 6, no. 2 (2004): 200.

442. Debate in 2004, at New York University.

443. Taken from M. Oppenheimer's *New York Times* column on a conversation he had with Antony Flew, "The Turning of an Atheist," *New York Times*, November 4, 2007.

Eventually it hit me: "Oh my goodness, these arguments are right after all!" By the summer of 2001 I would find myself trying to argue my wife's skeptical physicist brother-in-law into philosophical theism."[444]

100. **Neil Manson**, Professor of Philosophy at the University of Mississippi: *"The multiverse hypothesis is alleged to be the last resort for the desperate atheist."*[445]

---

444. https://edwardfeser.blogspot.com/2012/07/road-from-atheism.html.

445. Neil A. Manson, "The Much-Maligned Multiverse," in *God and Design: The Teleological Argument and Modern Science*, ed. Neil A. Manson (New York: Routledge, 2003), 18.

14.

What Do Scientists Believe In?

The previous chapter explored the challenges and surprises that some of the greatest scientists of the twentieth century faced as they grappled with the implications of their own discoveries. This leads us to the related question of belief among scientists in general.

Conventional wisdom holds that modern scientists are not very religious, or at least far less religious than the general population. Some people consider this a sign that science leads naturally to unbelief, an indirect proof that God does not exist.

To address this common perception, this chapter will study belief among scientists in depth, analyzing a selection of surveys and studies on the question.

The first thing to note is that the claim is made up of two distinct underlying questions that we will consider separately:

1. Do materialist scientists today outnumber the scientists who believe in *something*? Do they represent an overwhelming majority?

2. If unbelieving scientists are indeed an overwhelming majority, is this a consequence of their scientific knowledge or of other factors?

I. Let us begin by examining the major studies on religious belief among scientists

1. A 2009 study by the Pew Research Center,[446] "Scientists and Belief," has the largest sample size and is the most recent of studies on the

---

446. Pew Research Center, "Scientists and Belief," November 5, 2009, http://www.pewforum.org/2009/11/05/scientists-and-belief/.

issue. It shows that a majority of American scientists (51%) believe in *something* while a minority (41%) are atheists. Seven percent of scientists polled declined to answer the question. If we compare these results with rates of belief among the general population, it is undeniable that a greater proportion believe in *something*—believers made up 95% of those surveyed.

It is worth noting that younger scientists, those under the age of thirty-four, are much more likely to claim some kind of belief (66%) than older researchers. In fact, only 46% of researchers over sixty-five claimed some form of belief.

This study is so valuable because it is recent (2009) and takes into account researchers working in every scientific field. Additionally, the survey was carried out in the United States, the undeniable world leader in scientific research and a place where scientists are particularly numerous. In contrast to other studies, the survey offered a wide selection of possible responses, including, for example, belief in a creative spiritual being[447] that is not a personal God.[448] The survey provided space for respondents to express their doubts and even allowed them to skip questions they did not wish to answer.

2. **A 2003 study by geneticist Baruch Aba Shalev** analyzed the beliefs of all Nobel Laureates since the prize's inception. Entitled "100 Years of Nobel Prizes,"[449] the study showed that 90% of Nobel prize winners identified with some religion and that two-thirds of them have been Christians.[450] Further, 35% of the recipients of the Nobel Prize in Lite-

---

447. Many studies make the mistake of excluding the category of scientists who, like Einstein, do not believe in a personal god but do believe in some higher created spirit. Einstein claimed not to believe in the God of the Bible, but he frequently asserted in speech and in writing his conviction that some kind of higher being was necessary to explain the order present in the Universe. See Chapter 15.

448. Generally speaking, a "personal God" is understood to be a God like that of the Bible, who we can talk to, who listens to us, and who may respond to our needs and wants.

449. Baruch Aba Shalev, *100 Years of Nobel Prizes* (New Delhi: Atlantic Publishers, 2003).

450. It's important to keep in mind that this study involved only a general identification and no

rature have been atheists, as compared with only 10% among scientists. Incidentally, this goes to show that, at least according to this survey, there are more atheists in the field of literature than in the sciences.

Between 1901 and 2000, the 654 Nobel Laureates belonged to twenty-eight different religions. A majority of them (65.4%) identified as Christian. More precisely, Christians won 72.5% of the prizes awarded in chemistry, 65.3% in physics, 62% in medicine, and 54% in economics. Jews were awarded 17.3% of the prizes in chemistry, 26.2% in medicine, and 25.9% of the prizes in physics. Atheists, agnostics, and freethinkers were awarded 7.1% of the prizes in chemistry, 8.9% in medicine, and 4.7% in physics. Muslims have won thirteen prizes in total, or 2% of the total prizes awarded (including three in the scientific categories).

This second survey is of interest because while its scope is very broad in terms of time and space, it concerns only Nobel Prize recipients. Since Nobel laureates represent the world's scientific elite, their opinions and beliefs carry more weight than those of the general population. As the Nobel Prize was created in 1901, the study spans the entire twentieth century.

The questionable aspect of this survey is the uncritical identification made between culture and belief, which in no way guarantees that the belief is actually embraced by the individual concerned.

3. **In a 1989 survey of heads of research units in the exact sciences at the French National Centre for Scientific Research,**[451] 110 of the researchers surveyed described themselves as believers, 106 as unbelievers, and 23 as agnostic. 70% believed that science can neither prove nor disprove the existence of God. The percentage of believers in this study is similar to the 2009 study cited above, around 50%.

---

account of more relative details about the beliefs of individual respondents.

451. Study cited by Georges Minois in *L'Église et la science, histoire d'un malentendu—de Galilée à Jean-Paul II* (Paris: Fayard, 2014), II:1151–1159, II:1287.

4. **James H. Leuba's studies.** Two studies undertaken in 1914 and 1933 by the American psychologist James H. Leuba[452] showed disbelief increasing over time among people identified as what the study terms "greater scientists," while the percentage had remained stable among scientists in general since 1914—around 60.7%. What were his criteria for making this judgement? There is no answer.

These first four studies show relatively homogeneous results. We will now consider two others that draw a diametrically different conclusion: that believers make up a very small percentage of scholars.

5. **A 1998 Nature study,** entitled *"Leading Scientists Still Reject God,"*[453] reported that among the American scientists of the National Academy of Scientists, only 7% describe themselves as believers, 20% as agnostics, and all the rest as atheists.

What the study does not tell us is that this academy was founded in 1867, in the context of a bitter polemic between science and religion. Since its foundation, all members have been hand-picked for their commitment to the academy's particular ideals. The percentage of its scholars who are believers is so markedly different from that found in the other studies that it is obviously unrepresentative. Even the title of the study, "Leading Scientists Still Reject God," displays a marked bias that reinforces our doubt about the survey's objectivity.

6. **A survey of 1,074 members of the British Royal Society carried out by E. Cornwell and M. Stirrat** by email in 2007 presents a similar case. 86% of respondents categorically denied believing in a personal God, while only 3% said they believed. As with the previous survey, this study has the major defect of being confined to one academy whose limited number of members has always been self-selecting.

---

452. J. H. Leuba, The Belief in *God and Immortality* (Boston: Sherman, French & Co., 1916). J. H. Leuba, *God or Man? A Study of the Value of God to Man* (New York: Henry Holt & Company, 1933).

453. Edward J. Larson and Larry Witham, "Leading Scientists Still Reject God," *Nature* 394, no. 313 (1998), https://www.nature.com/articles/28478.

The resulting figures are a statistical anomaly, making this study as questionable as the last and deserving of similar skepticism.

As we see it, the results of the last two studies are compromised by bad sampling. Their sample sizes are limited and unrepresentative, since the scholars surveyed belong to academies with a self-selecting, closed membership.

The conclusion that emerges from the first four surveys, whose sample size is in fact large enough to be statistically significant, is that the proportion of scientists who believe in *something* remains high. According to the first study, believers even constitute a majority in the United States, the world's leading producer of scientific research. Therefore, the claim that scientists who believe are now only a small minority is completely unfounded and inaccurate.

II. Let us now examine the fact that the proportion of believers among scientists is smaller than the proportion of believers among the general population, and let's see what factors might cause this difference

What are we to make of the fact that the percentage of scientists who are believers is lower than that of believers in the general population? What factors might be at the root of this difference? If we go by the results of the Pew Research Center's poll, the most recent and broadest survey of the scientific community, we find that, as in most studies of this type, the number of respondents who believe in *something* is lower among American scientists (51%) than among the rest of the US population (95%). This difference could be chalked up to several causes, including:

- Science itself
- The socio-economic class of scientists
- The historic conflicts between science and religion
- The very recent nature of the discoveries supporting the existence of a creator God (the thermal death of the Universe, the Big Bang, fine-tuning, the improbability of life arising by chance, etc.)

1. The influence of socio-economic status on belief

A recent survey from the Pew Research Center was categorical: wherever you are in the world, the richer you are, the less religious you are.[454] This phenomenon, already mentioned in the introduction to this book, should come as no surprise. Material comfort, the security of social systems, and medical progress have made turning to any God to solve mankind's problems seem unnecessary.

If we look at the Pew Research Center's study in detail, the correlation between standard of living and disbelief is extremely well established.

Since scientists tend to be well off, it might be expected that they are less religious than others.

The correlation between belief and standard of living is firmly established, but there has not yet been any empirical evidence establishing a correlation between belief and scientific education.

It is therefore impossible to claim at this stage that science makes people less religious. If scientists are indeed less religious than the rest of the population, it is possible that this is simply attributable to their high standard of living.

2. Correlation is not causation

Hasty correlations often lead to false conclusions. The question of correlation is a complex one. Here we are dealing with a well-established and clear correlation between a low-status, high-belief group and a high-status, low-belief group, and scientists belong to the latter. It is therefore unsurprising that, due to this factor alone, scientists tend to be less religious than the average person.

---

454. George Gao, "How do Americans stand out from the rest of the world?" Pew Research Center, March 12, 2015, https://www.pewresearch.org/fact-tank/2015/03/12/how-do-americans-stand-out-from-the-rest-of-the-world/.

WHAT DO SCIENTISTS BELIEVE IN?

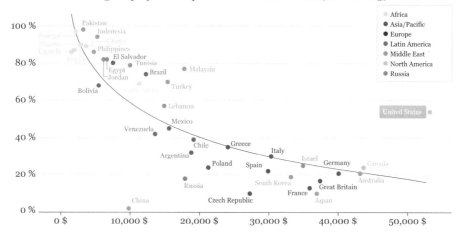

Wealthier Nations tend to be less religious, but the United States is a notable exception

% who claim religion plays an important role in their life (2011–2013)

Estimated per capita GDP in 2013 (PPP, current international dollar)

Note: The curve represents the logarithmic relationship between the per capita GDP and the percentage of persons who state that religion plays a significant role in their lives. Source: Spring 2011, 2012, 2013. Survey on global attitudes. Figures for per capita GDP are taken from the FMI's database of perspectives on the world economy, April 2014.

PEW RESEARCH CENTER

Source: https://www.pewresearch.org/fact-tank/2015/03/12/how-do-americans-stand-out-from-the-rest-of-the-world/

If we aren't satisfied with that answer, what would it take to prove a causal relation between advanced scientific knowledge and lack of belief in God? We would start by determining more precisely how the "standard of living" parameter functions as a factor in disbelief. This would require a similar survey of individuals in a variety of professional sectors who enjoy a comparable standard of living but have no special scientific knowledge. We could study professors, lawyers, writers, and actors, for example.

If it turned out that the people of this category—enjoying a social status and high standard of living comparable to leading scientists but without their advanced scientific knowledge—believed in God at a higher rate, then we could conclude that there is indeed a causal link between greater scientific knowledge and a decreased belief in God. Inversely, if individuals of socially equivalent status but lacking advanced scientific training turn out to have similar rates of unbelief to those of the scientists, then science can be eliminated as a factor influencing unbelief.

Unfortunately, to our knowledge none of these surveys have been conducted, nor is it at all certain that they would be conclusive. What would happen if it turned out that professors, philosophers, actors, and writers were just as unbelieving as scientists? The popular myth of causality between science and disbelief would definitively fall apart.

We are not claiming that there is no causal link between science and belief, as we do not have the data to prove this. What we do claim is that based on the evidence available today, it is impossible to draw a conclusion from this correlation. The limited data available to us does tend to make us doubt it. As noted above, Nobel Prize-winning scientists are much more likely to be believers than the winners of the top prize in literature.

3. Long-standing conflicts between science and religion

Although most of the great scientific discoveries of the modern era were made by Christian scientists (scientists who remained Christian even after they made their discoveries, like Copernicus, Galileo, Newton, and Kepler), their findings, perhaps inevitably, sparked lively controversy. But these controversies were often instrumentalized by Christianity's detractors.

As a result, the history of modern science is usually told as a battle between scientists struggling to spread their discoveries and religious people attempting to maintain their own interests and worldview. Perhaps influenced by this version of scientific history, many scientists of our time have rejected religion in general and by extension the very idea of the existence of God. It is obviously impossible to put a precise figure on the impact of this distorted historical narrative, but it is important to keep in mind nonetheless.

4. Many scientific discoveries favoring the existence of God are very recent

New and convincing scientific proofs for the existence of a creator God have only emerged very recently—most of them only within the last generation:

- The thermal death of the Universe was not definitively proved until 1998.
- The necessity of a beginning to the Universe (Borde-Guth-Vilenkin theorem), dates only to 2003.
- The discovery of the fine-tuning of the Universe dates only from the 1980s.
- The complexity of DNA and the intricacy of the smallest living cells—coupled with the improbability of life arising through pure chance—was discovered only in the early 1960s.

The controversies generated by these modern discoveries resemble the uproar provoked by Galileo and Darwin in their own times and are probably just as inevitable.

How long did it take for Darwin's discoveries to be accepted? Perhaps one hundred or one hundred and fifty years? No doubt this new proof for the existence of God will also take time to produce its effect.

According to the first Pew study cited above, the percentage of younger American scientists who believe in *something* (66%) is strikingly higher than the percentage of scientists over sixty-five (only 46%) who do. The survey may illustrate the early stages of a shift in opinion among scientists.

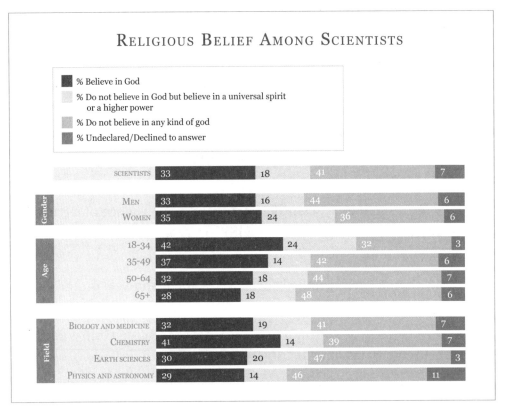
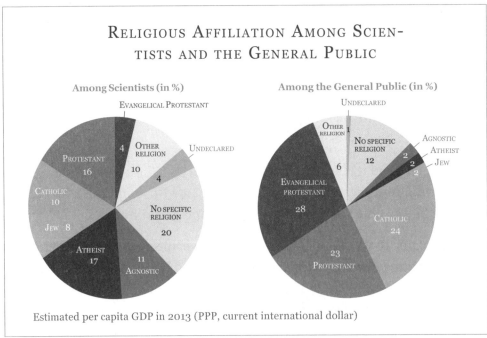

5. Another thing to consider when drawing a causal relationship between belief and scientific knowledge

Baruch Aba Shalev's study revealed that Nobel Laureates in science are more religious than Nobel Laureates in literature. The fact is of utmost importance because Nobel Prize winners are held to be remarkable, regardless of the field in which they excel. We can therefore learn from this that, among remarkable individuals, science is more of a factor in belief than disbelief.

Conclusion

The first assumption noted at the beginning of this chapter, that there are not many believers among scientists today, is at best unfounded and probably inaccurate.

The second assumption is equally baseless: there is no verifiable evidence to support the claim that scientists are less religious because of their scientific knowledge. It is likely that we need to look to other factors, such as their socio-economic status, or the fraught history of science itself, to explain this tendency.

Two more surveys of scientists on the question of religious belief

- In a 2002 report analyzing forty-three studies conducted since 1927, **Mensa** (an association of "gifted" people with an IQ over 132, some 2% of the population) concludes: *"The higher one's intelligence or education level, the less one is likely to be religious or hold 'beliefs' of any kind."*[455] Once again, we encounter a logical fallacy: a correlation is observed and wrongly presented as causation, without any justification.

---

455. *Mensa Magazine*, UK Edition, February 2002, 12.

- **A study published by CIRST** (*Centre interuniversitaire de recherche sur la science et la technologie*, based in Montréal, Canada) and carried out by two sociologists (Kristoff Talin, research fellow at CNRS, and Yves Gingras, Professor of History at the University of Québec in Montréal)[456] concludes: "*The more religious one is, the less competent in science.*" The two authors carried out a comparative analysis of religious practice, belief, and level of scientific advancement with samples taken internationally. For them, the results seem clear: the more people identify with and practice a religion, the weaker their scientific skills are: "*As a general rule, more religion translates into less science.*" This CIRST study commits the same error as did the previous one; it confounds correlation and causation.

How can we account for the significant differences between the conclusions of these various studies?

How is it possible that in dealing with a question as apparently simple as "Does science make people believe less in God?," we find studies with results at opposite ends of the spectrum? On one hand, we find 51% of US scholars believing in *something*, according to the Pew Research Center study, and 90% of the Nobel-Prize-winning scientists self-identified as believers in Baruch Aba Shalev's dubious study. With only 3% of scholars reported as believers in Cornwell and Stirrat's study, we get the impression that on this topic it is possible to read everything and its opposite.

These divergent results likely stem from the sensitive nature of the subject itself. The problem of the existence of God, which underlies the relationship between science and belief, is difficult to approach dispassionately. This translates into the choice of methods, whether conscious or not, that skew the polls' results according to the presuppositions of the surveyors. As we have already seen, multiple parameters are subject to bias here:

---

456. Translated from a report in *Le Monde*, April 21, 2020.

- Poorly worded questions
- Correlation hastily interpreted as causation
- Non-representative or even clearly biased samples
- Statistics based on averages between groups that are too different

As we come to the end of these two chapters devoted to the opinions of scientists in general on this matter, a question springs naturally to mind: what were the beliefs of the scientists who played a major role in the great scientific advances of the twentieth century? Two such cases prove particularly interesting: that of Einstein—without question the greatest physicist of the last century—and Gödel—whom John von Neumann called the greatest mathematician and logician since Aristotle.

As we shall see in the two chapters to follow, these two men were no strangers to the question of God. What's more, their reflections on the question of the existence of God appear to have been directly linked to their scientific work, and this has been the case for many other scholars in recent decades.[457]

---

457. Several great atheist or agnostic scientists became deists purely on the basis of twentieth-century scientific discoveries. In *The Return of the God Hypothesis*, Stephen Meyer gives a few examples, including Fred Hoyle, who was convinced by the discovery of the astonishing adjustments needed for the genesis of carbon in the stars (see p. 164) and famous Caltech astronomer Allan Sandage (p. 31) as well as his colleague Owen Gingerich (p. 31), but also physicist Georges Greenstein (p. 176) and biologist Dean Kenyon (p. 204).

15.

What Did Einstein Believe In?

Einstein is widely celebrated as one of the greatest scientists of the twentieth century. As an intellectual authority who captured a unique public respect and interest, his statements on the existence of a creator God have always carried a great deal of weight. Though many have tried to claim him for their own camp, Einstein's statements on the question can seem vague or even contradictory, resisting easy classification. We will need to take a closer look.

In piecing together any coherent view Einstein may have had on this question, it is worth distinguishing between Einstein the scientist and Einstein the man. In his role as a scientist, he often recognized the necessity for an *"infinitely superior"* being responsible for the creation of the world. But Einstein the man never acknowledged any personal God or followed any particular religion. Since this is actually a common situation among scientists, the case of Einstein may be illustrative of more general trends.

We can divide his statements on the subject into two groups: Einstein, who had no religious training,[458] makes a clear distinction between the God behind the origin of the Universe, whom he believes in—because the fact of that God's existence stems, so to speak, from his scientific work—and the God of the Bible, whom he personally does

---

458. Thus, Richard Dawkins, in his best-seller on atheism, *The God Delusion*, gives a biased seven-page argument (working from a collection of quotations from Einstein in Max Jammer's *Einstein and Religion*), that Einstein was not a believer but that he had a kind of vague *"pantheistic reverence."* *"Einstein sometimes invoked the name of God,"* he writes, but only *"in a purely metaphorical, poetic sense."* Dawkins concludes: *"The metaphorical or pantheistic God of the physicists is light years away from the [. . .] God of the Bible." The God Delusion* (London: Bantam, 2006), 14-19. This is a biased and inaccurate interpretation of Einstein's statements. Conversely, the German philosopher Eric Gutkind sought to use Einstein's words in defense of the Jewish religion in the 1950s, drawing a well-known and often-quoted letter of reprimand.

not believe in. What might otherwise seem incoherent begins to make sense when interpreted with this in mind. And so we can classify his statements into two large groups: those in which he affirms his belief in a creator God on the one hand and, on the other, those in which he rejects religion.[459]

Quotations from Einstein display his belief in a creator God

- *"But also, everyone who is seriously involved in the pursuit of science becomes convinced that some spirit is manifest in the laws of the universe, one that is vastly superior to that of man. In this way the pursuit of science leads to a religious feeling of a special sort, which is surely quite different from the religiosity of someone more naïve."*[460]

- *"I consider the comprehensibility of the world (to the extent that we are authorized to speak of such a comprehensibility) as a miracle or as an eternal mystery [. . .]. There lies the weakness of positivists and professional atheists who are elated because they feel that they have not only successfully rid the world of gods but 'bared the miracles.' Oddly enough, we must be satisfied to acknowledge the 'miracle' without there being any legitimate way for us to approach it. I am forced to add that just to keep you from thinking that— weakened by age—I have fallen prey to the priests."*[461]

- *"My religiosity consists of a humble admiration of the infinitely superior spirit that reveals itself in the little that we can compre-*

---

459. *"In fact, my first religious training, in any form, was the Catholic catechism. Of course, only because the primary school I started at was Catholic"* [*"En fait, ma première formation religieuse, de quelque forme que ce soit, a été le catéchisme catholique. Bien sûr, uniquement parce que l'école primaire où j'ai commencé était catholique."*] (Interview by Gustav Bucky in Peter A. Bucky and Allen G. Weakland, *The Private Albert Einstein* (Kansas City: Andrews McMeel Publishing, 1993).

460. Albert Einstein, "Letter to a Young Girl," printed in *Dear Professor Einstein: Albert Einstein's Letters to and from Children*, ed. Alice Calaprice (Amherst: Prometheus Books), 129.

461. A. Einstein, *Letters to Solovine*, translated by Wade Baskin, with an introduction by Maurice Solovine (New York: Philosophical Library, 1987), 132-133.

hend of the knowable world. That deeply emotional conviction of the presence of a superior reasoning power, which is revealed in the incomprehensible universe, forms my idea of God."[462]

- "I want to know how God created this world. I am not interested in this or that phenomenon, in the spectrum of this or that element. I want to know His thoughts, the rest are details."[463]

- "Everything is determined, the beginning as well as the end, by forces over which we have no control. It is determined for the insect, as well as for the star. Human beings, vegetables, or cosmic dust, we all dance to a mysterious tune, intoned in the distance by an invisible piper."[464]

- "I'm not an atheist. I don't think I can call myself a pantheist. [. . .] What separates me from most so-called atheists is a feeling of utter humility toward the unattainable secrets of the harmony of the cosmos..." "The fanatical atheists [. . .] are like slaves who are still feeling the weight of their chains which they have thrown off after hard struggle. They are creatures who—in their grudge against the traditional 'opium of the people'—cannot hear the music of the spheres."[465]

- "To sense that behind anything that can be experienced there is a something that our mind cannot grasp and whose beauty and sublimity reaches us only indirectly and as a feeble reflection, this is religiousness. In this sense I am religious."[466]

---

462. Albert Einstein, letter of August 1927, quoted by Alice Calaprice in *The Expanded Quotable Einstein* (Princeton, NJ: Princeton University Press, 2000), 204. Peter A. Bucky and Allen G. Weakland, *The Private Albert Einstein* (Kansas City: Andrews McMeel Publishing, 1993).

463. In Esther Salaman, "A Talk with Einstein," *The Listener* 54 (1955): 370–371.

464. Ronald W. Clark, *Einstein: The Life and Times* (New York: William Morrow, 2007), 422.

465. Walter Isaacson, "Einstein and Faith," *Time* 169 (April 5, 2007). https://time.com/archive/6680629/einstein-faith/.

466. Einstein wrote and recorded his "Credo" in 1932 to benefit the German League of Human Rights. English translation at: https://einstein-website.de/en/credo/.

- *"On the other hand, I maintain that the cosmic religious feeling is the strongest and noblest motive for scientific research."*[467]

- *"A contemporary has said, not unjustly, that in this materialistic age of ours the serious scientific workers are the only profoundly religious people."*[468]

- *"[This scientist's] religious feeling takes the form of a rapturous amazement at the harmony of natural law, which reveals an intelligence of such superiority that, compared with it, all the systematic thinking and acting of human beings is an utterly insignificant reflection."*[469]

Statements on God scattered throughout his entire career and discussions with his peers:

- *"Coincidences are God's way of remaining anonymous."*[470]
- *"Science without religion is lame, religion without science is blind."*[471]
- *"The discovery of a fundamental verified law of nature is an inspiration of God."*[472]

Einstein often spoke of the "*marvellous Spinoza,*" and even explicitly approved of this philosopher's notion of God: "*I believe in Spinoza's God, Who reveals Himself in the lawful harmony of the world, not in a God who concerns Himself with the fate and the doings of mankind.*" It was again Spinoza's God that Einstein had in mind when he exclaimed to Niels Bohr in 1927: "*God does not play dice with the universe!*"

He also occasionally made reference to Leibniz, notably when he explained about Max Planck that "*nobody who has really gone deeply

---

467. Albert Einstein, *Ideas and Opinions* (New York: Crown Publishers, 1960), 39.

468. Ibid, 40.

469. Albert Einstein, *The World as I See It* (New York: Philosophical Library, 1949), 21.

470. Albert Einstein, *Discours et entretiens* (1879-1955).

471. Albert Einstein, *Ideas and Opinions*, 46.

472. Princeton, 1932, as reported by Henry Margenau in *Cosmos, Bios, Theos* (Chicago: Open Court, 1992), 62.

into the matter will deny that in practice the world of phenomena uniquely determines the theoretical system, in spite of the fact that there is no logical bridge between phenomena and their theoretical principles; this is what Leibnitz described so happily as a 'pre-established harmony.'"[473]

Quotes in which Einstein rejects the idea of a personal God and organized religion

We must begin with his famous letter to Gutkind:

- "Still, without Brouwer's suggestion I would never have gotten myself to engage intensively with your book because it is written in a language inaccessible to me. The word God is for me nothing more than the expression and product of human weakness, the Bible a collection of honorable, but still purely primitive, legends which are nevertheless pretty childish. No interpretation, no matter how subtle, can change this for me. For me the Jewish religion like all other religions is an incarnation of the most childish superstition." **Later:** "Now that I have quite openly stated our differences in intellectual convictions it is still clear to me that we are quite close to each other in essential things, i.e; in our evaluations of human behavior."[474]

- "It was, of course, a lie what you read about my religious convictions, a lie which is being systematically repeated. I do not believe in a personal God and I have never denied this but have expressed it clearly. If something is in me which can be called religious then

---

[473]. Albert Einstein, "Motive des Forschens," in "Zu Max Plancks sechzigstem Geburtstag. Ansprachen, gehalten am 26. April 1918," *Der Deutschen Physikalischen Gesellschaft* (Karlsruhe: C. F. Müller), 29–32. English translation: "Principles of Research," trans. Sonja Bargmann, in *The Collected Papers of Albert Einstein, Volume 7: The Berlin Years*, English translation supplement (Princeton, NJ: Princeton University Press), 44. Available online at https://speakola.com/ideas/albert-einstein-principles-of-research-1918.

[474]. Handwritten letter from Einstein to Eric Gutkind. Sold at Christie's for $2.89 million in 2018. Image, transcription, and translation available at: https://lettersofnote.com/2009/10/07/the-word-god-is-the-product-of-human-weakness/.

- *it is the unbounded admiration for the structure of the world so far as our science can reveal it."*[475]

- *"I am a deeply religious nonbeliever. This is a somewhat new kind of religion."*[476]

- *I have never imputed to Nature a purpose or a goal, or anything that could be understood as anthropomorphic. What I see in Nature is a magnificent structure that we can comprehend only very imperfectly and that must fill a thinking person with a feeling of humility. This is a genuinely religious feeling that has nothing to do with mysticism."*[477]

- *The idea of a personal God is quite alien to me and seems even naive."*[478]

- *"If this being is omnipotent then every occurrence, including every human action, every human thought, and every human feeling and aspiration is also His work; how is it possible to think of holding men responsible for their deeds and thoughts before such an almighty Being? In giving out punishment and rewards He would to a certain extent be passing judgment on Himself. How can this be combined with the goodness and righteousness ascribed to Him?"*[479]

- *"No, clearly not. I do not believe that a man should be restrained in his daily actions by being afraid of punishment after death or that he should do things only because in this way he will be rewarded after he dies. This does not make sense."*[480]

---

475. Letter from Einstein to Joseph Dispentiere, March 22, 1954. Quoted by Helen Dukas, in Albert Einstein: The Human Side—New Glimpses from His Archives, ed. Helen Dukas and Banesh Hoffmann (Princeton, NJ: Princeton University Press, 1981), 43.

476. Albert Einstein, letter to Hans Muehsam, March 30, 1954, *Einstein Archive* 38-434.

477. 1954 or 1955, quoted in Helen Dukas and Banesh Hoffmann, eds. *Albert Einstein: The Human Side—New Glimpses from His Archives*, ed. Helen Dukas and Banesh Hoffmann (Princeton, NJ: Princeton University Press, 1981), 39.

478. Albert Einstein, letter to Beatrice Frohlich, December 17, 1952, *Einstein Archive* 59-797.

479. Albert Einstein, *Out of My Later Years* (New York: Philosophical Library, 1950), 32.

480. Interview by Gustav Bucky in Peter A. Bucky and Allen G. Weakland, *The Private Albert Eins-*

Einstein's is an interesting case. He received no religious education, professed no personal belief, and at times was even openly hostile to religion. But as a scientist he contemplated the extraordinary harmony of the cosmos and called it *"the intelligence of the Universe."* In the end, his position might best be described as a form of deism: *"My religion consists of a humble admiration for the illimitable superior spirit that reveals itself in the slight details we are able to perceive with our frail and feeble minds."*[481] Einstein, one of the greatest scientists of the twentieth century, was led to believe in a creator God thanks to science, and in spite of his prejudices, which were completely hostile to the idea. And as we have seen, the reasons he gave for his belief were entirely scientific.

---

tein (Kansas City, MO: Andrews McMeel, 1993).

481. Quoted in Peter Bucky, *The Private Albert Einstein* (Kansas City: Andrews and McMeel, 1992), 86.

16.

What Did Gödel Believe In?

A clap of thunder in the sky of mathematics

Kurt Gödel (1906-1978)

David Hilbert, one of the greatest mathematicians of the twenty-first century, had a list of unsolved mathematical problems. Of the twenty-three on his list, one seemed to him to be of special importance: to demonstrate that mathematics is a complete and coherent system. Though seemingly insignificant, the result would be of enormous philosophical importance. Theoretically, at least, one could demonstrate the truth or falsity of any logical proposition using math. Hilbert boldly called this the "final" solution to the problem of logic.[482] One can clearly see the ideology that motivated this research—the usual idea of "fencing in" the realm of reality, of saying, *"Okay, we've covered this question. Move on, there's nothing more to see here. We have exhausted what reality has to show us; we have worked it all out with our equations."*[483] These ideas lay at the heart of logical positivism and of the material dialecticism that then reigned in the sciences.

---

482. Called a "finitistic" proof in some translations.

483. To firmly denounce what he called the German *ignorabimus* (Latin: *we shall never know it*), a sort of intellectual shyness or even defeatism in the face of new knowledge, Hilbert came up with a catchy alliterative slogan that is engraved on his grave: "*Wir müssen wissen, wir werden wissen*" (We must know. We will know).

But on October 7, 1930, Hilbert's project fell apart at Königsberg, Kant's hometown, where a group of mathematical elites had gathered at a conference to discuss the theme *"The Epistemology of the Exact Sciences."* At the end of the conference, a shy young man in wire-rimmed glasses, thin and small of stature, spoke up. He was named Kurt Gödel, and he had just completed his thesis at the University of Vienna under the direction of Hans Hahn. He said just one thing: *"One can even give examples of propositions… whose contents are really true but are unprovable in the formal system of classical mathematics."*[484]

Witnesses say that Gödel spoke loudly enough to be heard by everyone present. But what he said did not seem to sink in.

There is nothing more formal than the notion of mathematical truth. Something is true only if it can be demonstrated to be so. Did this young man just say that there might be mathematical propositions that are both true and unprovable? We must have misheard him—that simply can't be possible. No one reacted. No one asked Gödel what he meant. Only one participant understood what had just happened: John von Neumann, a towering intellectual figure of the twentieth century, who would go on to design the first computer and take part in the Manhattan project that developed the atomic bomb.

In a private conversation after the conference, von Neumann asked Gödel if his remarks had implications for simpler mathematics, like arithmetic.

Gödel replied, *"Of course undecidable propositions about integers could be so constructed, but they would contain concepts quite different from those occurring in number theory like addition and multiplication."*[485]

---

484. Translation of Gödel's comments adapted from John W. Dawson Jr., "Discussion on the Foundation of Mathematics," *History and Philosophy of Logic* 5, issue 1 (1984): 126.

485. For details on von Neumann and Gödel's conversation, see Hao Wang, "Some Facts about Kurt Gödel," *The Journal of Symbolic Logic* 46, no. 3 (September 1981): 654–655.

Even Gödel had not yet seen the full implications of his theory, but he soon surprised even himself with the discovery of his second theorem: no reasonably rich mathematical system can contain a proof of its own consistency.

This implies that every coherent logical system is inevitably incomplete (which is why we call it an "incompleteness" theorem).

Von Neumann immediately grasped that Hilbert's program was doomed to fail. Logic can not be based on itself. Neither can mathematics, nor even arithmetic.

When Gödel published his proof the following year, a veritable tsunami swept over the mathematical elite. Leading mathematician Hermann Weyl called it a "*debacle*," a "*catastrophe*." As Palle Yourgrau describes it: "*The two-thousand-year-old axiomatization inaugurated by Euclid—the paradigm of captured rationality—had been shattered, and the blow had been struck, annoyingly, just when... Hilbert had succeeded in perfecting the very idea of a formal system of axioms. Not only the results but the very methods employed in Gödel's theorem were so unexpected that it was years before mathematicians and logicians began to grasp their full significance.*"[486]

A rendezvous with success

Gödel's article cannot be grasped at first glance. Even mathematicians at the highest levels had difficulty understanding it. John von Neumann championed it in these terms: "*Kurt Gödel's achievement in modern logic is singular and monumental—indeed it is more than a monument, it is a landmark which will remain visible far in space and time.*"[487] Gödel's name began to circulate at mathematics conferences.

---

486. Palle Yourgrau, *A World Without Time: The Forgotten Legacy of Godel and Einstein* (London: Hachette, 2009), 69.

487. P. R. Halmos, "The Legend of John von Neumann," *The American Mathematical Monthly 80*, no. 4 (April 1973): 382–394.

As for Hilbert, he was devastated. When the profound implications of the incompleteness theorem dawned on him, he realized that the blond-haired boy he had first crossed paths with three years before—and taken for half-mad—had sunk his dream of demonstrating and solving every question through mathematics alone.

Albert Einstein and Kurt Gödel at Princeton

Kurt Gödel received the first Albert Einstein Prize in 1951

Without intending it, Gödel rapidly became an international celebrity within the mathematical community. Among the many other invitations he received was an offer to join the Institute for Advanced Study at Princeton, a sort of safe haven for geniuses that offered scientists fleeing the Nazi regime a welcome refuge in the United States. Gödel settled permanently at Princeton in 1940, but only after he had escaped the Nazis in a remarkable adventure, taking the Trans-Siberian with his wife across the whole USSR and catching a boat from Japan to San Francisco. There he met a scientist named Einstein, who was to become one of his closest friends.

For more than fifteen years, their friendship was punctuated by the unchanging ritual of daily walks together through the university park. Einstein's particular esteem for Gödel is reflected in his comment to economist Oskar Morgenstern: *"My own work no longer meant much,"* that he came to the Institute merely *"... to have the privilege of walking home with Gödel."*[488]

Unfortunately, the conversations between these two great geniuses of the twentieth century are lost to us forever. Since they saw each other every day, they never exchanged letters. We do have a few snippets of his talks with Einstein that Gödel recorded in letters to his mother.

What does the famous "incompleteness theorem" tell us?

English mathematician Marcus du Sautoy, of Oxford University, sums up Gödel's insight in these terms: *"Gödel showed that any system of mathematics must contain true statements that are unprovable."*[489]

Gödel begins with a logical system, S, that can be shown to be "coherent"—that is to say, excluding false propositions like $2 + 2 = 5$. Next, he manages to construct a proposition (of the type now called G in his honor) that says, *"There is no demonstration of G that belongs to system S."* This is true for the simple reason that if we can construct from S (a system we've already shown to be coherent) a true proposition that says that it itself cannot be demonstrated from within system S, this is because it's impossible to do: the S system would be incoherent otherwise!

The proposition is therefore true but not demonstrable within system S. Though it might be demonstrated in an expanded system S' we would immediately run into proposition G', true in its turn but not demonstrable within system S'. G' is demonstrable in the even larger system S", but there we find proposition G", which can't be demonstrated within system S". And on and on it goes (we call this "recurrent reasoning"). Thus within any coherent system we find many true but unprovable propositions.

---

488. Rebecca Goldstein, *Incompleteness, the Proof and Paradox of Kurt Gödel* (New York: W. W. Norton and Company, 2005), 33.

489. Marcus du Sautoy, "What We Cannot Know," *IAI News* 54 (March 2, 2017). https://iai.tv/articles/what-we-cannot-know-auid-787?ts=1617959881.

As Gödel himself said, his system is based on the idea that the notion of truth in mathematics is larger than the notion of demonstrability.[490] This conclusion is a hard pill to swallow for some mathematicians even today.

What are the consequences of Gödel's theory?

Though today Gödel is much less well known to the general public than Einstein is, his peers considered him a singular genius. The "*Gödelian revolution*" is considered one of the greatest transformations in the history of mathematics and logic. The words of French mathematician and philosopher Pierre Cassou-Noguès clearly express the consensus on Gödel: "*Kurt Gödel was one of the greatest logicians in history. His incompleteness theorem, published in 1931, is perhaps the most significant mathematical proposition of the twentieth century.*"[491]

But more than half a century after Gödel's death, the debate on the direct and indirect consequences of his theorem rages on. Since the notion of truth is much broader than the notion of demonstrability in mathematics (as we've seen), the most generally accepted view is that the theory confirms the "Platonic" position on mathematics. This position, which holds that mathematical objects exist independently of the human mind, has been summarized very well by the great mathematician and physicist Roger Penrose (2020 Nobel laureate in Physics), who is passionate about Gödel's work and its applications: "*According to Plato, mathematical concepts and mathematical truths inhabit an actual world of their own that is timeless and without physical location. Plato's world is an ideal world of perfect forms, distinct from the physical world, but in terms of which the physical world must be understood. It also lies beyond our imperfect mental constructions; yet, our minds do have some direct access to this Platonic realm through an "awareness" of mathematical forms, and our*

---

490. Hao Wang, *Kurt Gödel* (Paris: Armand Collin, 1990), 201.

491. Translated from Pierre Cassou-Noguès, *Les démons de Gödel: Logique et folie* (Paris: Le Seuil, 2007), back cover.

ability to reason about them. We shall find that whilst our Platonic perceptions can be aided on occasion by computation, they are not limited by computation. It is this potential for the 'awareness' of mathematical concepts involved in this Platonic access that gives the mind a power beyond what can ever be achieved by a device dependent solely upon computation for its action."[492]

It is ultimately an ontological question, not a philosophical one. If mathematical forms exist "*beyond our imperfect mental constructions,*" it stands to reason that there is another world, or at least another dimension, in which these objects exist without anyone around to think of them.[493]

The last sentence of Penrose's remark deserves our particular attention. He says that the direct access the human mind has to the "*Platonic realm*" gives it a power superior to that of computers, which by definition have no access to it.

In two works that completely scandalized those who defended a materialist conception of the human mind, Penrose claimed to demonstrate that Gödel's theorem alone proves that the human mind is superior to any machine.[494] The second work includes, in passing, the most accessible proof of Gödel's theorem that has ever been proposed.[495]

Without going into too much detail about Penrose's claim, which is still being debated in the fields of logic, mathematics, and artificial intelligence, the argument is as follows: if the human mind worked like a computer, we would use a set of S axioms to determine the truth or falsity of mathematical propositions, just as today's artificial

---

492. Roger Penrose, *Shadows of the Mind: A Search for the Missing Science of Consciousness* (Oxford: Oxford University Press, 1994), 50–51.

493. The debate between mathematician Alain Connes and neurologist Jean-Pierre Changeux, a committed atheist, reveals what materialists have at stake in their attempt to refute this Platonic position. See their *Matière à pensée* (Paris: Odile Jacob, 1989).

494. Roger Penrose, *The Emperor's New Mind: Concerning Computers, Minds and The Laws of Physics* (Oxford: Oxford University Press, 1989).

495. Roger Penrose, *Shadows of the Mind*, especially the chapter entitled "The Gödelian Case," 64–116.

intelligence can automatically prove theorems. But this system would necessarily contain its own proposition G, a proposition not provable within the system but which a human observer immediately understands to be true.

For Penrose, this implies that *"human mathematicians do not use a computational procedure to find the truth of proposition G."* We might say that we understand the truth of this proposition through our transcendent access to mathematical truths. But might not the human mind run up against the same barrier? That is what Gödel himself believed. While he forcefully and repeatedly asserted that the human mind was immaterial, he thought that his theorem alone was not enough to demonstrate its immaterial nature.

While Gödel did not publish much during his lifetime, he left behind thousands of pages in small notebooks. Like Leonardo da Vinci, he wrote in code, using *Gabelsberger*, a special form of shorthand that only few people in the world can now decipher. John and Cheryl Dawson learned this language with the express purpose of sharing Gödel's ideas with the world.

Forty years before Penrose wrote, Gödel noted down in his mysterious shorthand in one of his little notebooks: *"My incompleteness theorem makes it likely that mind is not mechanical, or else mind cannot understand its own mechanism."*[496] In other words, at least one of the following statements has to be true:

- Either the human mind is a reality independent of the material world,
- Or mathematical objects have a reality outside the material world.

For Gödel, both propositions were absolutely true, but what his theorem actually demonstrates is that at least one of them is true. In his eyes, that fact assured the defeat of materialism.[497]

---

496. Hao Wang, *A Logical Journey: From Gödel to Philosophy* (Cambridge, MA: MIT Press, 1996), 186.

497. Most mathematicians are not aware that Gödel drew this implication from his arguments, but this has been demonstrated by Pierre Cassou-Noguès in his analysis of Gödel's notebooks: Pierre Cassou-Noguès, *Les démons de Gödel* (Paris: Seuil, 2007), 121–126.

As he put it: "*My theorem only shows that the mechanization of mathematics, that is to say removing the mind and abstract entities from the equation, is impossible.*"[498]

Mind, are you there?

Thus mind was the central focus for Gödel, leading him to wider claims about the nature of matter: "*Matter and mind are two different things. [. . .] It is a logical possibility that the existence of mind [separate from matter] is an empirically decidable question. [. . .] The mere possibility that there may not be enough nerve cells to perform the function of the mind introduces an empirical component into the problem of mind and matter.*"[499]

Gödel extended his non-materialist conception of mind to his view of the nature of life and of evolution: "*I don't think the brain came in the Darwinian manner. In fact, it is disprovable. Simple mechanism can't yield the brain.*"[500] He believed that Darwinism, which he called "*simple mechanism,*" would one day be refuted rationally: "*More generally, I believe that mechanism in biology is a prejudice of our time which will be disproved. In this case, one disproof, in my opinion, will consist in a mathematical theorem to the effect that the formation within geological times of a human body by the laws of physics (or any other laws of a similar nature), starting from a random distribution of the elementary particles and the field, is as unlikely as the separation by chance of the atmosphere into its components.*"[501]

In fact, Gödel was very coherent, attempting to do for neurology and biology what he had done for logic: construct a theory that demonstrates the insufficiency of the materialist approach.

---

498. Translated from Pierre Cassou-Noguès, *Les démons de Gödel* (Paris: Seuil, 2007), 122.. Gödel's use of the word "only" shows his modesty and prudence.

499. As quoted in Hao Wang, *Kurt Gödel: A Logical Journey* (Cambridge, MA: MIT Press, 1996), 191.

500. Ibid., 192.

501. Ibid., 192.

Regarding neurology, Gödel speculated that there are simply not enough neurons in the human brain to provide sufficient storage capacity for all the mind's operations. In the realm of biology, he argued that even the immensity of geological time was not sufficient for every kind of existing living being to have developed, in all their complexity, from the first living cell through a Darwinian process of mutation driven by chance and natural selection.

Although he was a theoretician, Gödel also ran experiments. In letters to his mother, he mentions more than two hundred tests he performed on his wife Adele, who had a very strong sense of intuition. She was able to guess playing cards without seeing them, with a much better result than could be marked up to blind chance.[502] Gödel also wrote extensively on the sudden and instant character of intuition (as opposed to the more laborious process of acquiring knowledge), which seemed to him further proof of the nonmaterial character of the human mind.

But if the human mind is not material, if it is not identical with the brain, can it survive after death? Gödel answered this question in the affirmative.

Between July and October 1961, Gödel wrote four letters to his mother Marianne, who had remained behind in Vienna. The pair did not know if they would ever see each other again in their lifetimes, and she asked him if they would meet again after death.

Gödel, who was hyperrational,[503] privileging logic and reason above all else, was not the sort of person to distort his thinking merely to console someone, even his own mother. To him, that would have seemed a reprehensible betrayal of logic and reason.[504] For this reason, we can be sure that his message to his mother (written in simple and accessible language) contains his true thoughts on the matter. Here,

---

502. Cassou-Noguès, *Les démons de Gödel*, 36.

503. Or at least who wanted to be so, though hyper-rationality can lead to madness.

504. Gödel's candor may be one of the reasons Einstein sought out his company. Gödel would speak his mind to him as if he were any other person, while others always spoke to Einstein with deference.

in a letter composed the 23 of July, 1961, one can find his reasoning in full:

> *If the world [Welt] is rationally constructed and has meaning, then there must be such a thing [as an afterlife]. For what sense would there be in creating a being (man), which has such a wide realm of possibilities for its own development and for relationships to others, and then not allowing it to realize even a thousandth of those [possibilities]? That would be almost like someone laying, with the greatest effort and expense, the foundations for a house, and then letting it all go to seed again. But does one have reason to suppose that the world is rationally constructed? I believe so. For it is by no means chaotic or random, but, as science shows, everything is pervaded by the greatest regularity and order. Order, however, is a form of rationality [Vernunftigkeit]. How would one envision a second [another] life? About that there are naturally only guesses. However, it is interesting that it is precisely modern science that provides support for such a thing. For it shows that this world of ours, with all the stars and planets in it, had a beginning and most probably will also have an end (that is, it will literally come to "nothing"). But why, then, should there exist only this one world—for just as we one day found ourselves in this world, without knowing why and wherefrom, so can the same thing be repeated in the same way in another world too.*[505]

It is interesting that, though it was too early for Gödel to speak of the famous fine-tuning of the Universe, he nevertheless affirms that the Universe had a beginning. He also claims that the regularity and order ruling the Universe show that it is rationally arranged—implying a meaning behind the order. Gödel observes that each human being has much more potential than can be tapped and brought to fruition in the course of a single lifetime, and so it would be illogical for there to be no life after death. To his mind, there is no reason that this irrational exception should exist in an otherwise rational world.

---

505. Kurt Gödel, letter of July 23, 1961, in Hao Wang, *A Logical Journey* (Cambridge, MA: MIT Press, 1996), 106.

And where's the Universe in all this?

Gödel's conception of the world as rational and rules-based made an irrational inconsistency like the absence of life after death seem impossible to him.

So we might well be astonished that he did not try to apply his incompleteness theorem to the Universe. If the Universe is a coherent system based on the laws of mathematics, doesn't that imply that it is incomplete—that its source can only be found outside of itself?

Gödel did show some interest in the Universe, especially during the years 1949–1950, when he even gave Einstein as a birthday present a new solution to the equations of Relativity called "Gödel's Universe," in which it would be possible to travel in time.[506] But he does not seem to ever have taken a serious interest in the question. Others have picked up the thread for him, either explicitly or implicitly pointing to Gödel's theorem to argue that man will never comprehend the Universe in its entirety. As astrophysicist Trinh Xuan Thuan puts it: *"Will the Universe one day be revealed in all its glorious reality? [. . .] In this respect, we may mention the work of the Austrian mathematician Kurt Gödel, who showed in 1931 that, within a mathematical system, there will always be certain propositions that cannot be proved. Just as it is impossible to prove everything in mathematics, the human mind will never be able to comprehend the whole of the Universe. The Universe will remain forever inaccessible. The melody will remain secret forever."*[507]

In the conclusion to his *Mind of God*, celebrated physicist and cosmologist Paul Davies definitely has Gödel's theorem in mind: *"But in the end a rational explanation for the world in the sense of a closed and complete system of logical truths is almost certainly impossible. We are barred from ultimate knowledge, from ultimate explanation,*

---

506. His solution treats a rotating universe where time travel would be possible—known as "Gödel's universe"—not the Universe as we know it.

507. Trinh Xuan Thuan, *The Secret Melody* (Oxford: Oxford University Press, 1995), 274.

by the very rules of reasoning that prompt us to seek such an explanation in the first place."

To pass beyond these limits that Gödel's theorem draws around our conception of the Universe, Davies goes on to encourage us to turn to mysticism: *"If we wish to progress beyond, we have to embrace a different concept of 'understanding' from that of rational explanation. Possibly the mystical path is a way to such an understanding. I have never had a mystical experience myself, but I keep an open mind about the value of such experiences. Maybe they provide the only route beyond the limits to which science and philosophy can take us, the only possible path to the Ultimate."*[508]

Celebrated astrophysicist Stephen Hawking too was struck by the implications of Gödel's thought. After spending most of his scholarly career trying to construct a *"Theory of Everything,"* he changed his mind. Reflecting on Gödel's theorem, he ultimately stated that such a theory was impossible to reach.[509]

And where is God in all this?

In his deliberately illegible papers, Gödel left a fourteen-point "creed." The first point affirms—as should come as no surprise—that the world is rational. The second says that human reason can theoretically be developed more highly than it has been yet, and the thirteenth says: *"There is a scientific (exact) philosophy and theology, which deals with concepts of the highest abstractness; and this is also most highly fruitful for science."*[510]

---

508. Paul Davies, *The Mind of God*, 231.

509. *"What we need is a formulation of M-theory that takes account of the black hole information limit. But then our experience with supergravity and string theory, and the analogy of Gödel's theorem, suggest that even this formulation will be incomplete. Some people will be very disappointed if there is not an ultimate theory that can be formulated as a finite number of principles. I used to belong to that camp, but I have changed my mind. I'm now glad that our search for understanding will never come to an end, and that we will always have the challenge of new discovery. Without it, we would stagnate. Gödel's theorem ensured there would always be a job for mathematicians."* Steven Hawking, "Gödel and the End of Physics" (lecture, Texas A&M, College Station, TX, March 8, 2002), https://lappweb.in2p3.fr/~dinis/Hawking.pdf.

510. Hao Wang, *A Logical Journey* (Cambridge, MA: MIT Press, 1996), 316.

This last point leads us to understand why Gödel grappled with the ultimate question: the existence of God. He tried to demonstrate this existence through simple logic, without falling back on philosophy or theology. He started from St. Anselm's famous ontological argument, which comes close to being a tautology and can be summarized thus: *"Since God is by definition a being who possesses all qualities, he must possess existence, since existence is a quality."*

During St. Anselm's lifetime (in the eleventh century), the monk Gaunilo objected: *"We can imagine that in the middle of the ocean there exists the most beautiful island possible, richer than all other islands. But that doesn't mean the island exists!"*

Despite this objection, the argument has continued to be taken seriously and developed by philosophers like Leibnitz, Descartes, and Spinoza.[511] Emmanuel Kant and David Hume, on the other hand, exerted a considerable amount of effort attempting to refute it. Gödel, a great admirer of Leibnitz, could not fail to follow in the footsteps of someone he considered a master thinker.[512]

Though Gödel shared his work on the existence of God with several colleagues, his proof was not published until 1987, nine years after his death.

The famous "Ontological Proof": 12 lines of formal logic, 5 axioms, 3 definitions and 4 theorems[513]

No one questioned its formal significance. Christoph Benzmüller (of the Free University of Berlin) and Bruno Woltzenlogel Paleo (of the Vienna University of Technology), both mathematicians researching artificial intelligence, responded by publishing an article with the provocative

---

511. Leibniz even recorded an account of a discussion he had with Spinoza on the subject.

512. In his famous letters to his mother, Gödel explains that he is attempting to develop Leibniz's approach, holding that reason can tackle life's toughest questions. His mother doubted that reason was equal to the task of answering questions like the existence of God or of life after death.

513. It takes at least two pages to list it all. See for example Hao Wang, *A Logical Journey*, 114-116.

title "*Formalization, Mechanization and Automation of Gödel's Proof of God's Existence.*"[514] With the help of computers, the two researchers were able to verify that Gödel's proof was formally correct—at least mathematically—according to higher order modal logic.[515]

"*With the help of computers, we can verify the coherence of a logical proposition very quickly,*" explains Christoph Benzmüller, who confirms Gödel's conclusions: "*God, by the most widely used metaphysical definition, necessarily exists. You cannot conceive of a world in which he does not exist.*"[516]

Is the debate closed, then? Mathematician Piergiorgio Odifreddi, of Cornell University, has cautioned us not to be so hasty. For this logician, Gödel's ontological proof resembles the idealistic arguments of Berkeley, which, according to Hume, cannot be contradicted but do not convince anyone. As with any mathematical demonstration, the strength of the conclusion depends on the axioms upon which it rests. If we accept the truth of the axioms, the conclusion inexorably follows. Of course not all materialists will accept, for example, axiom three: "*The property of being God-like is positive.*"

What is certain is that Gödel was completely theistic. As he himself put it: "*I went home with Einstein almost every day and talked about philosophy, politics, and the conditions of America. Einstein was democratically inclined. His religion is much more abstract, like that of Spinoza and Indian philosophy. Mine is more similar to church religion. Spinoza's God is less than a person. Mine is more than a person, because God can't be less than a person. He can play the role*

---

514. The article may be accessed online here: https://www.researchgate.net/publication/255994541_Formalization_Mechanization_and_Au-tomation_of_Godel's_Proof_of_God's_Existence#fullTextFileContent.

515. Even the popular French monthly *Science & Vie*, little given to spirituality, ran an article about this computer test on their front page in April 2020, stating in particular that "*the results are unequivocal [. . .] God must exist.*" Gödel's theorems (lightly adapted) are validated: "*The claim that 'God exists' is a true proposition in the logical and mathematical sense.*"

516. *Science & Vie*, special edition, no. 1235, "Pourquoi on croit en Dieu: Les mathématiques ont enfin la réponse," August 2020, 64–73.

of a person."[517] Since Gödel presented himself as a *"baptized Lutheran,"*[518] we can assume that this last sentence is perhaps a reference to the Incarnation.

He did distance himself, however, from organized religions, distinguishing them from *Religion*, as in this key passage from another letter to his mother: *"I believe there is a lot more sense in religion—though not in the churches—than one usually thinks, but from earliest youth we (that is, the middle layer of mankind, to which we belong, or at least most people in this layer) are brought up prejudiced against it [religion]—from school, from poor religious instruction, from books and experiences."*[519]

Though Gödel was quite critical of individual religions, he nevertheless viewed religion in a positive light, as the last point of his creed shows: *"Religions are, for the most part, bad—but Religion is not."*[520] He thought of his efforts to rationalize religion as *"no more than a vivid representation and adaptation to our present way of thinking of certain theological doctrines, that have been preached for 2000 years—though mixed with a lot of nonsense, to be sure."*[521]

And so Gödel's Religion with a capital *R* seems to be the Christian religion, since he mentions two thousand years and not the two thousand five hundred years that Greek philosophy has been around. But

---

517. Hao Wang, *A Logical Journey* (Cambridge, MA: MIT Press, 1996), 88.

518. Other researchers have leveraged Gödel's theorem to attempt to prove the existence of God. According to Antoine Suarez, since mathematics must be "thought" by someone and since Gödel's theorem demonstrates, at the very minimum, that mathematics must exist outside the human mind, we must posit a "super mind" capable of thinking mathematics. Juleon Schins, of the Delft University of Technology, argues that Gödel's results *"firmly establish the existence of something that is unlimited and absolute, fully rational and independent of the human mind."* He continues: *"What would be a more convincing pointer to God?"* Juleon Schins, "Mathematics: A Pointer to an Independent Reality" and Antoine Suarez, "The Limits of Mathematical Reasoning: In Arithmetic there will always be Unsolved Solvable Problems," in *Mathematical Undecidability, Quantum Nonlocality and the Question of the Existence of God*, ed. Alfred Driessen and Antoine Suarez (New York: Springer, 1997), 55, 58.

519. Hao Wang, *A Logical Journey* (Cambridge, MA: MIT Press, 1996), 107, 523. Ibid.

520. Ibid., 316.

521. Ibid., 108.

he draws a distinction between a purified Christianity and establishment churches.

Conclusion

The man his peers considered *"the greatest logician since Aristotle"*[522] demonstrated that Truth ultimately lies beyond demonstration and showed that we cannot do away with spirits and abstract entities. This lends great credibility to those who claim to have been in direct contact, beyond any demonstration, with the "world of mathematical truths."

His theory also suggests that there is some way to get in touch with the spiritual world. As we have seen, Gödel believed that the human mind, though associated with a machine (the brain), is not itself material. He argued that the laws governing the evolution of life are more complex and subtle than is currently understood, and that it makes sense to believe in an afterlife. He even attempted to demonstrate the existence of God.

But the greatest strength of Gödel's claims is his deconstruction of the positivistic and reductionistic theses that shaped what he called the *"spirit of the time."* As the tenth point of his credo tells us, *"materialism is false."* So Gödel joins the ranks of great scientists—Newton, Faraday, Maxwell, Pasteur, and the mathematician Bernard Reimann (1826–1866)[523]—for whom science practiced at the highest level leads towards God, not away from him. This is yet further support that corresponds perfectly with the conclusions of other rational investigations into the question of the existence of God.

---

522. Both Robert Oppenheimer and John von Neumann referred to Gödel using these words.

523. Bernhard Riemann (1826–1866), another great mathematical genius, was the founder of "Riemannian geometry," which gave rise to a revolution that permitted—among other things—Einstein's formalization of the theory of General Relativity. He is the author of the "Riemann hypothesis," the most difficult of the seven mathematical puzzles of the millennium, which is numbered as one of the most difficult twenty-three problems ever (as identified in 1900 by David Hilbert, who described it as the most important problem in mathematics and even the most important problem of all). See Marcus du Sautoy, *The Music of the Primes* (London: Fourth Estate, 2003). Riemann frankly admitted his faith in *"an eternal, personal, omniscient, almighty and benevolent God."* *Gesammelte Mathematische Werke: Fragments Philosophiques*, 1876, (Leipzig: B. G. Teubner Verlag, 1876).

EVIDENCE FROM
OUTSIDE THE SCIENCES

Introduction

We have now brought the scientific portion of our book to a close. We began by studying the logical consequences of the claim that there is no creator God. Then we showed how the great discoveries of the twentieth century do not square with these implications.

Today, three implications of the claim that God does not exist are challenged by science: the idea that the Universe had no beginning, that the fine-tuning of the Universe is nothing more than chance, and that the emergence of life from inert matter is a natural, comprehensible, and replicable phenomenon. But science has contested all three of these ideas. Therefore the thesis they flow from no longer seems tenable.

We could stop there. But that would be a pity, and our study would be incomplete if we concluded our book after this first part, however essential it is. Reason is much broader than the field of empirical science, and the thesis that God does not exist has implications in other fields of human knowledge that are just as powerful and interesting. They leave us with clear alternatives to choose from. Let's consider a few of these:[524]

- Miracles can't happen.
- Prophecy and divine revelation are impossible.
- There is no enigma that can't be resolved through "normal" materialist explanations.

But, as we shall see, these implications cannot stand the test of reason.

In the following chapters we submit four enigmas for your consideration.

---

524. See Chapter 3.

- 3,000 years ago, the Hebrews were the only people in the world who knew that the sun was nothing other than a luminous body, not a god. Where do the extraordinary cosmological and anthropological truths revealed by the Bible come from?

- Everyone on our planet arranges their calendars and signs their contracts based on the birth year of a man named Jesus of Nazareth. Who could he be? A wise man? A failed opportunist? Or more than that?

- Is the astonishing survival of the Jewish people over thousands of years nothing more than chance?

- Does the extraordinary event that happened in Fátima in 1917 and which was foretold months in advance by illiterate children have a plausible materialist explanation?

Those of you, our readers, who feel repulsed by this kind of talk can stop here or skip the next four chapters. We understand the reaction—but you don't know what you're missing!

17.

The Humanly Inaccessible Truths of the Bible

Is it possible for a small, obscure, and impoverished people to hold great truths unknown to the rest of the world?

In the ancient world, there was a small nation of nomads who occupied a modest territory of semi-desert, at the mercy of their powerful idolatrous neighbors. Their nation had no scientists, no observatories, no great cities, no natural resources, no ports, and no monumental architecture. But they did have a precious book. This people? The Hebrew people. This book? The Bible, which contained a goldmine of knowledge about humanity and the cosmos. Most of the Bible's claims about the nature and origin of the cosmos would remain unverifiable and inaccessible to human reason throughout antiquity and for many centuries thereafter. But over the last century or so, modern science has proved them true.

What truths did the Hebrews hold? They knew that the Sun and the Moon are nothing more than "lights," luminous bodies in the sky. They knew that the Universe was created from nothing, that it had a beginning, and that it would someday come to an end. They were convinced that mankind was made out of the same matter as the rest of the physical world and that neither stars, rivers, springs, nor forests were inhabited by deities.

It's easy for us to be indifferent to these beliefs. They seem so obvious. But in antiquity, they were all revolutionary ideas. The Jews' more prestigious neighbors—the Sumerians, Egyptians, Assyrians, Babylonians, Persians, Greeks, and Romans, with all their scholars, pyramids, observatories, and great libraries—held diametrically opposed convictions. They lived in a world of idols, while their more modest neighbors, thanks to the truths of the Bible, lived free of these superstitions.

What explains the extraordinary uniqueness of the Hebrew people? Where did these many truths come from, that were so counterintuitive and contrary to the common beliefs of their day, truths that the Hebrews had held for centuries? How did they manage to keep their faith in these truths intact through all the ups and downs of history?

We first need to clarify the line of questioning we'll use to examine some of the Bible's truths:

1. What is a humanly inaccessible truth?
2. What are the humanly inaccessible truths that the Bible gave to the Hebrews?
3. Are we sure we are reading the same Bible the Hebrews had 2,500 years ago?
4. Are we certain that we know how they understood the Bible's claims?
5. How advanced were the Hebrews, and what was their ability to tackle complex questions on their own?
6. How advanced were their neighbors, and to what extent were they able to impose their beliefs on the Hebrews?

I. What is a humanly inaccessible truth?

In general, we can consider a piece of knowledge to be out of reach if it meets the following criteria:

- The scientific or philosophical tools necessary to attain it through reason did not exist at that time.
- It is counterintuitive.[525]

---

525. "Counterintuitive" means "contrary to what our senses lead us to believe." For example, it is intuitive to think that the Sun revolves around the Earth, as this is what our senses tell us. The truth is the opposite, so the fact is "counter-intuitive." In the same way, the claim that the Universe was created *ex nihilo* appears counterintuitive, as our senses present us with a universe

- It is eventually revealed to be true.
- It is at odds with the beliefs of neighboring peoples who are more advanced and more numerous.
- The group who holds it is small, undeveloped, and weak.

If these criteria are met, the historical existence of a factual belief is a mystery, because those who hold it could not have obtained it by their own means or through the intervention of their neighbors.

II. What out-of-reach truths did the Hebrews know about the cosmos and humanity thanks to the Bible?

The Hebrews knew that:[526]

1. The Sun and the Moon are nothing but luminous bodies.
2. The Universe was created from nothing and had a beginning at a particular point in time.
3. Since the Universe will end, time is unidirectional, not cyclical.
4. Mankind comes from matter.
5. There are no gods in the stars, rivers, springs, or forests.
6. All people come from the same stock, which is the basis of their equality; kings, pharaohs, and emperors are only human.
7. The world was not created all at once but rather developed in stages.
8. In the process of creation, mankind appeared last.
9. There were other humanoid beings that eventually went extinct.
10. The astrology and magical rites that dominated the lives of all the ancient peoples, from Babylon to Rome, were nothing but superstitions.

Two things about this list immediately stand out. First of all, the Hebrews' worldview is correct on too many points to be the result of chance. In fact, there are so many surprising accuracies that we do not have room to study all of them here. Second, this list contains a

---

that appears stable and unlimited in space and time.

526. On these topics, see Nathan Aviezer, *Au commencement* (Genève: Éditions MJR, 1990).

number of fundamental, important teachings regarding man and his place in the cosmos. These truths had major real-life implications that inspired the Hebrews to adopt a way of life different to that of their neighbors, namely:

- The absence of human sacrifice.
- A very minimal practice of slavery that was carefully regulated to protect the rights of slaves.
- Better conditions for women.
- A greater respect for human life.
- The condemnation of astrology and superstition.

III. Is our Bible the same as the one the Hebrews had?

We must answer this question to put to rest a lingering suspicion. We might reasonably doubt whether the texts of the Bible have been altered over the course of time and if the truths in question were introduced only at a later date.

Fortunately, we can easily put this fear to rest. First of all, in the pre-exilic books—those written before the sixth century BC, which are the ones we are most concerned with here—there are only minor differences between the Jewish and Christian versions of the Bible read today. It would have been impossible for one group to alter the texts without the other immediately noticing. Consequently, the texts we have are the same ones the Hebrews read more than two millennia ago.

The discovery of the Dead Sea Scrolls in 1947 settled the case beyond doubt. The excavated texts are almost identical to the texts we read today.

Most experts admit today that the texts in question have not been significantly altered since the sixth century BC.

We can be fairly certain, therefore, that the texts and its teachings have not been modified in the last 2,500 years and that they are the very ones that the Hebrews of that time put their trust in.

IV. Are we sure that they interpreted the Bible's claims in the same way we do today?

We run very little risk of misinterpreting these claims because they recur in many places in the Bible, worded in various ways. The claim that the Sun and Moon are only luminous objects, for instance, is presented in various forms in several different passages.

Further, contemporary non-Jewish authors sometimes expressed their astonishment at the Bible's claims, which proves that these ideas went against the shared worldview of antiquity.

Finally, in some cases we know of, neighboring peoples persecuted the Hebrews for these beliefs. They were punished by the Romans, for example, for refusing to participate in the cult of the emperors.

V. How advanced were the Hebrews?

From Abraham to Jesus Christ, the Hebrews were always a small nation. Their kingdom, when they had one, was tiny and wielded little political influence.

Aside from Kings David and Solomon, whose actual importance is controversial, they had no great kings, no great scholars, no philosophers, no generals, no explorers, and no great conquerors.

Socrates, Plato, Pythagoras, Euclid, Archimedes, Parmenides, Thales, Ramses, Darius, Cyrus, Alexander, Caesar—not a single one of these great figures of antiquity came from the Hebrew nation.

The Hebrews were frequently conquered and oppressed by their more powerful neighbors. They were enslaved in Egypt, long held subject to the Canaanites, forced into exile by the Assyrians, crushed by the Babylonians, and deported en masse to Babylon. After their liberation, they became vassals of Persia, only to be invaded by the Greeks and finally occupied by the Romans, who destroyed their country, scattered them to the four corners of the world, and ended their presence in Palestine for almost two thousand years.

At best, they could boast of a handful of prophets, men who were usually on the run from their own co-religionists, who were infuriated by their rebukes and prophecies of doom.

The Hebrews had no pyramids, observatories, libraries, or great ports—not even impressive buildings (with the late exception of the Temple of Herod, built when their numbers and influence were at their peak). But by that time, the texts of the Bible that concern us here had already been written a long time ago.

The Hebrews possessed none of the usual qualities or accomplishments associated with high civilization or advanced intellectual activity. Needless to say, this was not fertile ground for cultivating great discoveries.

In the face of such a discrepancy between the Hebrew people's modest resources and the incredibly advanced and prescient truths they possessed about the world and mankind, it is difficult to believe that they made these discoveries on their own.

VI. How advanced were their neighbors?

Conversely, since the neighboring peoples were more numerous, more advanced, and more powerful, it stands to reason that they would influence their weaker neighbors and impose their beliefs upon them.

Indisputably, the Hebrews' neighbors surpassed them on almost every level. There is simply no comparison between the Egyptian pyramids and the buildings of this group of shepherds living 1,000 years before Christ, or between the skill and refinement of the objects discovered in Tutankhamun's tomb and Hebrew pottery of the time.

When it comes to technology and competence in various fields of knowledge, there is no possible point of comparison. Pythagoras, Thales, Plato, Socrates, Homer, and the wise men of Babylon—the genius of Darius, Cyrus, and Alexander—find no equal among the few notable historical figures of the Hebrew people.

Yet the Jews clung stubbornly to beliefs that held nothing in common with those of their more advanced neighbors. Their convictions were not at all borrowed from the mental universe of other nations, and thus it is all the more extraordinary that this little people had the courage to hold onto beliefs that their more brilliant, powerful, and numerous neighbors deemed absurd and sacrilegious.

This anomaly is what leads us to speak of humanly unattainable truths, since normally these peoples would have imposed their deities and beliefs on the weaker group.

But actually the opposite happened: the star of the pagan deities slowly died, their cults fell into disuse and then oblivion, while the certainties of the Hebrews were maintained until, millennia later, they were proven to be perfectly accurate.

VII. Some of these truths "fall from the sky"

1. The Sun and Moon are nothing more than luminous bodies

The Hebrew people had a unique perspective on the world around them. They did not consider the Sun or Moon to be gods, nor did they ascribe any special power to them. Significantly, archaeologists have never found a Hebrew temple dedicated to the Moon or to the Sun. The Bible calls them "lights," that is to say, luminous objects.

Of all the ancient sacred texts known to us, this is the only one that does not place the Sun and Moon on a divine pedestal.

Let's take a look at the main passages in the Bible where this unique perspective is especially clear:

- *"And God said, 'Let there be lights in the dome of the sky to separate the day from the night; and let them be for signs and for seasons and for days and years; and let them be lights in the dome of the sky to give light upon the earth'"* (Gen 1:14–15).
- *"God made the two great lights—the greater light to rule the day and the lesser light to rule the night—and the stars"* (Gen 1:16).

- *"You have made the moon to mark the seasons; the sun knows its time for setting"* (Ps 103:19).
- *"The heavens are telling the glory of God; and the firmament proclaims his handiwork"* (Ps 19:1).
- *"When I look at your heavens, the work of your fingers, the moon and the stars that you have established"* (Ps 8:4).
- *"By the word of the Lord the heavens were made, and all their host by the breath of his mouth"* (Ps 32:6).
- *"Who alone stretched out the heavens and trampled the waves of the Sea; who made the Bear and Orion, the Pleiades and the chambers of the south"* (Job 9:9).
- *"Who made the great lights... the sun to rule over the day... the moon and stars to rule over the night"* (Ps 135:7–9).
- *"Lift up your eyes on high and see: Who created these? He who brings out their host and numbers them, calling them all by name"* (Isa 40:26).

In the thought systems of all great ancient empires, the Sun and Moon were divine beings:

- The **Sumerians** believed that Enlil, the king of the gods, begot Nanna, the divine personification of the Moon, who was the father of Utu, the Sun god.
- This conception of a second sun god is also found among the **Akkadians**, **Assyrians**, and **Babylonians**, in the person of Ilu, who begot Sin (the Moon), the mother of the god Shamash (the Sun), who begot Truth and Law. The Sun and Moon were worshipped at sanctuaries across Mesopotamia from time immemorial.
- The most important deity of the **Egyptians** was the sun god Ra, who had the head of a falcon and whose light brought life to the Universe. He issued forth from the primordial ocean (Nun)—some said the god Ptah created him—and he made the world and the other gods. The most important cult centers in Egypt were dedicated to his worship, notably at Heliopolis, the "city of the Sun," near modern-day Cairo.
- The **Persians** associated the god Mithras with the Sun, whose Persian name "Cyrus" was taken up by Cyrus the Great, founder of the

Persian Empire. Mah was the moon goddess.
- Even the most enlightened philosophers among the **Greeks** considered the Sun and the stars to be part of the divine world. The philosopher Aristotle himself, in his treatise *De caelo*, writes that the Sun and the Moon have souls and that they are alive.
- For the **Romans**, the whole starry realm was permeated with divinity, and Jupiter, the master of the Olympian gods, borrowed his radiance from the Sun.

The table below provides a summary of the deities associated with the Sun and Moon worshipped by the great ancient peoples. The Hebrews are conspicuously absent, the only people who saw the Sun and Moon as nothing more than two created objects.

Civilization	Sun	Moon
Sumerians	Utu	Nanna
Egyptians	Ra	Thot or Khonsou
Assyrians	Shamash	Sin
Babylonians	Shamash	Nanna
Persians	Mithra	Mylissa
Greeks	Helios and Apollo	Selene and Artemis
Romans	Sol and Phoebus	Luna and Diana
Phoenecians	Melqart	Astarte
Aztecs	Huitzilopochtli	Coyolxauhqui
Incas	Inti	Mama Quilla
Jivaroans	Etsa	Nantu
Hindus	Surya	Chandra or Soma
Chinese	Xihe	Heng-Ngo or Chang'e
Japanese	Amaterasu	Tsukiyomi
Celts	Belenos	Sirona

Table showing the names of the gods of the Sun and the Moon in the different civilizations

From a twenty-first-century vantage point, it seems natural to believe that the Sun and Moon are just two luminous bodies, because science has made us thoroughly familiar with the ideas of space, stars, and a solar system governed by physical laws. Before modern science, however, this would have been counterintuitive and contrary to human sense perception. The ancient peoples simply couldn't imagine a cosmos capable of keeping the stars and other celestial bodies turning regularly in the skies without the intervention of some divine power. A **representation** of the cosmos according to physical laws as we understand them today was unimaginable and unthinkable, since they simply lacked access to the technology and advanced observation necessary to arrive at our modern model of the solar system. Much to the contrary, the Hebrew people's neighbors—as well as peoples far separated from them in space and time, like the Incas—naturally assumed that the Sun and Moon were gods, as we have seen.

The cosmology of the chosen people was therefore completely iconoclastic from the point of view of other nations. It reportedly provoked this scandalized and, in retrospect, amusing comment from the philosopher Celsus in his *On the True Doctrine: A Discourse Against the Christians* (c. 178 AD): *"The first thing about the Jews which may well cause amazement is that although they worship the heaven and the angels in it, yet they reject its most sacred and powerful parts, the sun, moon, and the other stars, both the fixed stars and the planets. They behave as though it were possible that the whole could be God but its parts not divine."*[527]

Knowledge that was objectively inaccessible in biblical times

Without telescopes, clocks, or advanced mathematics, it would have been utterly impossible to conclude that the Sun and Moon are nothing more than two luminous objects. The works of Pythagoras, Thales, Euclid, and Hipparchus were certainly remarkable initial efforts to-

---

527. Origen, *Contra Celsum*, ed. and trans. Henry Chadwick (Cambridge: Cambridge University Press, 1980), book V.6.

wards our understanding of celestial bodies, but their work was kept to their own inner circle and preserved the divine character of the stars. At any rate, they made their discoveries long after the Hebrews had recorded, in the Bible, their belief that the stars were not divine.

Centuries Later: Modern science puts an end to the divinity of the stars

Heirs to the knowledge contained in the Bible, Christians have always known that the Moon and Sun are nothing but objects. Over time, increasingly precise scientific observation has confirmed this central biblical claim. In 1609, Englishman Thomas Harriot made the first observations of the Moon through a telescope, just a few months before Galileo made his observations.

Thus, the Hebrews were right dozens of centuries before their time, long before anyone had the necessary scientific instruments at their disposal to prove it.

2. The Universe had an absolute beginning and was created from nothing by a God who is exterior to it: a biblical notion contrary to all cosmogonies

The Jews claimed that the Universe had a beginning at a particular point in time, a creation from nothing by a God who stood outside of it. This striking metaphysical claim is repeated throughout the Bible:

- *"In the beginning God created the heavens and the earth"* (Gen 1:1).
- *"Then God said, 'Let there be light'; and there was light"* (Gen 1:3).
- *"I beg you, my child, to look at the heaven and the earth and see everything that is in them, and recognize that God did not make them out of things that existed. And in the same way the human race came into being"* (2 Macc 7:28).
- *"I am the Lord, who made all things, who alone stretched out the heavens, who by myself spread out the earth"* (Isa 44:24).
- *"Thus says God, the Lord, who created the heavens and stretched*

them out, who spread out the earth and what comes from it"* (Isa 42:5).
- *"By the word of the Lord the heavens were made, and all their host by the breath of his mouth"* (Ps 32:6).
- *"O Lord, Lord God, Creator of all things"* (2 Macc 1:24).
- *"He who lives forever created the whole universe"* (Sir 18:1).
- *"Long ago you laid the foundation of the earth, and the heavens are the work of your hands"* (Ps 101:26).
- *"Before the mountains were brought forth, or ever you had formed the earth and the world, from everlasting to everlasting you are God."* (Ps 89:2).

The Jews' neighbors, in contrast, believed that gods were part of this world, arising out of a primordial matter that has always existed:

- The **Sumerians**, the **Akkadians**, the **Assyrians**, and the **Babylonians** shared a common understanding of their origins. The goddess Nammu, the primordial sea, gave birth to the sky and the Earth. From her arose the two principles of the world and the other gods: the female, Tiamat (saltwater), and the masculine, Apsu (fresh water).
- The **Egyptians** believed that the Universe emerged from a mound or egg that arose from the primordial watery abyss, named Nu, which was also chaos. Egyptian cosmogony varied slightly from place to place, and evidence for three or four great creation myths has been discovered at Heliopolis, Memphis, and Thebes.
- The **Persians** thought that in the beginning there were two hostile primordial principles separated by an empty space. The evil principle declared war and created demons. In response, the good principle created angels and men.
- Among the **Greeks** two alternate versions coexisted. According to Hesiod's *Theogony*, in the beginning was the primordial element Chaos, an immeasurable whole containing all the elements that make up today's world. Four entities separated from it: Gaia (the Earth), Eros (sexual Desire as a primordial creative force), Erebus (the Darkness of the Underworld), and Nix (Night). The orphic

tradition is slightly different: the water and the elements gave rise spontaneously to the earth, from which a monster called Chronos arose. He created Ether, Erebus, and Chaos, then engendered an egg from which Eros hatched. Eros in turn generated the Moon, the Sun, and the Night, with whom Chaos fathered Uranus and Gaia. From them come all the gods.[528]

- The **Romans** adopted Greek mythology.

The ancient cosmogonies largely lack the concept of an absolute beginning to the Universe. After all, it is not a very intuitive idea.

To the naked eye, space seems infinite in all directions, immobile and motionless, with neither beginning nor end.

Not until the middle of the twentieth century did people begin to doubt what had been held as a certitude for so many centuries.

Even the great Einstein refused at first to admit evidence to the contrary, and he arbitrarily modified his own equations to make them yield a stationary Universe that conformed with his own belief. Einstein's fudged figures are evidence of just how counterintuitive the notion of a beginning to the Universe is to human reason.

But despite all odds, the Hebrews have constantly maintained this truth about the beginning of the Universe.

This second truth, therefore, can also be classed as humanly unattainable knowledge: No rational explanation can satisfactorily explain the existence of this knowledge, at that time, and amongst these people.

3. The Universe is progressing towards its end, following the unidirectional arrow of time

Most ancient civilizations, whether in Mesopotamia, Asia, America, or Greece, conceived of time as a cycle. They thought about it by analogy with the cycles of nature: day and night, the four seasons, birth and

---

528. https://en.wikipedia.org/wiki/Cosmogony, consulted November 2019.

death. As Mircea Eliade has pointed out, *"the myth of the eternal return"*[529] is at the foundation of the religious and philosophical beliefs of these peoples.

In this view of time, a person is not confined to a single point in a linear history. Everything always begins again. The eternal cycles of the cosmos are celebrated in cyclic rites that annually abolish the time that has passed and inaugurate a fresh new cycle.

In this system, the individual human life or action has little value. Belief in reincarnation, often associated with a cyclical understanding of time, devalues the importance of the present life, since this is only one of a potentially infinite number of lives that one will live. Fatalism and passivity, then, often go hand in hand with faith in the eternal return.

At the opposite end of the spectrum, materialists affirm that the Universe is eternal. For them, it is inconsistent to speak of the end of the Universe, since an end presupposes a beginning, which in turn implies a creation and therefore a creator.

The biblical doctrine charts a unique path between these two extremes, one that confers on every human action a unique and unprecedented value. It sets the world on a linear timeline: unidirectional, limited, and unrepeatable. Hebrews and Christians share this peculiar way of thinking of time as linear and unidirectional.

- *"Long ago you laid the foundation of the earth, and the heavens are the work of your hands. They will perish, but you endure; they will all wear out like a garment. You change them like clothing, and they pass away"* (Ps 101:26–27).
- *"For it is he who gave me unerring knowledge of what exists, to know the structure of the world and the activity of the elements; the beginning and end and middle of times, the alternations of the*

---

529. Mircea Eliade, *The Myth of the Eternal Return: Or, Cosmos and History* (Princeton: Princeton University Press, 1971), 141. The Greeks had the concept of palingenesis (παλιγγενεσία), which is close to the idea of "rebirth."

solstices and the changes of the seasons, the cycles of the year and the constellations of the stars" (Wis 7:17–19).
- "*His powerful spirit looked into the future and he predicted what was to happen before the end of time*" (Sir 48:24).
- "*For the heavens will vanish like smoke, the earth will wear out like a garment, and those who live on it will die like gnats; but my salvation will be forever, and my deliverance will never be ended*" (Isa 51:6).
- "*Declaring the end from the beginning, and from ancient times things not yet done*" (Isa 46:10).

The Bible rescues man from the infinite and hopeless cycle of the eternal return.

Not until the end of the twentieth century, when scientists proved the impossibility of a "Big Crunch" (or "Great Collapse") followed by a new "Big Bang," did science definitively prove that time is not circular. Rather, it is developing inevitably towards its end.

Like the others, this third truth is contrary to the worldview of the Hebrews' neighbors, absolutely counterintuitive, and humanly inaccessible with the skill and learning of the time. Centuries later, it has been confirmed by scientific observation.

Though we can now explain how this truth was ahead of its time, we still can't explain how the Hebrew people obtained it.

4. The human body is made up of nothing but physical matter

Thanks to the Bible, the Hebrews knew that the human body is nothing other than matter. Only the soul, a spiritual reality that God infuses into man, gives him a different nature.

- "*Then the Lord God formed man from the dust of the ground, and breathed into his nostrils the breath of life; and the man became a living being*" (Gen 2:7).
- "*By the sweat of your face you shall eat bread until you return to the ground, for out of it you were taken; you are dust, and to dust you shall return*" (Gen 3:19).

- *"Remember that you fashioned me like clay"* (Job 10:9).
- *"You turn us back to dust, and say, 'Turn back, you mortals'"* (Ps 89:3).
- *"All human beings come from the ground, and humankind was created out of the dust"* (Sir 33:10).
- *"When you take away their breath, they die and return to their dust"* (Ps 103:29).
- *"And the dust returns to the earth as it was, and the breath returns to God who gave it"* (Eccl 12:7).

By contrast with the Hebrews, most peoples believed that humans were created out of the same divine stuff as the gods, although in greatly diluted form. In fact, many mythologies trace man's origins back to the corpses of creator gods. In these traditions, there is something divine about the human body. A brief tour of the mythological landscape illustrates this fact:

- The **Sumerians**, **Akkadians**, and **Assyrians** believed that man was made by the god Marduk from clay soaked in the flesh and blood of a sacrificed god. This gave Marduk's creature a share in the divine intelligence.
- For the **Babylonians**, man and the Universe were created from the dismembered corpse of the conquered and sacrificed goddess Tiamat.
- In the **Egyptian** world, texts found on sarcophagi (2000–1780 BC) have the god Atum-Ra, the creator of most of the other gods, say, *"Mankind came into being from the tears which flowed from mine eye."*[530]
- Among the **Greeks** we find several myths that tell of the creation of man. The most obscure of them claims that he was born from the teeth of the serpent Ophion after he was trampled by the Earth goddess Gaia. The better-known story is that of Prometheus, whose name means "foresight," and his brother Epimetheus, "hindsight."[531] Prometheus shaped man from water and mud, but man was not

---

530. Alan W. Shorter, *The Egyptian Gods: A Handbook* (San Bernardino: Borgo Press, 1994), 8.
531. David Leeming, *The Oxford Companion to World Mythology* (Oxford: Oxford University Press, 2005), s.v. Epimetheus.

able to protect himself against the animals created by Epimetheus, who had given them strength, speed, feathers, fur, wings, shells, and other advantages. Not wanting to leave his creation powerless, Prometheus stole fire from heaven and gave it to man.[532] According to a third myth, the first generation of humans sprang from the womb of the mother goddess Gaia,[533] and Zeus created a second and third generation of humans.

- The **Romans** were little concerned with these questions and merely adopted the Greeks' ideas.
- According to the **Vedic** mythology of the Hindus, all beings arose from the division of the primitive being Purusha: "*In the beginning this was Self alone, in the shape of a person (purusha)... He was so large as man and wife together. He then made this his Self to fall in two (pat), and thence arose husband (pati) and wife (patnî).*"[534]
- In the **Brahmanic** version, Brahma creates man by dropping his seed as he pursues his daughter Sarasvati or, according to another version, Sandhya, the daughter of Shiva.
- Among the **Tamils**, the earth (Prithvi) is represented by the cow, and the fruits of the earth are symbolized by her milk. She gives birth to Manu, the ancestor of humankind, in the form of a calf.
- The **Norse** believed that the enormous giant Ymir, formed out of primordial ice and fire, was the first living creature. While he slept, giants leaped out of his body, and a man and a woman emerged from his armpits.
- Among the **Mayans**, the gods created four men and four women from yellow and white corn. The men were very wise, and the gods feared that they would become their equals. To prevent this, they blew steam in their eyes to cloud their wisdom.

---

532. Hesiod, *Theogony* 507–545.

533. In Hesiod's cosmogony, she is the personification of the Earth.

534. See Hoành-Sơn Hoàng-Sỹ-Quý, "Le mythe indien de l'Homme cosmique dans son contexte culturel et dans son évolution," *Revue de l'histoire des religions* 175, no. 2 (1969): 133–154. Also see *The Upanishads*, Part 2 (SBE15), trans. F. Max Müller (Oxford: Clarendon Press, 1875), Brahmana 4.1, 85.

- The **Aztecs** held that Quetzalcoatl, in the form of the dog god Xolotl, went to Mictlantecuhtli, the underworld, and stole some dried-out bones. He sprinkled them with his own blood to give life to men.
- The **Shinto** belief of Japan teaches that the sky and earth were born from a primordial egg that generated the first divine couple, Izanagi and his sister Izanami. After entering the realm of the dead in a failed attempt to bring the dead Izanami back to life, Izanagi ends up taking on human form.
- In **Korea**, the two goddesses Gung-hee and So-hee each gave birth to two divine men and two divine women, who generated a dozen infants, the ancestors of humans.

These mythological accounts attempt to explain where man came from and what he is made of. But science has now answered these questions. We know that the human body is made up of matter only, shaped through a long process of increasing complexity.

The human body is just an intelligent machine. Only the soul, if we believe it exists, is of a different nature.

This conception of the human body is now commonly accepted. Yet many find it hard to accept fully because it is humiliating to the human ego.

Once again, we are faced with a mystery: the Hebrews possessed a lot of knowledge which was counterintuitive. To put it another way, they understood major truths which had important metaphysical implications. How curious...

5. Nature and the elements are not inhabited by divine beings

For most ancient peoples, inexplicable natural phenomena were a cause of great fear and worry. Most explained them as signs of divine activity. So for the Greeks, Poseidon made the sea move and the ground shake, Zeus sent down lightning bolts, the rainbow traced the path followed by Iris (the messenger of the gods), and the streams were inhabited by naiads.

The Bible's demythologizing of natural forces is original and revolutionary: for the Hebrews, seas, forests, springs, hills, mountains, rivers, storms, and lightning were not divine beings:

- *"Their idols are silver and gold, the work of human hands. They have mouths, but do not speak; eyes, but do not see. They have ears, but do not hear; noses, but do not smell. They have hands, but do not feel; feet, but do not walk; they make no sound in their throats"* (Ps 113B:4–7).
- *"We know that 'no idol in the world really exists,' and that 'there is no God but one'"* (1 Cor 8:4).

The process of demythologization that the Bible initiated did not always go smoothly. The entire history of the Hebrew people, in fact, was marked by the struggle, and frequent failure, to avoid falling into the idolatrous beliefs and practices of their neighbors, and the superstitious belief that natural phenomena (like thunder and lightning) were manifestations of the divine.

And now, more than three thousand years later, idols have all but vanished from the world and no one believes that lightning is hurled to earth by Zeus, Thor, or Indra. Far before science told us why, the Bible had separated the creator God from the realities of his Universe, reducing the latter to the status of simple phenomena. Thanks to this revealed truth, the Jews led a life free from the fear induced by superstition.

6. Humanity traces its origin back to one man and one woman, implying that there is no innate hierarchy among men

This thesis of the shared origin of every human is today demonstrated by the fact that the entire human species shares the same chromosomes. The Hebrews already called all men "sons of Adam" (Num 23:19, Deut 32:8, Ps 13:2), and the same doctrine is professed in many places in the Bible:

- *"So God created humankind in his image, in the image of God he created them; male and female he created them"* (Gen 1:27).

- *"You made Adam, and for him you made his wife Eve as a helper and support. From the two of them the human race has sprung"* (Tob 8:6).
- *"I also am mortal, like everyone else, a descendant of the first-formed child of earth"* (Wis 7:1).

The Hebrews are likely the only ancient people who never claimed divinity for a mortal being, in stark contrast to the legends and foundation myths of the other ancient civilizations, in which gods and mortals intermingle and seduce one another and heroes, kings, pharaohs, and emperors are demigods in their own right. Built on these mythological foundations, these societies unsurprisingly preserved rigid hierarchies among men, from the god-emperor at the top to the abject slave at the bottom.

- The **Mesopotamian** myth of Enki and Ninmah centers on the creation of humans. It begins with the creation of the world and its initial population by the gods, who mingle and multiply until they have to produce their own food to survive. Finding the task extremely unpleasant, they complain to the goddess Namma, who asks her son Enki to make substitutes to work in place of the gods. Enki crafts a mold and his mother fills it with clay, shaping human beings, who are then brought to life by a group of goddesses, Ninmah first among them.[535]
- The **Babylonians** divinized their kings and thought their first king (Cronus, husband of the goddess Rhea) had been a god, like his ancestor Nimrod. In Babylonian society, as in neighboring India, people were defined by the caste to which they belonged by birth. The great Babylonian myths reworked older creation stories to present a coherent vision of the earliest times, from the creation of the gods to the creation of the world and finally the creation of man. This allowed them to develop their vision of the relationship between man and the gods, with royalty serving as a link between the two.[536]

---

535. See W. G. Lambert, *Babylonian Creation Myths* (Winona Lake, Indiana: Eisenbrauns, 2013), 337.

536. See Roy E. Gane and Constance E. Gane, "Cosmic Conflict and Divine Kingship in Babylonian

- Among the **Persians** we find genies, supermen, and a relatively complicated creation story: After *creating the world, Ohrmazd created the first man, Gayomard, out of clay, and the primeval ox. But Ahriman attacked the world and killed the ox and the man. Out of the ox's body arose various species of good animals, and from the man arose various precious metals and a seed that sprouted into the rhubarb, from which the primordial human couple would spring.*[537]
- In **Egypt**, the pharaoh was the incarnation of Horus. Therefore, he belonged to the lineage of the gods.
- For the **Greeks**, men interacted on earth with the full retinue of Mount Olympus: gods, demigods, titans, and divine heroes. The *Iliad* and the *Odyssey*, for example, describe the exploits of the demigod Achilles, son of the nereid Thetis and the mortal Peleus. On the other end of the spectrum, they considered non-Greeks "barbarians" who belonged to a subhuman class because they could not speak Greek. These barbarians were treated harshly in the Greco-Roman world, although in Rome during the late imperial period, the master's right to kill or mistreat his slaves was restricted by law.
- The **Roman** emperors always wanted to be seen as gods. Many had themselves proclaimed gods posthumously through "apotheosis." It all started with Augustus: when he was declared the heir of the deified Julius Caesar, he proclaimed that he was also the son of Apollo. From that point on, the Roman emperors were considered living gods.

Rather than deifying some and denigrating others, the Hebrews held this truth that was so foreign to their contemporaries: since all men, however different they may be, have the same origin, there is no way to hierarchize them by deifying kings and heroes while denying the humanity of slaves or members of certain castes.

Religion and Biblical Apocalypses" *Faculty Publications* 208 (2016). https://digitalcommons.andrews.edu/pubs/208.

537. See Carlo Cereti, "GAYŌMARD," *Encyclopædia Iranica*, online edition (2015), available at http://www.iranicaonline.org/articles/gayomard (accessed on 16 July 2024).

Mankind's common descent: a single pair of ancestors

Scientists have been debating the question of mankind's origins for a long time now. Even up to the 1930s, some scientists held the Enlightenment-era idea that the various races came from completely separate lineages.

But the discovery of chromosomes and the sequencing of the human genome has probably closed this debate, even if there is still some room for discussion.

Various studies of the genome,[538] focusing on the Y-chromosome (transmitted only from father to son) have demonstrated that all humans descend from one and the same father, now known universally as "Y-chromosomal Adam."[539]

Likewise, studies of matrilineal ancestry carried out on mitochondrial DNA (transmitted only through the mother) have established that all humans descend from one and the same mother, who is known as the "Mitochondrial Eve."[540] As strange as it may seem, some have argued that "Y-chromosomal Adam" and "Mitochondrial Eve" lived in different epochs, but this hypothesis has also been disputed since 2013.[541]

---

538. https://en.wikipedia.org/wiki/Y-chromosomal_Adam.

539. Notably, the forty-million-dollar 2005 Genographic Project, led by geneticist Spencer Wells and made possible by collaboration between IBM and *National Geographic*, analyzed more than a million genomes. Entitled "The Search for Adam," it concluded that there was a single mother and father who lived between 60,000 and 150,000 years ago. Since then, this date has been advanced and pushed back by various studies, notably that of Fulvio Cruciani, "A Revised Root for the Human Y Chromosomal Phylogenetic Tree: the Origin of Patrilineal Diversity in Africa," published in *The American Journal of Human Genetics* 88, no. 6 (19 May 2011)—with a wide range spanning 5,000 years and the first appearance of *Homo sapiens* at perhaps 300,000 years. The debate was reignited in 2013 with the analysis of the very unique DNA of Albert Perry, an African American man, who supposedly came from a line very different from ordinary people, dating back to a cross with a different line than most humans 340,000 years ago, before *Homo sapiens*.

540. https://en.wikipedia.org/wiki/Mitochondrial_Eve.

541. G. David Poznik et al.,"Sequencing Y Chromosomes Resolves Discrepancy in Time to Common Ancestor of Males Versus Females," *Science* 341, issue 6145 (August 2013), 562–565, https://science.sciencemag.org/content/341/6145/562.abstract. Studies in this direction abound. In addition to the article already cited by Fulvio Cruciani, see that of Douglas L. T. Rohde of MIT, "On

Facts only recently discovered by modern science were revealed to the Hebrews three thousand years ago and transmitted from generation to generation through the Bible.

What can we conclude about these truths? Did they fall from the sky?

We can agree on at least one point: thanks to the Bible, the Hebrews enjoyed knowledge unique to their time, and this knowledge, which was counter-intuitive and opposed to that of their neighbors, was only proven correct much later.

In the twentieth century, when all these truths of the Bible and science met, some scientists were amazed, especially after the discovery of the Big Bang.

Let's re-read some quotations from two Nobel Prize winners already seen previously:

"The best data we have are exactly what I would have predicted, had I nothing to go on but the first five books of Moses, the Psalms and the Bible as a whole" (Arno Penzias, 1978 Nobel Prize in Physics).[542]

"Certainly if you are religious, I can't think of a better theory of the origin of the universe to match with Genesis" (Robert Wilson, 1978 Nobel Prize in Physics).[543]

Now it is up to you, the reader, to judge this anomaly and form an opinion.

---

the Common Ancestors of All Living Humans" (November 11, 2003), which used mathematical modeling to calculate that the earliest common ancestor of all humans alive today would have lived between 2,000 and 5,000 BC.

542. Malcolm Browne, "Clues to the Universe's Origin Expected," *The New York Times*, March 12, 1978, https://www.nytimes.com/1978/03/12/archives/clues-to-universe-origin-expect- ed-the-making-of-the-universe.html.

543. Robert Wilson, cited by F. Heeren, *Show Me God* (Nashville: Day Star Publications, 1997), 157.

There is not much wiggle room between two alternatives: it is either the result of pure chance or of a divine revelation.

In conclusion, it is amusing in retrospect to read the following remark made by the Roman philosopher Celsus around the year 175, now that the scientific discoveries of the twentieth century have given authoritative support to the Bible's insights. As Origen reports, Celsus claimed: *"There is an ancient doctrine which has existed from the beginning, which has always been maintained by the wisest nations and cities and wise men,"* including among these *"the Egyptians, Assyrians, Indians, Persians, Odrysians, Samothracians, and Eleusinians"* but not, pointedly, the Jews, whose cosmogony Celsus calls *"utterly stupid."*[544]

---

544. Origen, *Contra Celsum*, ed. and trans. Henry Chadwick (Cambridge: Cambridge University Press, 1980), book I.15.

18.

The Alleged Errors of the Bible, Which Are Not Errors

Going through the pages of the previous chapter, we were surprised by the enigma posed by the advanced cosmological and anthropological knowledge in the Bible. At the conclusion of our analysis of this paradox, only two viable hypotheses remained: the presence of these truths is the result either of extraordinary chance or a divine revelation which should constitute evidence for the existence of God.

The Bible's detractors reject this second hypothesis, pointing to a number of errors they believe that they find in the biblical texts. If it turns out that there are falsehoods mixed in with the truths we've seen, the divine revelation hypothesis loses much of its credibility. A book inspired by God cannot mix the two. Moreover, the errors the critics put forward are multiple and are the subject of a large number of articles, books and specialized websites.

This chapter aims to assess the legitimacy of these concerns. This will sometimes entail thinking outside our usual framework to better understand the underlying objectives, the constraints involved, and the methods employed to overcome these constraints. At the end of this effort, the reader will realize that the errors put forward in fact only result from a misunderstanding of the goals, constraints and processes of the Bible.

The Bible's objectives

The purpose of the Bible is to tell us who God is, who we are, how we must live in order to re-join him and, finally, how he has prepared a people capable of transmitting these truths to the world and of welcoming the Messiah announced since the dawn of time.

As we can see, these objectives are numerous and important, and we have no need to look for more.

The Bible is therefore not a work of history, as we understand that discipline today, nor is it a treatise on the nature of God, nor a book of science, geography, or philosophy. The Bible no more fits into any of these genres than it qualifies as a work of fiction.

The Bible which, in fact, is not a single book but an entire library, is a unique genre. For this reason, it resists easy classification. Trying to read or interpret it according to the constraints of a particular genre leads inevitably to misinterpretation.

Constraints and challenges

Let's put ourselves for a moment in the place of the God of the Bible. Wishing to make himself known to mankind, he decided to reveal himself through a book to the Hebrews, a small tribe whose culture at that time was little developed. Their lack of advanced culture made transmitting through them a message intended for peoples of all times a difficult task.

Not least was the obstacle posed by the Hebrews' very limited stock of vocabulary and general knowledge.

In addition to the difficulties posed by the Hebrew language and culture, the stories of Genesis originated in a society that had no means of writing down texts. This meant that divine revelation had to be imparted to them in an easy-to-remember format.

Finally, teachings had to be presented in such a way that they could be disseminated over time and space, survive for thousands of years, adapt to other languages and cultures, and endure the difficulties inherent in translation to other languages.

All these complicating factors constituted a considerable problem requiring intelligent and imaginative solutions.

Let's look at some of these constraints in more detail to get a better idea of what was involved.

Constraints imposed by language, literature, and writing

The vocabulary of the Hebrew language was extremely limited, notable for the absence of important concepts and their corresponding lexicon. According to Wikipedia, in Biblical times, Hebrew had around 8,000 words, a tiny vocabulary compared to Greek in the same era (120,000 words), modern French (100,000 words), or modern English (200,000 words).

Further, the Hebrews were under the influence of Aramean culture, which in Biblical times had already built up a system of knowledge and beliefs. It wasn't possible to start with a completely blank slate.

Finally, the scarcity of writing materials and means for preserving written texts in the ancient Near East encouraged composition techniques that aided memorization and oral transmission of the teachings, allowing them to be preserved through ages when writing was uncommon.

Constraints imposed by the objectives of timelessness and the universality of biblical revelation

The Bible's aim of being universal across time and space imposes three additional constraints:

- It must employ communication techniques that preserve its meaning through translations across languages and allow it to endure millennia of cultural change.
- Its message must remain comprehensible to people from vastly different cultures and historical periods, from ancient Hebrews, Greeks, and Romans to the people of the Middle Ages and the Chinese of the third millennium.
- It must be accessible to all individuals within these diverse cultures—young and old, shepherds and scientists alike—so that each can learn from it.

Achieving these requirements while overcoming these constraints demands both extraordinary ingenuity and specific methods.

This short preamble helps explain why critics of the Bible have so often misunderstood it.

Now we will consider five of the supposed errors in more detail.

I. The errors most commonly ascribed to the Bible

A quick internet search pulls up a good number of sites that claim to expose the errors contained in the Bible. The criticisms tend to fall under three main categories: cosmology, anthropology, and the historicity of the Bible's great stories. We've made a list of the most common ones:

"Errors" in cosmology

1. According to the Bible, the world was created in six days, but it actually took 13.8 billion years.

2. The Earth began to take shape on the very first day, even before the stars. But in reality our planet did not appear until nine billion years after the Big Bang and the first stars.

3. Light was created on the first day. But we know that the stars, the source of the Universe's light, did not form until later.

4. The Bible says the Earth is flat.

Anthropological "errors"

5. God created man in a single instant from a bit of earth. In fact, man descends from animals, through the primates, as the result of a slow process of evolution that took several billion years.

1. Supposed "error" number one: The world was created in seven days

According to the Bible, the world was made in seven days. But modern science tells us that it actually took 13.8 billion years for the Universe

THE ALLEGED ERRORS OF THE BIBLE, WHICH ARE NOT ERRORS

to form and bring about the appearance of mankind. This is by far the most frequently cited "error."

Here is a sampling from websites making this claim:

- *"Creation as literally depicted in Genesis is indeed supported by faith (and needs to be, since it is not supported by anything else, certainly not the Pope, nor the Roman or Anglican hierarchies). Evolution, on the other hand, is supported by evidence. Any science teacher who denies that the world is billions (or even millions!) of years old is teaching children a preposterous, mind-shrinking falsehood."*[545]

- *"First absurdity: it is clearly stated that everything from the origin of the Universe to today was created in 6 days. Yet we know for certain that the Earth is much older than that: it formed more than 4 billion years ago. This can be found in any science textbook, any book, or on any website. I know some Christians interpret the term 'day' as a period equivalent to millions or billions of years; but in that case, 1) why use the term 'day' instead of a more precise term? 2) If so, each 'day' or period would have the same duration. However, science now tells us that the Universe appeared approximately 14 billion years ago."*[546]

- *"Bible says in Genesis Ch. No. 1 Verses No. 3 and 5,...'Light was created on the first day.' Genesis, Ch., 1 Verses, 14 to 19... 'The cause of light – stars and the sun, etc. was created on the fourth day'. How can the cause of light be created on the 4th day – later than the light which came into existence on the first day? – It is unscientific."*[547]

In this first case, the error lies with the critics and their ignorance of the realities and constraints of the Hebrew language. At the time the Bible was written, the word "billion" and the numerical concept associated with it simply did not exist. The concept "billion" was invented in the

---

545. Richard Dawkins, "A Scientist's View," *The Guardian*, March 8, 2002, https://www.theguardian.com/uk/2002/mar/09/religion.schools1.

546. Translated from the French: https://forum.hardware.fr/hfr/Discussions/Sciences/1000-absurdites-bible-sujet_44793_1.htm.

547. https://ihsaan.wordpress.com/2020/03/25/bible-errors/.

fifteenth century by a pair of French mathematicians.[548] Thus, it would have been completely impossible for the author of the Bible to inform us about a duration of 13.8 billion years.

This first example perfectly illustrates the gap of understanding between critics who adopt a literal—we might even say fundamentalist—reading of the Bible and the complexity of reality and the constraints of language.

We know that the writer of Genesis could not have revealed the age of the Universe in billions of years, and if we reflect a little, we can agree that the precise number is of no interest at all, from either a metaphysical or physical point of view. Moreover, it's important to remember that we have only known the age of the Universe for about fifty years now, and our estimate could very well change tomorrow.

Unable to reveal the historical age of the Universe, we can understand that the biblical author, working under these conditions, chose to reveal to men only what was important to them, thus delivering them from false idols and vain terrors.

The essence of his message is that the Universe had a beginning and will have an end; that God is outside the Universe; and that between this beginning and this end, all things were made by him alone, progressively over time (as symbolized by the seven days of creation), and man appeared last.

In choosing "seven days" to represent the creation of the world, revelation made a leap from the cosmological to the anthopological plane. It set up a rhythm of life around a week of seven days: six days of work followed by one day of rest.

In ancient times, calendars varied: Mesopotamians had a seven-day week, Romans an eight-day one, and the Greeks ten. Eventually, the biblical calendar became the norm. The seven-day week is convenient from an

---

548. The concept of the "billion" was invented in the fifteenth century by Jehan Adam and Nicolas Chuquet. The word itself was not coined until the next century by the French mathematician Jean Trenchant, who wrote of "*miliars*" in 1558. The Sanskrit *kalpa* cannot be considered an exception since it was not mathematically calculated.

astronomical point of view because it makes up roughly a quarter of a lunar month. It also likely reflects a harmonious rhythm for human work.

At the same time, the Bible raises the number seven to a symbol of perfection and completion, an expression of the fullness of creation. Once this basic signification is established, it is reused with the same meaning throughout the rest of the Bible. It is the first in a long series of numbers with special significance—three, seven, ten, twelve, forty—scattered throughout the text.

Faced with a lexical constraint, revelation has redirected its course like a river, flowing through different anthropological and symbolic channels while counting on man's intelligence and interpretative capacity to achieve its goal. The great thinkers of the Church realized early on how they should interpret this biblical text. Origen, one of the very first Christian exegetes, wrote in around 230 AD: *"Now what man of intelligence will believe that the first and the second and the third day, and the evening and the morning existed without the sun and moon and stars? And that the first day, if we may so call it, was even without a heaven? And who is so silly as to believe that God, after the manner of a farmer, 'planted a paradise eastward in Eden,' and set in it a visible and palpable 'tree of life,' of such a sort that anyone who tasted its fruit with his bodily teeth would gain life; and again that one could partake of 'good and evil' by masticating the fruit taken from the tree of that name? And when God is said to 'walk in the paradise in the cool of the day' and Adam to hide himself behind a tree, I do not think anyone will doubt that these are figurative expressions which indicate certain mysteries through a semblance of history and not through actual events."*[549]

As we can see, there is no "error" in the biblical account of the world's creation, but rather a technical obstacle that divine revelation bypassed and in doing so revealed a much more important message.

---

549. Origen, *On First Principles*, trans. G. W. Butterworth (Notre Dame: Ave Maria Press, 2013), III.1. St. Augustine also warned against a too-literal reading of biblical texts. Unfortunately, his advice has not always been followed.

2. A second "error" in the Bible: "In the beginning, God created the heavens and the earth"

The Bible opens with a well-known phrase of poetic simplicity: "*In the beginning, God created the heavens and the earth*" (Gen: 1:1).

Detractors seize upon this text to disprove Genesis, noting that the Earth did not come into existence at the beginning of the Universe—it didn't appear until nine billion years after the Big Bang. Let's hear what they have to say:

- *The Stupidities of the Bible – Error no. 2: The Earth existed before the stars.*
  Genesis 1:1-16: 'In the beginning, God created the heavens and the earth. [. . .] God called the dry ground land, and he called the gathered waters seas. And God saw that it was good. [. . .] God made the two great lights—the greater light to govern the day and the lesser light to govern the night. He also made the stars.'"

- *"The reality: The oldest known star is called HD 140283, also known as the Methuselah star. It is a subgiant star in the Milky Way, located about 190.1 light-years from Earth. It formed shortly after the Big Bang. HD 140283 is currently one of the oldest known stars in the Universe. As a reminder, the Universe is 13.7 billion years old. According to current knowledge, the Earth is 4.54 billion years old."*[550]

- *"The stars, created on the fourth day, were useless as lighting bodies for a long period of time. A light-year is a unit of distance. It is the distance that light can travel in one year. Light moves at a velocity of about 300,000km each second. So in one year, it can travel about 9.5 trillion kilometres. In other words, one light-year is equal to 9,500,000,000,000 kilometres. The distance which separates Proxima Centauri from the earth is given by multiplying 9.5 trillion kilometres by 4.22."*

---

550. Translated from the French: http://antiobscurantisme.over-blog.com/2016/09/les-stupidites-de-la-bible.html.

- *"If God created the stars on the fourth day, their light didn't reach earth the same night, so at the beginning they were useless. Some creationists tried to find a solution to this problem in the idea that God would have created the universe already mature and the light would have been created in transit. This isn't a good explanation for the lack of purpose of the stars created in the fourth day of creation."*[551]

- *"In relation to the creation of the celestial bodies on the fourth day of creation, the author or authors of the book of Genesis could have made such an error only because they didn't know how the universe really works. At the same time, God surely knew how the universe works so either He didn't inspire the book of Genesis or He deliberately has misled us about the issue. The first option is the correct one because God cannot lie and be perfect at the same time. Taking into consideration the huge distances between the stars and the earth the light from the stars couldn't have reached the earth and become visible in the night after the fourth day of creation. In this case, it isn't clear why God declared that their creation was good. How good was the creation of the stars, in the fourth day, if their light was unable to reach Earth, in the same day? It wasn't that good."*[552]

- *"The idea that the Earth and the Universe were created in six days a few thousand years ago is completely unscientific. The age of the universe is calculated as being around 13.8 billion years while Earth is calculated to have formed 4.5 billion years ago. The difference between these two dates is significantly greater than six days. For those biblical literalists who argue for a 7,000 year old earth, there are multiple lines of evidence against a recent creation and—by extension—against a six day creation."*[553]

---

551. Gabriel Baicu, "Contradictions in the Bible: Creation of the Stars, Sun and Moon," 20 September 2017, https://www.contradictionsinthebible.net/chapter-2-an-absurd-order-of-creation-effects-before-causes/viii-creation-of-the-stars-sun-and-moon.

552. Ibid.

553. https://rationalwiki.org/wiki/Biblical_scientific_errors#Six_day_creation.

These criticisms are themselves riddled with errors.

The first mistake is one we've already seen: expecting the Bible to quantify the billions of years that separated the creation of the Universe from the creation of planet Earth. As we've established, this would have been impossible.

The second mistake is conflating the word "earth" in the biblical text with "planet Earth." These critics write "Earth," with a capital E, without realizing that this changes the meaning of the text.

The original Hebrew word is *ha'arets*, a term that corresponds to the ground beneath the feet of the writer, or the land he inhabits. It could not imply "planet Earth," a concept about which neither the author of Genesis nor his contemporary readers could have had the slightest idea.

The third mistake is their word-by-word analysis of the text, when the phrase "the heavens and the earth" is a common Hebrew expression meaning "the whole." The true sense of the phrase is this: "In the beginning, God created all that exists."[554]

But if we dare to go deeper than this first and primary meaning, we see that the phrase from Genesis is composed of the three dimensions that make up the Universe:

- "in the beginning" = time
- "God created the heavens" = space
- "and the earth" (the ground, the country) = matter

So the sentence would go something like this: *"God began by creating time, space, and matter."* This is exactly the scientific description of the

---

554. "The expression 'the heavens and the earth' is a semitism: 'to create the heavens and the earth' is a way of saying 'creating the whole Universe.' In the Our Father, 'thy will be done, on earth as it is in heaven,' means 'may your will be done everywhere, in all that exists.' In chapter seven of the book of Isaiah, when the prophet tells Ahaz to call for a sign, 'let it be deep as Sheol or high as heaven,' he is urging him to ask for a sign of cosmic significance, manifest in all creation. The expression 'in the beginning, God created the heavens and the earth' means therefore that God created everything that exists." Christophe Rico, director of POLIS: The Jerusalem Institute of Languages and Humanities.

beginning of the Universe that we have known since Einstein; that is to say, for current science, space, matter and time are inseparable and appeared simultaneously.

Whatever one thinks of this last surprising and perhaps audacious interpretation, it is clear that, in reality, this second cosmological "error" is not a mistake at all. It is even, perhaps, rather an extraordinary insight into reality that has only recently been confirmed by science.

3. "Error" number three: "According to the Bible, light is created even before its source, the sun and the stars"

In its very first verses, the book of Genesis tells of the creation of light: *"Then God said, 'Let there be light'; and there was light"* (Gen 1:3). This happens even before God created the stars and the Sun. Critics have seized upon these passages as an example of a cosmological error.

- *"Are the creationists right? Is Genesis accurate in the order and timing of the events it describes? Is Genesis a historical narrative that accurately describes the appearence of life? A cursory examination shows that it is not."*
- *"Notice here that there is no Sun yet, it not having yet been created. This account is thus contradicted by science on several grounds. Since a "day" is itself based on the earth's rotation near the Sun, there could have been no "day" until AFTER the sun appeared. Nor is there any cosmic source of "day light" other than the sun. Scientifically, we know that the sun actually condensed first, and was already burning nuclear fuel when the earth first began to appreciably accrete. The Genesis account, which has the earth and the "waters" formed before the Sun, is simply wrong."*[555]

Once again, critics commit the error of interpreting the text in an overly literal way and even fall into some scientific errors of their own. Scientific discoveries of the twentieth century have effectively demonstrated that

---

555. Creation Science Debunked, "Is Genesis Scientifically Accurate?", http://www.huecotanks.com/debunk/genesis.html.

light did indeed emerge from the primordial Universe about 380,000 years after the Big Bang, long before the stars existed, at a time when the decreasing density of the Universe allowed the original photons to stop their ceaseless erratic motion and begin to radiate in a straight line.

In the early Universe, therefore, light did appear before the emergence of the stars. Not only is the Bible not in error, but we can even be tempted to see in it an astonishing additional cosmological revelation.

4. A fourth "error": "The Bible teaches that the earth is flat"

Let us make it clear from the outset: this regularly repeated accusation is really surprising because there isn't any passage in the Bible which teaches that the Earth is flat.

The biblical passages cited to prove this claim are notably few in number and entirely unconvincing. They are everyday expressions we still use, like "*the ends of the Earth,*" "*the four corners of the Earth,*" and "*from one end of the Earth to the other.*"

When we use these expressions today, it does not mean we think the Earth is flat; nor, when a meteorologist on TV announces the hour of sunrise and sunset, does anyone think the Sun is actually rising or setting, or that it revolves around the Earth.

Here are a few examples of the biblical passages that the Bible's critics typically use to support their argument:

- "*There are dozens of references in the Bible to 'the ends of the Earth' and 'the four corners of the Earth.' A spherical planet cannot have ends or corners, but a flat one can. A few examples:*
    - *Deuteronomy 28:64 - And the Lord shall scatter thee among all people, from the one end of the earth even unto the other;*
    - *Job 28:24 - For he looketh to the ends of the earth, and seeth under the whole heaven;*
    - *Job 38:13 - That it might take hold of the ends of the earth, that the wicked might be shaken out of it?*

THE ALLEGED ERRORS OF THE BIBLE, WHICH ARE NOT ERRORS 375

- *Isaiah 11:12 - And he shall set up an ensign for the nations, and shall assemble the outcasts of Israel, and gather together the dispersed of Judah from the four corners of the earth."*[556]

- *"Flat earth clearly IMPLIED in Scripture: The devil took Jesus to a very high mountain & showed Him ALL the kingdoms of the world. Matt 4:8."*[557]

As we can see, there is not a single passage in the Bible that teaches that the Earth is flat. Where it seems to do so, it is only using a common expression. All these arguments are therefore ridiculous. It's sad to see that people keep repeating these absurd ideas.

Did the Bible cause medieval Christians to believe in a flat Earth? A persistent but baseless legend...

We often think of the Middle Ages as a time of great ignorance, with a Church hostile to science working to keep people in ignorance. Reviewing a few historical facts will help us recognize that this was not the case.[558]

---

556. https://atheist.fandom.com/wiki/Flat_Earth.

557. Denis O. Lamoureux, "The Bible and Ancient Science 2: Ancient Geography/Geology," https://sites.ualberta.ca/~dlamoure/wlas2.pdf.

558. http://ancientresource.com/.

With the exception of Lactantius, a *"writer celebrated in other ways but very little in mathematics"* (as Copernicus put it),[559] the scientists of the medieval West never questioned the roundness of the Earth. The majority followed Plato's ancient dialogue, the *Timaeus* (which they had in Latin thanks to Cicero), and the erudite commentaries upon it by Calcidius, a Neoplatonic philosopher of the fourth century. In his *Commentary on Ephesians*, the famous biblical translator Saint Jerome (347–420) criticized those who deny the spherical nature of the Earth. Isidore of Seville, bishop and scholar, compared the Earth to a ball in his *Etymologies* (ca. 590–636). Much later, Thomas Aquinas, Doctor of the Church, though cautious when it came to astronomy, considered the Earth a sphere and described the Universe as a series of concentric interlocking spheres.

The first great sea expeditions, like that of Marco Polo in 1270, added the weight of observation and experience to these theories. More concretely, the royal orb, an object that represented temporal power, is clear evidence that the medieval West considered the Earth to be round. First appearing at the beginning of the fifth century, the orb, a hand-held globe topped by a cross, was used as a royal emblem at the coronation of most European monarchs.

Coin minted in Constantinople: the Virgin crowns Emperor Romanus III Argyrus (968–1034), who carries the orb and cross in his left hand

---

559. Nicholas Copernicus in his dedication to Pope Paul III in the 1543 edition of *On the Revolutions of the Celestial Spheres*. The work is available in *Great Books of the Western World*, trans. Charles Glenn Wallis (Chicago: William Benton, 1995), 16:497–838.

The cross symbolizes Christ, the savior of the world (represented by the spherical globe). This small object tells us a lot about medieval conceptions of the Earth.[560]

Let's avoid anachronistic platitudes

Let's back up a bit. In antiquity, the Earth was widely believed to be flat. This view was a completely reasonable deduction at that time from sense observation of the Earth and celestial bodies. In fact, curvature to the horizon can only be seen with the naked eye from altitudes of more than forty-nine thousand feet. It was only natural for man before the aid of advanced technology to conclude that the Earth is flat or that the Sun revolves around it.

We had to wait for the genius of the Greeks, starting with Pythagoras, to realize that our senses can deceive us and that the Earth is actually spherical.

If the Bible is so visionary on other cosmological points, why does it have nothing revolutionary to say about the shape of the Earth? As we have seen, of course, it never said the Earth was flat, but why didn't it tell us the Earth is round?

The reason is simple: the Bible pursues purely supernatural ends, and therefore corrects only those errors that prevent us from understanding who God is and who we ourselves are.

It needed to show us that the Sun and Moon are not gods, because this is necessary for belief in one God alone. Similarly, it had to teach us that the Universe had a beginning and will have an end, because it was created by an all-powerful God. But since the roundness of the Earth has no metaphysical implications, there was no reason to correct erroneous beliefs about it—for the purposes of the Bible they're irrelevant.

---

560. https://en.wikipedia.org/wiki/Myth_of_the_flat_Earth.

Revealing who God is, who we are, and how we enter into relationship with him: these are the real aims of the Bible.

5. Fifth Error: God made man out of earth in a single day (Genesis 2:7), while in reality man was not the product of a separate creation but descended from primates and hominids, following a process of evolution over millions of years

When they read that God made the first man out of mud in one day,[561] the Bible's adversaries scoff yet again. Man is just a highly evolved primate who did not benefit from any special treatment, was not created separately, and is just the last link in a long chain of evolution.

Which of these two stories is true?

In the chapter on biology, we saw that all humankind probably descends from a single male and a single female ancestor.

But where did this first pair of ancestors come from, and when did they live? This is the real, metaphysically essential question.

It is common knowledge that *Homo sapiens* (who appeared an estimated 300,000 years ago) descended from *Homo erectus* (1.9 million years ago), *Homo habilis* (2.5 million years ago), *Australopithecus* (4 million years ago), and ultimately from the earliest hominids (7 million years ago).

And before that? Man is a mammal, and mammals arose after a long process of evolution that began with fish and successively gave rise to amphibians, reptiles, and birds.

And further back than that? We are today in the exceptionally fortunate position of being able to respond satisfactorily to this question: science was able to unravel this mystery only a generation ago. According to the most commonly accepted theory, if we follow the chain of life back link-by-link to our true ancestor, to a common point uniting all living beings, we find

---

561. In Hebrew *adamah*: earth, or matter.

the most remote ancestor of all life, called the Last Universal Common Ancestor, or LUCA. LUCA is supposed to have appeared on Earth in a rather extraordinary way 3.8 billion years ago, in the form of unicellular organisms, the first one that knew how to self-replicate and which then began to evolve on their own up to the appearance of man. From there, so the story goes, a process of evolution eventually ended with man. Very well. But what did LUCA descend from? This too is something we know today: LUCA, the first living being, the first organism capable of reproducing itself, was a mere collection of proteins and macromolecules—in other words, of molecules, atoms, and particles.

So there you have it. Our true ancestor, our ultimate origin, our first parent, is matter. Looking at the marvelous technology of the human hand or eye, it is hard to believe that it is nothing more than cleverly arranged matter. But that's what it is.

We're closing in on our answer

If a traveler steps off the train at the end of a long journey and someone asks him where he came from, he won't name the last of the many stops he made along the way. He will always answer with his starting point. The same is true if we ask where man comes from. Hominids are nothing but the final stage on our journey. Our starting point—where we come from—is matter. Only a soul, if it exists, breathed into us by God at conception and spiritual in nature, differentiates us from a simple hominid.

Once again, in stating that man is made from earth—in other words, from matter—the Bible makes no mistake. In fact, it perfectly answers the question of the origin of man in a way that has proved clairvoyant.

It's ultimately more scientifically correct to say that we come from matter than to say that we descend from primates. Metaphysically, it's much more significant.

As for the length of this trajectory, must the Bible have identified that 3 billion years separated LUCA from our first ancestors rather than merely a single day? We have already addressed this question earlier: a "billion" was unknown in antiquity and, further, this information is of no interest metaphysically.

Preliminary conclusions

And so not a single one of these commonly referenced "errors" turns out to be an error, after all. In fact, each one turns out to contain surprising cosmological and anthropological truths.

We have examined five alleged errors in the Bible, which are, in fact, not errors at all. Addressing all such claims would require an entire book, which is beyond our scope here. We selected these five because they are among the most frequently mentioned and criticized today. By exploring these examples, we aim to demonstrate how such criticisms often lack depth and substance.

It's been said that to err is human. If we can't find a single error in the Bible's thousands of pages, that may suggest that its inspiration lies elsewhere, beyond the human realm.

The end of this chapter on the Bible naturally leads us to consider its main objective: the arrival of the Messiah. The question now arises: Who could Jesus be?

19.

Jesus: Who Could He Be?

No one has the right to dismiss this question: "Who could Jesus be?" Four astonishing or inexplicable facts demand our attention:

- **Fact 1: The entire population of the Earth—over eight billion people—uses the year of his birth as year one of their calendar, even those who are not Christian and have never heard his name.** All the contracts in the world, all the legal acts in the world, all the publications in the world, date events by the year of his birth, even after many attempts to obscure this reference in our common calendar. The French revolutionaries chose 1793 as year one of their new calendar, but their renumbering lasted a mere twelve years and never caught on outside France. Mussolini similarly tried to set 1925 as year one of a new Italian calendar, but his attempt also failed. Jews, Muslims, and the Chinese do maintain their own traditional calendars, of course, but in each case the use of these calendars is restricted to the bounds of their own group. For most international or cross-cultural interactions, they have to use the birth of Jesus as year one. We can think of the year of Jesus' birth as an absolute and universal meridian, a sort of "temporal equator" that divides human history into two broad time periods, a "before" and an "after."

- **Fact 2: He is a unique phenomenon: more than 20,000 books have been written about him in the last century alone**, and hundreds of new ones are published every year. The Bible is the most widely distributed and translated book in the history of the world. Today, 2.5 billion people—or one-third of the world's population—say they believe in the divinity of Jesus Christ.

- **Fact 3: By all accounts, Jesus ought to have remained an inglorious nobody.** *"If anyone were destined to be forgotten after his* *life on earth it would be this modest craftsman from Nazareth, who wielded neither sword nor pen, held no position or performed any function in his country. This carpenter with no fortune, no wife, no children or relations, declared himself the Messiah. The authorities of his country arrested him and sentenced him to an ignominious yet common form of death, and most of his supporters abandoned him at his execution. By all accounts, even his name should have been completely forgotten! And yet his name would come to occupy the first place among all the names in the history of the world! And a lowly craftsman from an obscure village in Galilee acquired such fame? Nothing could ever be less likely!*"[562] We all implicitly sense that the extraordinary story of Jesus of Nazareth exceeds the bounds of ordinary human comprehension.

- **Fact 4: Jesus asked his followers, and through them, indirectly asks us, an unsettling question: "But you, who do you say that I am?"** At first glance this question, "*Who do you say that I am?*" (Matt 16:15), seems simple, harmless, and open-ended, but it will actually end in a problem of implacable logic. But we soon find that there are very few logical ways to answer it. Moreover, we have enough information to eliminate most of the alternatives.

Here are the seven answers that have been given historically and which are the only logical answers to the question "Who could Jesus be? (They've tried it all!):[563]

> I. **He didn't exist; he is a myth created later on**
> Beginning in the eighteenth century, this thesis was supported by some atheists and authors like Bauer, Couchoud, and Onfray.

---

562. Translated from Ferdinand Prat, *Jésus Christ, sa vie, sa doctrine, son œuvre* (Paris: G. Beauchesne, 1938), 82.

563. We have not considered certain marginal beliefs, such as those of the Jehovah Witnesses, who today have 8 million members and for whom Jesus is the incarnate form of Saint Michael the Archangel. Nevertheless, the same arguments work just as well for the refutation of these theories.

II. He was a great sage

This thesis is the thesis of Renan and Jefferson, and it is currently held by many freemasons and a segment of the general public.

III. He was a madman

This theory has been held by a number of philosophers since the nineteenth century: first by David Strauss and Nietzsche, and later by physicians and psychiatrists such as Binet-Sanglé and William Hirsch.

IV. He was a failed opportunist

This thesis is the thesis of the Talmud and it can also be found in mainstream Judaism.

V. He was a prophet

This thesis is the thesis of the Muslims and it was also held by some of Jesus' contemporaries.

VI. He was the Messiah and an extraordinary man, but only a man

Thesis of the the Arians and Cathars.

VII. He was the Messiah and God made man

This is the thesis of Christians and of some Messianic Jews.[564]

Each of these seven possibilities will now be subject to a thorough scrutiny, and we will call reason, history, the Bible, and most importantly, the testimony of Jesus' enemies to the witness stand. The testimony of those who witnessed the trial and execution of Christ will also prove especially useful. Their disorganized and contradictory accusations furnish some of the most valuable data available. To judge the relative merit of these hypotheses, no particular skill, university degree, title, or expertise is required. If the more technical scientific chapters of this book required us to rely on the expert witness of scientists, in this chapter

---

564. Jews who both recognize Jesus (Yeshua) as the Messiah of Israel and continue to practice Judaism.

everyone can come to his own conclusion. No one can claim ignorance when asked the question "Who could he be?" We ask you to accompany us actively in each step of this chapter.

Dear reader, take your seat among the jurors—the hearing is about to start.

I. Jesus never really existed. He is a myth that was invented later

This theory is untenable, since the historical reality of Jesus is very well-documented, even if we only take into account the testimony of his detractors, who had no vested interest in proving his existence—quite the contrary. Since these valuable testimonies are numerous enough on their own to prove Christ's historical existence, we won't even have to cite a single apostle, saint, or church father.

1. Ancient authors and historians on the witness stand

The Jewish historian Flavius Josephus (37–97) reports the stoning of James the Less, the "brother" of Jesus, which took place in Jerusalem in the year 62: *"The younger Ananus, who, as we have said, had been appointed to the high priesthood [. . .] convened the judges of the Sanhedrin and brought before them a man named James, the brother of Jesus who was called the Christ, and certain others. He accused them of having transgressed the law and delivered them up to be stoned."*[565]

In the seventies AD, **Mara bar Serapion, a notable Syrian of the Stoic school,** writes in a letter to his imprisoned son about the misfortunes that befall people who persecute wise men: *"For what advantage did the Athenians gain by the murder of Socrates [. . .] or the Jews by the death of their wise king, because from that same time their kingdom was taken away?"*[566]

---

565. Josephus, *Jewish Antiquities* 9.20, trans. Louis H. Feldman, Loeb Classical Library 456 (Cambridge, MA: Harvard University Press, 1965), 197.

566. Letter from a Syrian named Mara bar Serapion to his son Serapion, in a manuscript from the library of the Syrian monastery in Wadi El Natrun, Egypt, acquired in 1843 by the British Museum (*The Nitrian Collection—Manuscript no. 14658*), available in William Cureton, *Spicilegium Syriacum: Containing Remains of Bardesan, Meliton, Ambrose and Mara Bar*

The Roman historian Suetonius (69–125) mentions the persecutions of Christ's followers in 50 AD: "*Since the Jews constantly made disturbances at the instigation of Chrestus, he [Claudius] expelled them from Rome.*"[567]

The Roman historian Tacitus (55–118) mentions Jesus by name: "*Nero supplied defendants and inflicted the choicest punishments on those, resented for their outrages, whom the public called Chrestiani. The source of the name was Christus, on whom, during the command of Tiberius, reprisal had been inflicted by the procurator Pontius Pilate; and, though the baleful superstition had been stifled for the moment, there was now another outbreak, not only across Judaea, the origin of the malignancy, but also across the City, where everything frightful or shameful, of whatever provenance, converges and is celebrated.*"[568]

Writing about these persecutions, **Suetonius** also says: "*During [Nero's] reign many abuses were severely punished and put down. [. . .] Punishment was inflicted on the Christians, a class of men given to a new and mischievous superstition.*"[569]

Pliny the Younger (61–114), the governor of Bithynia, gives a report to Emperor Trajan on his treatment of the Christians: "*Amongst these I considered that I should dismiss any who denied that they were or ever had been Christians when they had repeated after me a formula of invocation to the gods and had made offerings of wine and incense to your statue (which I had ordered to be brought into the court for this purpose along with the images of the gods), and furthermore had reviled the name of Christ: none of which things, I understand, any*

Serapion (London: F. and J. Rivington, 1855). For further information, see Kathleen E. McVey, "A Fresh Look at the Letter of Mara Bar Serapion to His Son," in *V Symposium Syriacum 1988, Orientalia Christiana Analecta* 236, ed. R. Lavenant (Rome: Pontificium Institutum Studiorum Orientalium, 1990), 257–272.

567. Suetonius, "Life of Claudius," 25.11, *Lives of the Caesars*, trans. K. R. Bradley and John Cardew Rolfe, rev. with a new introduction by K. R. Bradley, Loeb Classical Library 31 (Harvard: Harvard University Press, 1913, rev. 1997–1998).

568. Tacitus, *The Annals*, XV.44, trans. A. J. Woodman (Indianapolis: Hackett, 2004).

569. Suetonius, "Life of Nero," 16.3, *Lives of the Caesars*.

genuine Christian can be induced to do. [. . .] [Those who said that they were Christians] declared that the sum total of their guilt or error amounted to no more than this: they had met regularly before dawn on a fixed day to chant verses alternately amongst themselves in honour of Christ as if to a god."[570]

The Greek writer Lucian of Samosata (125–192) *"The Christians, you know, worship a man to this day, the distinguished personage who introduced their novel rites, and was crucified on that account, [. . .] and then it was impressed on them by their original lawgiver that they are all brothers, from the moment that they are converted, and deny the gods of Greece, and worship the crucified sage, and live after his laws."*[571]

The famous physician Galen (129–216): *"One might more easily teach novelties to the followers of Moses and Christ than to the physicians and philosophers who cling fast to their schools."*[572]

The philosopher Celsus (second century AD) also taunts Jesus: *"And [they] assert that a man who lived a most infamous life and died a most miserable death was a god."*[573]

Flavius Josephus, the great Jewish historian who sided with the Romans during the war of 70 AD, writes in his *Jewish Antiquities*: *"About this time there lived Jesus, a wise man, if indeed one ought to call him a man. For he was one who wrought surprising feats and was a teacher of such people as accept the truth gladly. He won over many Jews and many of the Greeks. He was the Messiah [the Christ]. When Pilate, upon hearing him accused by men of the highest standing among us, had condemned him to be crucified, those who had in the first place come to love him*

---

570. Pliny the Younger, *Correspondence*, 10.96.5–7. *The Letters of the Younger Pliny*, trans. Betty Radice (Harmondsworth, New York: Penguin Books, 1969).

571. Lucian, "The Death of Peregrine," 11–13, *The Works of Lucian of Samosata*, vol. 4, trans. H. W. Fowler and F. G. Fowler (Oxford: Clarendon Press, 1949).

572. Galen, *On the Variety of Pulses*, 3.3, in Richard Walzer, *Galen on Jews and Christians* (Oxford: Oxford University Press, 1949), 14.

573. Celsus, *Speech against Christians*, quoted in Origen, *Contra Celsum* VII.53, trans. Henry Chadwick (Cambridge: Cambridge University Press, 1980).

did not give up their affection for him. On the third day he appeared to them restored to life, for the prophets of God had prophesied these and countless other marvelous things about him. And the tribe of the Christians, so-called after him, has still to this day not disappeared.[574]

This passage (known as the *Testimonium Flavianum* or *Testimony of Flavius*) appears without modification in every Greek manuscript of the *Antiquities* that has come down to us. Since the seventeenth century, specialists have hotly debated the authenticity of this text, with some scholars claiming that the passage, or at least parts of it, was a later addition. Given that Flavius Josephus did not believe Jesus was the Messiah, the argument goes, he never would have written such a positive testimony. Upon closer consideration, however, we have to admit the passage does not really reflect the mind of a Christian. The text portrays Jesus as an extraordinary man ("*a wise man*"), not as the Son of God made man. A Christian would never have written, "He was the Christ," but rather, "*He is the Christ.*" In this passage, *Christos* is simply the name given to Jesus by many pagans, such as the Latin historians Tacitus and Suetonius (as we saw above).

Looking for evidence of authenticity outside the text, we find the passage quoted in full on several occasions in antiquity:

- Eusebius of Caesarea quotes this text in two separate books between 314 and 333 AD, each time in the same way (*Ecclesiastical History* I.11.7–8 and *Proof of the Gospel* III.105–106).

- St. Jerome (ca. 393 AD) quotes the text of the *Testimonium* almost word for word in his *De viris illustribus* (13).

The idea that the passage was interpolated[575] into Josephus' text seems difficult to sustain. Since Eusebius and St. Jerome were already quoting the text in the fourth century, the interpolation would have to have occurred between the year 90 (the approximate date Josephus wrote the *Antiquities*)

---

574. Josephus, *Jewish Antiquities*, vol. 8, books 18–19, trans. Louis H. Feldman, Loeb Classical Library 433 (Cambridge, MA: Harvard University Press, 1965), 18.63–64.

575. Interpolation: inserting into a text an element not in the original (a gloss, a variant, etc.).

and the very beginning of the fourth century (when Eusebius cites it). Those two intervening centuries were precisely the period in which Christians were persecuted. Manuscripts were generally kept in public or private libraries, under close guard, so mass falsification is a highly unlikely hypothesis. The alleged purpose of the interpolation was to "prove" that Jesus existed and worked miracles, but at that time no one questioned the historicity of the man called Jesus or the possibility of miracles. As we shall see below, even the Talmud testifies indirectly to his miracles. So there was less motivation to falsify the text than it might seem. More fundamentally, it is difficult to imagine how every single manuscript of Josephus in the empire at the time could have been falsified, and only with regard to this isolated passage.

But the coup de grâce to the hypothesis that Jesus was a myth is given by the Jews themselves, who were at once the most informed and most resolute adversaries of Jesus Christ. For example, the Babylonian Talmud says:[576] *"The mishna teaches that ... on Passover Eve they hung the corpse of Jesus the Nazarene after they killed him by way of stoning. And a crier went out before him for forty days, publicly proclaiming: Jesus the Nazarene is going out to be stoned because he practiced sorcery, incited people to idol worship, and led the Jewish people astray. Anyone who knows of a reason to acquit him should come forward and teach it on his behalf. And the court did not find a reason to acquit him, and so they stoned him and hung his corpse on Passover eve. Ulla said: And how can you understand this proof? Was Jesus the Nazarene worthy of conducting a search for a reason to acquit him? He was an inciter to idol worship, and the Merciful One states with regard to an inciter to idol worship: 'Neither shall you spare, neither shall you conceal him.'"* (Deuteronomy 13:9).[577]

---

576. *Tractate Sanhedrin* 43a. English from the William Davidson digital edition of the *Koren Noé Talmud*, with commentary by Rabbi Adin Even-Israel Steinsaltz, https://www.sefaria.org/texts/Talmud/Bavli.

577. Peter Schäfer, a respected academic and author of *Jesus in the Talmud* (2007), comments on the Babylonian Talmud (which dates from around the year 300), finding Jesus' condemnation to be logical: *"More precisely, I will argue—following indeed some of the older research—that they are polemical counternarratives that parody the New Testament stories, most notably the story of Jesus' birth and death. They ridicule Jesus' birth from a virgin, as maintained by the Gospels of Matthew and Luke, and they contest fervently the claim that Jesus is the Messiah*

In the Jerusalem Talmud, Jesus is presented as a *mamzer*,[578] born of an adulterous affair between a Jewish woman and a Roman centurion. Following a dispute with a rabbi, he was allegedly banished, broke with Judaism, engaged in idolatry, and deliberately misled the people of Israel. Trained as a magician in Egypt, he left that land with secret magical formulae hidden in a fold of his skin. According to a legend, he made clay birds and brought them to life but lost a magic contest to the rabbis. When he was sentenced to death for witchcraft, he had to be hung on a cabbage stalk because he had ordered all the trees to refuse to serve as his gallows (*Tractate Sanhedrin* 43b).

The Babylonian Talmud and the Jerusalem Talmud are major pieces of evidence: in presenting Jesus as a charlatan, they inadvertently attest to his existence better than any other source

For Christians, these stories are blasphemous, but they have immense merit because they validate the historical existence of Jesus.

---

and the Son of God. Most remarkably, they counter the New Testament Passion story with its message of the Jews' guilt and shame as Christ killers. Instead, they reverse it completely: yes, they maintain, we accept responsibility for it, but there is no reason to feel ashamed because we rightfully executed a blasphemer and idolater. Jesus deserved death, and he got what he deserved. Accordingly, they subvert the Christian idea of Jesus' resurrection by having him punished forever in hell and by making clear that this fate awaits his followers as well, who believe in this impostor. There is no resurrection, they insist, not for him and not for his followers; in other words, there is no justification whatsoever for this Christian sect that impudently claims to be the new covenant and that is on its way to establish itself as a new religion (not least as a 'Church' with political power). This, I will posit, is the historical message of the (late) Talmudic evidence of Jesus. A proud and self-confident message that runs counter to all that we know from Christian and later Jewish sources. I will demonstrate that this message was possible only under the specific historical circumstances in Sasanian Babylonia, with a Jewish community that lived in relative freedom, at least with regard to Christians — quite different from conditions in Roman and Byzantine Palestine." Peter Schäfer, *Jesus in the Talmud* (Princeton, N.J.: Princeton University Press, 2007), 9.

578. A person born from incest or another forbidden relationship.

Let's put all this in context. The writers of the Talmud saw themselves as the adversaries of Christianity and considered the Christian message a threat to Judaism. Having been the witnesses of the creation of this myth about an imaginary Messiah, whose message was sacrilegious, absurd, and disastrous for them—because it announced the fall of Jerusalem and its domination by pagans until the end of time—then first of all they would certainly have known it was a myth, and second, would not have let it develop without opposing it. They certainly would not have validated the existence of a nonhistorical, mythical Jesus.

It is even less likely that they would have furthered the spread of this detestable myth through the invention of counternarratives such as those reported in the Talmud.

And so the Talmud is probably the best witness to the historical existence of Jesus. It also validates the biblical account of his sojourn in Egypt, the miracles he performed, the enthusiastic crowds who gathered to hear him teach, and his condemnation and execution by the Romans on the eve of Easter.

2. Jesus, a myth? A fairly new strategy with an obvious motive

For seventeen centuries, not a single one of Jesus' adversaries, whether they were Jews or pagans, ever put forward the idea that Jesus was merely a myth—that is to say, an imaginary construction developed over time through oral tradition. Obviously, it would have been in their interest to do so, but there's not a word, not a sentence to that effect. This deafening silence alone should render highly suspect the idea that he is a myth.

So, where did this idea originate?

In the eighteenth century, Enlightenment thinkers and philosophers were looking for ways to reject God and Christianity. However, settling their score with religion meant dealing with Christ, who was the very incarnation of religion. They needed to find an explanation for Christ's existence, one that was sensible and convincing enough to oppose the dominant view of Jesus as the Messiah and Son of God.

There was not much room for them to maneuver, since the possible explanations are few and difficult to defend. This is why some decided to support the idea that he was a myth, some that he was a sage, and some that he was an opportunist, or an evil magician. But there was no coordination between them and their different versions testifying one against the other.

The idea that Jesus was a myth emerged in the eighteenth century (Bruno Bauer, Paul-Louis Couchoud). Although this thesis blatantly contradicts undeniable historical facts, some people still persist in defending it today, as does the French philosopher Michel Onfray in his *Atheist Manifesto*. This comes as no surprise, as it results from the absolute necessity of answering the question "Who could Jesus be?" in a way that is compatible with their denial of Christianity. If they have chosen the indefensible idea that Jesus is a myth, it is just because all the other options are even worse, as we shall see.

3. A myth? Since when?

For a myth to take shape, it has to emerge and spread well after the disappearance of the last eyewitnesses to the time and events. Therefore, the myth of Jesus should not have appeared before the year 100. But in addition to the many historical texts already cited, we now have a large collection of objective physical evidence that proves Christianity existed well before this date.

The discovery of Christian remains in the ruins of Herculaneum, a city destroyed in 79 by the eruption of Vesuvius, disproves the hypothesis that Jesus is a merely a myth

Tacitus and Suetonius made Nero's persecution of Christians in the sixties AD part of the historical record, and archaeological remains also furnish irrefutable evidence of a Christian presence in Italy from an early date. In 1938, a cross apparently belonging to a Christian slave was discovered during excavations of a villa in Herculaneum, a city that was completely buried by pyroclastic flow from the eruption of Mount Vesuvius in the year 79 (see picture below). A cross and a Christian inscription were also unearthed in nearby Pompei, a city covered by volcanic ash in the same eruption. The date Vesuvius erupted is well documented, and we know that the sites at Pompei and Herculaneum remained undisturbed between the eruption and the modern excavations in which these artifacts were uncovered. This means it's impossible that the myth of Jesus was created after the year 100.

The nail holes and lighter-colored area around the cross suggest the presence of lightweight wooden doors, or at least a frame for a veil, used to hide the cross from view. The cross, which was fixed on or engraved into the wall, may have been veiled out of respect for its mystery, or to hide it as a way of avoiding problems if strangers passed by.[579]

The room in the House of Bicentenary, as it was discovered on October 28, 1938.

4. Jesus an absurd and sacrilegious myth? That makes no sense!

Myths emerge as a response to a particular need: their purpose is usually to inspire people, galvanizing or unifying an oppressed nation by giving it pride and hope. None of this squares with the supposed myth of Jesus. Why create a mythical figure—sacrilegious for the Jews, absurd for the pagans—who ended his life as a resounding failure and suffered the ignominious death of a bandit? The idea involves too many improbabilities.

---

579. Photo and caption from http://www.eecho.fr/pompei-herculanum-vestiges-chretiens-avant-79/.

All the historical and archaeological evidence is unanimous: Jesus is not a myth. He definitely existed. The "Jesus myth" hypothesis, then, can be definitively discarded.

II. Jesus was a great sage

In 1863, Ernest Renan, the famous French historian, ended his work, *The Life of Jesus*, with this well-known declaration: "*All the ages will proclaim that, among the sons of men, there is none born who is greater than Jesus.*"[580] In other words, for Renan, Jesus was the greatest sage who ever walked the earth, but only a man like any other, and therefore not the Messiah or the Son of God.

Renan's book was a great success, no doubt partly due to the hint of scandal surrounding its publication. Any shocking revelation, claims that Jesus did not exist, that he was only a wise man, or even that he was married to Mary Magdalene—always arouse the interest of critics, excite curious readers, and ensure the author's enduring notoriety.

The publication of Renan's book was timed perfectly to lend encouragement and ammunition to the growing number of "freethinkers," who instantly seized upon it for justification. The "sage" theory had the advantage of being more plausible and less risky than the myth theory, packaged as it was in a respectable skepticism and hidden behind a veil of culture and resting on the authorship of a renowned historian.[581]

Nevertheless, the sage theory is the most absurd and the least defensible of all the possible accounts of Jesus' identity.

To realize this, we have only to read Jesus' words as recorded by the evangelists. No sage in the world would ever dare to say such things as the following:

---

580. Ernest Renan, *Life of Jesus* (New York: A.L. Burt Company, 1863), 393.

581. Renan was not the only one to defend the sage theory. Before him, in 1820, Thomas Jefferson, President of the United States, wrote *The Life and Morals of Jesus of Nazareth*, a work made up of fragments of Gospel texts cobbled together, excluding all the miracles and any words that made Jesus out to be more than a man.

1. Crazy words, words of an insane man

- *"Destroy this temple, and in three days I will raise it up"* (John 2:19).
- *"Your sins are forgiven"* (Matt 9:5; Mark 2:10; Luke 7:48).
- *"I am the bread that came down from heaven"* (John 6:41).
- *"I am the resurrection and the life"* (John 11:25).
- *"Before Abraham was, I am"* (John 8:58).
- *"Those who believe in me, even though they die, will live"* (John 11:25).

2. Infinitely pretentious words

- *"Apart from me you can do nothing"* (John 15:5).
- *"Heaven and earth will pass away, but my words will not pass away"* (Matt 24:35).
- *"All authority in heaven and on earth has been given to me"* (Matt 28:18).
- *"Whoever has seen me has seen the Father [God]"* (John 14:9).
- *"I am the light of the world"* (John 8:12; 9:5).
- *"I am the way, and the truth, and the life"* (John 14:6).
- *"Everyone who belongs to the truth listens to my voice"* (John 18:37).
- *"The queen of the South will rise up at the judgment with this generation and condemn it, because she came from the ends of the earth to listen to the wisdom of Solomon, and see, something greater than Solomon is here!"* (Matt 12:42).

3. Words impossible to live by

- *"Whoever comes to me and does not hate father and mother, wife and children, brothers and sisters, yes, and even life itself, cannot be my disciple"* (Luke 14:26).
- *"But if anyone strikes you on the right cheek, turn the other also"* (Matt 5:39).
- *"If any want to become my followers, let them deny themselves and take up their cross and follow me"* (Matt 16:24).
- *"You have heard that it was said, 'An eye for an eye and a tooth for

a tooth.' But I say to you, Do not resist an evildoer. But if anyone strikes you on the right cheek, turn the other also" (Matt 5:38–39).

4. Sacrilegious words

- "*Your sins are forgiven*" (Matt 9:5; Luke 7:48).
- "*Destroy this temple, and in three days I will raise it up*" (John 2:19).
- "*So the Son of Man is lord even of the sabbath*" (Mark 2:28).
- "*Very truly, I tell you, unless you eat the flesh of the Son of Man and drink his blood, you have no life in you.*" (John 6:53).

This last saying shocked his disciples so deeply that many turned away from him hurt, scandalized, and completely baffled. "*Because of this, many of his disciples turned back and no longer went about with him*" (John 6:66).

If Jesus is not the "God made Man" he claimed to be, then these are the words of a madman, a lunatic—not a sage.

5. A great sage at the head of a gang of liars and crooks?

We would logically expect a great sage to surround himself with people aspiring to wisdom, people who come to meditate on his words and be edified by his thoughts. At any rate, no "great sage" would seek the exclusive company of liars and lowlifes. But if this hypothesis is true, that is just what the great sage Jesus did.

The proof that it was the dirty dozen is that, immediately after his death, his disciples unanimously proclaimed his resurrection. The sage theory forces us to imagine that the disciples agreed to go and open Jesus' tomb on a moonlit night—a tomb watched by Roman soldiers who could expect to be executed if they failed to guard it—covertly stole his body, and buried it in a hiding place so secret that no one has found it in more than two thousand years. All this happened without anyone knowing and without the slightest leak of information. The notion is highly improbable. Since stealing corpses was punishable by death under both Jewish and Roman law, the operation would have been extremely perilous.

After hiding Jesus' body, the apostles then passed on the fiction that their master was risen, and every one of them, without exception, stood by the truth of this incredible fabrication until his death!

So, Jesus—the greatest sage the earth has ever seen—to paraphrase Renan—recruited only liars, lowlifes, and crooks? These charlatans put the pièce de résistance on their fables by inventing miracles that Jesus, merely a sage, could never have performed, and thought up a collection of sayings and entire discourses that he had never actually spoken. Even worse, these grifters spread out to disseminate their grotesque farce all over the world, even suffering voluntary martyrdom to safeguard their credibility. These imposter-disciples were crooks and liars completely lacking any sense of self-preservation. The entire scenario makes absolutely no sense.

And thus Renan's theory that Jesus was merely a wise man should be consigned to the reject pile. Its only function is to do away with the troublesome question of who Jesus is via a non-committal, non-confrontational hypothesis. "Jesus the sage" is a gentle character without any rough edges, a facile choice, but he is not the Jesus given to us in any historical or biblical source, and he makes no logical sense. As C. S. Lewis wrote:

> *"You must make your choice. Either this man was, and is, the Son of God, or else a madman or something worse. You can shut him up for a fool, you can spit at him and kill him as a demon or you can fall at his feet and call him Lord and God, but let us not come with any patronizing nonsense about his being a great human teacher. He has not left that open to us."*[582]

III. He was a madman, a crazy mystic

Was Jesus crazy? Even some of his contemporaries expressed doubts about his mental health. The Gospel writer Mark reports that some

---

582. C. S. Lewis, *Mere Christianity* (London: HarperCollins, 1952), 54–56.

people close to Jesus were worried about him because of his deeds and preaching: "*When his family heard it, they went out to restrain him, for people were saying, 'He has gone out of his mind'*" (Mark 3:21). At a time when undiagnosed mental disorders were thought to be cases of demonic possession, others believed that he was possessed: "*Many of them were saying, 'He has a demon and is out of his mind. Why listen to him?'*" (John 10:20); The Jews answered him, '*Are we not right in saying that you are a Samaritan and have a demon?*'" (John 8:48).

Almost twenty centuries later, philosophers like David F. Strauss and Friedrich Nietzsche called Jesus' mental health into question in their writings. Distinguished specialists at the time began to give their own diagnoses of his case. In his 1915 book *La folie de Jésus* (*The Madness of Jesus*), French doctor Charles Binet-Sanglé detected a religious paranoia in Jesus. The American psychiatrist William Hirsch concurred and declared Jesus a paranoiac, a diagnosis also given by the Soviet psychiatrist Y. V. Mints.

In fact, if we consider only the outrageous claims made by Jesus mentioned above, and assume Jesus was just a man and not the Son of God, it wouldn't be unreasonable to conclude that he was, in fact, a madman.

Since the hypothesis is supported by the authority of eminent specialists, it certainly has its attraction. But is it convincing?

1. Disconcerting but wise words

Though some of the sayings of Jesus seem crazy and disconcerting, others of his teachings are so deep and wise that they have been passed on for centuries and still have broad popular appeal among Christians and non-Christians alike. Here are three representative examples of Jesus' wise sayings:

- "*Give therefore to the emperor the things that are the emperor's, and to God the things that are God's*" (Matt 22:21). The relevance of this maxim is obvious. It has been used throughout the last two millennia, and especially in our own time, as a justification for the separation of church and state. The maxim's context is a trick question posed to Jesus by the Jewish leaders: *"Tell us, then, what you think. Is it lawful to pay taxes to the emperor, or not?"* (Matt 22:17). If Jesus answers "yes," he will anger devout Jews and zealots. If he answers "no," the Romans might arrest him as a rebel. His response is extraordinary for two reasons: on the one hand, he evades the trap; on the other hand, he makes the bold claim for all generations that it is possible to reconcile the tension between church and state.

- *"Let anyone among you who is without sin be the first to throw a stone at her"* (John 8:7). This warning has remained part of our received cultural wisdom for over two thousand years. Once again, Jesus gives a wise answer to a tricky question. A woman caught in adultery is brought to Jesus, and they ask him: *"Now in the law Moses commanded us to stone such women. Now what do you say?"* (John 8:5). A thorny question. If Jesus says, "Do not stone her," he pits himself against the law of Moses and risks being condemned. If he answers, "Stone her," he would have contradicted his own message of love and would have stirred up the mob against him. By responding as he does—forgive the turn of phrase—he kills two birds with one stone: he avoids the trap by turning the question against the woman's accusers, and at the same time transmits two profound teachings. Judgment and condemnation are reserved for God alone, and a sinner always deserves a second chance (*"Go your way, and from now on do not sin again,"* he later tells the woman).

- *"Father, forgive them; for they do not know what they are doing"* (Luke 23:34). Jesus has just been nailed to the cross. His enemies witness his agony, looking on with malicious glee and mocking him: *"He saved others; let him save himself if he is the Messiah of God, his chosen one!"* (Luke 23:35); *"He is the King of Israel; let him come down from the cross now, and we will believe in him"* (Matt 27:42). Their taunts are punctuated by merciless laughter. Despite his suffering, Jesus finds the strength to say aloud: *"Father, forgive them; for they do not know what they are doing"* (Luke 23:34). Even Christians find these strangely wise and merciful words disconcerting. A handful of early scribes were so scandalized by this verse that they even omitted it from their copies of the gospels.

We could give more examples of the wise sayings of Jesus, like the parable of the "good Samaritan," which has become a household word, or his teaching that *"the sabbath was made for humankind, and not humankind for the sabbath"* (Mark 2:27), or his counsel to build one's house (that is to say, one's life) on rock, not sand (Matt 7:24-27). Everyone probably has his own cherished references, and popular wisdom has amply borrowed from the wisdom of the Gospel: these phrases are part of our common cultural vocabulary. Therefore, even this brief selection presents a serious challenge to the idea that Jesus was insane.

2. Why didn't the Talmud portray Jesus as a madman?

If Jesus were merely a madman, or an illuminated visionary, why would the various versions of the Talmud lend him legitimacy by describing him as a dangerous schemer and a demonic magician? If Jesus had really been an unfortunate madman, his opponents wouldn't have needed to say anything more to completely and definitively discredit him.

3. Requiem for a madman

If Jesus were a sage, he would not have surrounded himself with crooks and liars. Likewise, if he were a madman, no one would have dreamed of following him around Galilee—no one except, perhaps, other equally crazy people who first made an incredible effort to stage a supposed resurrection and then traveled around the world to spread the news of it, in most cases even suffering death for their witness.

The whole Jesus story would be a comedy of fools!

The hypothesis that Jesus was crazy had to be taken seriously nonetheless. Some Christian apologists have felt the need to address the allegation directly. In *The Foundations of Christianity*, C. S. Lewis writes that identifying who Jesus is means facing a "trilemma," a three-way choice.

Either Jesus is God, or he is not. In the latter case, either he knows he is not but pretends to be, or he erroneously thinks he is because he is crazy.

We have investigated Jesus' possible madness, but this line of investigation is ultimately a blind alley.

In our imaginary courtroom, we have now heard testimony in support of three different hypotheses about the identity of Jesus. None of them withstands the force of evidence. Four witnesses remain to cross-examine.

IV. Jesus was a failed opportunist

Viewing Jesus as a seductive opportunist who was able to stir the crowds' excitement for a time before being halted in his tracks by an inglorious execution appears plausible at first sight.

This hypothesis has the advantage of corresponding with the claims of the Talmud, which was written by Jews who, unlike Renan and the philosophers of the eighteenth century, were eyewitnesses to Jesus' trial, condemnation, and death. It also coincides with some of the testimonies of the Latin authors we have cited. Finally, it fits with the temporal context, since the Messiah was prophesied to appear during this particular period, as we will explain later.

We should summarize this hypothesis before investigating its merits. Taking advantage of the feverish anticipation of a Messiah, an astute schemer who has learned magical arts in Egypt claims to be the expected Messiah. He seduces the crowds with "miracles" and tries to get them to revolt, intending to take power. This attempt is popular and dangerous enough to raise alarm among the authorities, and Jesus is arrested, executed, and buried. His disciples, who abandoned him and scattered at the time of his execution, secretly regroup, and within a matter of hours stage an elaborate and implausible second act for this magic show. After stealing his body and burying it in a secret place, they fabricate a monumental hoax, pointing to the now-empty tomb as evidence of Jesus' resurrection.

However romantic this story might be, it has a number of flaws.

1. Why didn't the high priests look for Jesus' corpse?

Jesus had been deemed a dangerous and subversive spiritual opportunist. He is finally dead, and his followers are dispersed. Against all expectations, his supporters return and decide to continue the ruse by themselves. They announce his resurrection and blame his death—that of the Messiah, the King of Israel—on the Jewish leaders. Consequently, after only a brief respite, the same Jewish leaders find themselves faced with a new threat to themselves and to their nation.

Why, then, didn't they look for his body? It would have been so simple. An absurd charade invented by Jesus' followers, this supposed resurrection gave them an ideal chance to put an end to the movement for good. All they had to do to finish off this incredible and dangerous masquerade was simply locate the body.

For several reasons, the theft and concealment of the body of Jesus would have been an impossible task. Jerusalem at that time was not a very large city, boasting only fifty thousand inhabitants. Moreover, on the nights after Passover, the moon was full, and thousands of pilgrims who had come for the festival were camping in and around the city. In these circumstances, it would have been next to impossible to move a body in complete secrecy. Jesus' supporters could probably have done no more than rebury the body close to the original tomb.

At that time, the Jewish leaders held almost all power in the area—religious authority, of course, but also temporal authority as delegated to them by the Romans. Why didn't they arrest some of his disciples, forcing them to confess the subterfuge and reveal the body's hiding place? That would have been the end of the story. This operation would have encountered little resistance, since almost all of Jesus' close supporters had fled, distressed and terrified. On this point the Gospel and the Talmud agree, and the Gospel even specifies that only a few women and a young man (the apostle John) were at the foot of the cross to witness Jesus' execution.

But no, nothing—no official search for the body. Even if the high priests did not immediately see the importance of putting a stop to this lie, once they had witnessed the emergence of Christianity and its diffusion throughout the Roman Empire, they still had time to react. They could have initiated a search at least until the first destruction of Jerusalem in the year 70. Many eyewitnesses, including the apostle John (perhaps the most important witness), lived until 100 AD.

Of course, they may very well have searched in vain and preferred to keep their failed efforts quiet. If so, they took that information with them to the grave.

Either way, the incoherent and incomprehensible attitude of the Jewish leaders does not plead in favor of Jesus as a religious opportunist.

2. Why would the disciples take the charade so far?

Jesus proclaimed himself the Messiah and then died. This fact alone should blow the narrative out of the water, since by definition the Messiah was destined by God to be the "Great King" and "deliverer of Israel." The disciples were stranded, left to their own devices. The Gospels say that only a few remained after most followers of Jesus had fled.

Are we really going to believe that this handful of cowards pulled off world history's biggest scam with only a few hours of preparation? That they managed to pass off an opportunistic charlatan as God on earth? That they secretly stole and hid his body and proclaimed that he had risen from the dead?

What did they stand to gain from this preposterous scenario? The only conceivable enticement would be the prospect of seizing power. Did the disciples premeditate a coup d'état, taking power in the name of Jesus by promoting the expectation of his return? In this case, why not stand together, on the spot? Why go to the ends of the world, each one on his own, without wife, without money, without children, telling the story of the resurrected Messiah to pagans who were ignorant of the very concept of a Messiah?

And for what kind of outcome? Except for John, every one of the apostles was killed for spreading the new belief: Peter was crucified at Rome in 64; Paul beheaded there in 67; John exiled to Patmos; James the Great decapitated by Herod in 41; Andrew crucified in Patras in 45; Bartholomew martyred in Arabon, Armenia, in 47; Simon the Zealot martyred in Mauritania around 60; Matthew burned to death in Upper Egypt in 61; James the Less thrown from the pinnacle of the Temple and then stoned in 62; Jude hanged and pierced with arrows in Armenia in 65 by order of King Sanatrouk; Matthias stoned and crucified in Ethiopia and buried in Biritov in 80; Thomas skinned alive and pierced by a spear in Meylafour south of Madras, India, in 72; Philip hanged by the feet and then crucified at Hierapolis in Phrygia around 95; Luke martyred at Thebes in 76; and Mark killed in Alexandria in 68.

From a human point of view, this macabre list would have made many change their mind. The hypothesis that Jesus was only an opportunist simply does not hold water.

3. Are these the words of an opportunist?

We have already quoted some of the wise sayings of Jesus. Are we supposed to believe that these words, which for Renan represented the highest wisdom ever spoken by any sage, were uttered by a petty revolutionary and his followers? This does not square with the ambitions of an opportunist.

Opportunists and revolutionaries need the support of at least a portion of the population to carry out their schemes. What benefit could Jesus or his "gang" expect from promoting teachings that are impossible to follow? No one who wants to take power can afford to drive the people—including his own supporters—away, yet that is essentially what Jesus and his followers did. His message was clearly not conducive to self-serving political goals.

4. Why was the apostles' preaching so successful?

Only thirty years after Jesus' crucifixion, the Christian religion had already spread everywhere. In Rome, Christians were wrongly accused of starting the great fire of 64 AD, and the emperor had them thrown to the lions.

Who can imagine for an instant that a bunch of poor Jewish adventurers preaching outrageous doctrines in distant and hostile lands could have peacefully changed the face of the ancient world, without violence, without money, and, in most cases, with very little education? It's unthinkable!

How did these men come up with the ludicrous idea of perpetrating a worldwide hoax, even to the point of a violent death, and where did they find the courage and obstinacy to carry out this irrational project? Paul has something to say on this point: *"Are they ministers of Christ? I am talking like a madman—I am a better one: with far greater labors, far more imprisonments, with countless floggings, and often near death. Five times I have received from the Jews the forty lashes minus one. Three times I was beaten with rods. Once I received a stoning. Three times I was shipwrecked; for a night and a day I was adrift at sea; on frequent journeys, in danger from rivers, danger from bandits, danger from my own people, danger from Gentiles, danger in the city, danger in the wilderness, danger at sea, danger from false brothers and sisters; in toil and hardship, through many a sleepless night, hungry and thirsty, often without food, cold and naked. And, besides other things, I am under daily pressure because of my anxiety for all the churches"* (2 Cor 11:23–28). If they were all just opportunists in search of temporal gain, it did not do them much good.

Later, in 190, Tertullian described the rapid spread of Christianity: *"We are but of yesterday, yet we have filled every place among you—cities, islands, fortresses, towns, marketplaces, camp, tribes, town councils, the palace, the senate, the forum; we have left nothing to you but the temples of your gods. [. . .] Even unarmed and without any uprising, merely as malcontents, simply through hatred and withdrawal, we could have fought against you. [. . .] Without a doubt, you would have been exceedingly frightened at your loneliness, at the silence of your surroundings, and the stupor, as it were, of a dead world."*[583]

---

583. Tertullian, *Apology* 37.4–7 *The Fathers of the Church: Tertullian Apologetical Works and Minucius Felix Octavius*, trans. Rudolphus Arbesmann, Emily Joseph Daly, and Edwin A. Quain (Washington, D.C.: Catholic University of America Press, 2008), 96.

We are not dealing here with the realm of adventure, of power grabs and acts of defiance against destiny and society. What's taking place is of another order and belongs to another dimension. Therefore, the opportunist hypothesis falls short, like all its predecessors.

V. He was a prophet

At first glance, this is an attractive view—after all, most of Israel's prophets were persecuted or killed. Jeremiah was persecuted by his own people, Daniel was thrown into the lions' den by the king of Babylon, Isaiah was sawn in half (according to some traditions), and John the Baptist was beheaded at the order of Herod Antipas. Jesus makes one more in a long line of persecuted prophets, being criticized violently during his life and finally put to death on a cross. Like the other prophets, he brought a message of radical conversion. Some of his contemporaries were not slow to recognize the prophetic tenor of his life and message: "*Herod the ruler[...] was perplexed, because it was said by some that John had been raised from the dead, by some that Elijah had appeared, and by others that one of the ancient prophets had arisen*" (Luke 9:7–9).

It's worth mentioning that the Koran, the holy book of the Islamic religion (which counts nearly 1.5 billion followers), also teaches that Jesus was a prophet.

But on closer examination, this theory comes up against some insurmountable difficulties.

- A prophet in the Jewish tradition is supposed to be wise and clear-sighted. A Jewish prophet would not have surrounded himself with criminals and bandits who dug him up, reburied him, and passed him off as "God made Man."
- A Jewish prophet would never have spoken the absurd and sacrilegious words we considered earlier.
- A prophet who is not God cannot rise from the dead. Therefore, everything we said above about the implausible charade of a vanished body and the falsified resurrection account applies here, too.

More generally, all the objections noted with regard to the theory that Jesus was a sage apply here as well. We can safely conclude that the prophet theory is untenable without trotting out all those arguments a second time.

Muslims escape one set of difficulties by creating others. They argue that while the apostles were not liars, they *were* mistaken: it wasn't Jesus who died on the cross, but a look-alike that God substituted for him. Further, to explain Christ's many unlikely, challenging, or impossible teachings, they argue that the Gospel was somehow "falsified," though they never specify who did the falsifying or when, how, or why it was done.

VI. Jesus is the Messiah and an extraordinary man, but only a man

This theory recognizes the truth of the New Testament and acknowledges Jesus as the Messiah but denies his divine nature. Historically this was a central tenet of Arianism, born in the fourth century, which later also formed the basis for Cathar beliefs. So let's take it one proposition at a time: "Jesus is the Messiah," and "He was only a man."

The first proposition: "Jesus is the Messiah"

Biblical chronology and a great number of Old Testament prophecies support the idea that Jesus is the Messiah.

The date of Jesus' birth coincides with the period in which the Jews expected the Messiah to appear. The date of his coming had been the subject of several prophecies throughout Jewish history, and foremost among them was the prophecy of the prophet Daniel: *"Seventy weeks are decreed for your people and your holy city: to finish the transgression, to put an end to sin, and to atone for iniquity, to bring in everlasting righteousness, to seal both vision and prophet, and to anoint a most holy place. Know therefore and understand: from the time that the word went out to restore and rebuild Jerusalem until the time of an anointed prince, there shall be seven weeks, and for sixty-two weeks it shall be built again with streets and moat, but in a troubled time* (Dan 9:24–26).

But how do we decrypt this sibylline message? When we read, "*seven weeks, and for sixty-two weeks,*" we might easily interpret it to mean sixty-nine weeks, but actually the "weeks" mentioned here are "weeks of years," as explained elsewhere in the Bible. That is to say, each "week" corresponds to seven years. The phrase "*from the time that the word went out to restore and rebuild Jerusalem*" refers to the edict of Artaxerxes, published in 457 BC. The "anointed" one is the Messiah, who, according to the calculation given by this prophecy, should have appeared 483 years later (69 x 7)—around the year 27.

Even if there is some variation among the possible interpretations of this prophecy, there are other reasons we can be certain that the advent of the Messiah was expected around this period. For example, even though Herod the Great, whom the Romans had installed on the throne of Judea at this time, was of Idumaean, not Jewish, descent, he knew that the Messiah was supposed to be the King of Israel. Thus, when he learned of the Messiah's birth in Bethlehem, Herod was so afraid of losing his grip on power that he ordered a mass infanticide of newborn children, which we now call the Slaughter of the Innocents.

The chronology of several prophecies foretelling the Messiah's coming pointed to this time period. The history is too long to discuss in great detail here, but the strong convergence of the ancient prophecies explains why multiple self-proclaimed Messiahs emerged just before and after the lifetime of Jesus. Between the years 1 and 135 AD, no fewer than six other contenders for the title of Messiah declared themselves: Judas the Galilean, Simon, Athronges, Theudas, Menahem, and Simon Bar Kokhba.

Many years after the death of Jesus, the Jewish people continued to wait for their Messiah. The wait went on and on. Rabbi Akiva, the greatest Jewish wise man of his time, recognized Simon Bar Kokhba as the Messiah. The Jews minted coins in his honor and supported him to the point of starting an absurd and hopeless war against the Romans, during which tens of thousands of poorly armed men attacked the greatest military power of the age. Predictably, they were crushed by the Romans in the year 135. Only their blind faith in Simon Bar Kokhba as the Messiah can explain the madness of waging such asymmetric warfare.

In light of these events, we can see how deeply convinced the Jews of Jesus' day were that the Messiah's hour had arrived. The wait was finally over.

Even the Jews who did not recognize Jesus have admitted that the Messiah was supposed to come on a very specific date, even as that date began rapidly receding into the past. Many years after Jesus' death, the Jewish people continued to wait for their Messiah, and the wait goes on. As the Talmud says: *"That is the course that history was to take, but due to our sins that time frame increased. The Messiah did not come after four thousand years passed, and furthermore, the years that elapsed since then, which were to have been the messianic era, have elapsed."*[584]

This especially interesting passage contains two important observations and a very telling interpretation. The first observation is that there were multiple prophecies predicting the date of the Messiah's coming. The second is that their predictions corresponded roughly to the lifetime of Jesus, since according to the compilers of the Talmud (who worked in the late first century, after the destruction of the Temple in 70 AD), the date for his coming had already passed. The Talmud's authors go on to attribute the delay of the Messiah's arrival to the chosen people's lack of piety. Of course, another interpretation is also possible: the Messiah did indeed come as announced by the prophecies, but went unrecognized except by a minority—he was Jesus of Nazareth.

Let us now examine the second part of the thesis: "Jesus is the Messiah, but he was only a man, certainly extraordinary, but just a man"

Admittedly, the Old Testament's predictions about the Messiah had led many people to expect that he would be a temporal king—an extraordi-

---

584. *Sanhedrin* 97b.

nary man who would be God's spiritual son, but only a man, neither a literal son of God nor God himself.

It is clear even in the gospels that the Jews of Christ's time expected this kind of King Messiah. In fact, it is precisely because Jesus did not meet their expectations of a worldly Messiah that many turned away from him.

However, this thesis of "Jesus is the Messiah, but is only a man" becomes, after his death, perfectly untenable, since the core definition of the Messiah is that he would be a king destined by God to reign over Israel. Instead, to the great disappointment of those who hoped that he would restore the temporal power of Israel, Jesus dies on the cross, abandoned by everyone. He cannot therefore in any case only be "a human Messiah."

In the final analysis, this theory, which is almost an iteration of the previous one (Jesus as prophet), can be refuted by the same arguments.

Despite a period of great success, the Arian heresy eventually lost ground in the Christian world because all the characteristics of divinity attributable to God alone (eternity, forgiveness of sins, etc.) were also found in Jesus. He cannot be a simply human Messiah.

Interlude

Let's go back to our imaginary courtroom, where our witnesses have been giving opposing versions of the facts. The audience shifts restlessly in their seats, each one leaning over to murmur to his neighbor. The air is full of rustling and excitement. There is a strong feeling that we have come to a decisive and crucial moment in the proceedings. One after the other, each of the witnesses have discredited one other. Some of their testimonies nearly had us convinced, but, at the last moment, a fact, an objection—and everything collapsed. Who was Jesus, really? We're on to the seventh hypothesis. Order in the court! Let us proceed.

1. In spite of themselves, Jesus' adversaries have shed very clear light on the case. The Talmud made him out to be an opportunist; Renan saw him as a sage, and a few modernists considered him a myth. But we have seen that the Talmud shoots down Renan, Renan undermines

the Talmud, and both discredit the mythologists. Thanks to them, we did not have to call upon a single Christian author.

2. Before we take a look at the final possibility, let's consider the option of creating a hybrid theory from the alternatives we've considered. Any such hybrid thesis would be hard to defend because it would contain the flaws of its constituent parts. We simply cannot get around the following facts:

a. Jesus existed.

b. If Jesus is only human, his crazy words or impossible exhortations are absolutely incompatible with his wise sayings. Trying to find a logical explanation to justify these different teachings is like trying to square a circle. It's a logical impossibility. The problem is inherent in the fact that Jesus claimed to be God made man (which, incidentally, is something unique in the history of mankind) and that the discourse resulting from this claim is absolutely absurd from a rationalist point of view.

c. There is no logical explanation or sense to be found in the actions of the apostles after his death. Their response is unprecedented. When opportunists or minor warlords are killed, their supporters always disperse. We have cited six false messiahs who appeared around the same time as Jesus: in all six cases, their supporters fled after their leader died and never spoke of their false messiahs again.

On this subject, the words of Gamaliel, a very wise doctor of the law and member of the Sanhedrin, are instructive. Shortly after Jesus' death, his apostles were arrested and questioned by the Sanhedrin's court: *"But a Pharisee named Gamaliel, a teacher of the law, who was honored by all the people, stood up in the Sanhedrin and ordered that the men be put outside for a little while. Then he addressed the Sanhedrin: 'Men of Israel, consider carefully what you intend to do to these men. Some time ago Theudas appeared, claiming to be somebody, and about four hundred men rallied to him. He was killed, all his followers were dispersed, and it all came to nothing. After him, Judas the Galilean appeared in the days of the census and led a band of people in revolt. He too was killed, and all his followers were scattered. Therefore, in the present case I advise you: Leave these men alone! Let them go! For if their purpose or activity is of human origin, it will fail. But if it is from God, you will not be able to stop these men; you will only find yourselves fighting against God.' His speech persuaded them. They called the apostles in and had them flogged. Then they ordered them not to speak in the name of Jesus, and let them go"* (Acts 5:34–40).

d. The success of the apostles' preaching has so far defied rational explanation. How was the Roman Empire peacefully converted by a handful of foreigners working far from their homes—without the help of families, weapons, or money—all of whom were put to death for the hoax they had invented? The story makes no sense.

VII. Jesus is the Messiah and God incarnate

So we come to the last remaining hypothesis. You will notice that it has been taking shape as we have gradually cleared away all the other options. It takes commitment to study it in an intellectually honest way, but everything begins to become clear as we make the effort.

1. The madman's words start to make sense

In fact, if this is the case, sayings that previously seemed disconcerting, even shocking, start to make sense.

If Jesus is the Son of God, then he exists from all eternity and can say:

The Shroud of Turin, which may have been the actual burial cloth of Jesus

- *"Before Abraham was, I am"* (John 8:58).
- *"Heaven and earth will pass away, but my words will not pass away"* (Matt 24:35).

He also has power over everything, even over death and sin, and can quite naturally say:

- *"Your sins are forgiven"* (Matt 9:5).
- *"I am the resurrection and the life"* (John 11:25).
- *"For the Son of Man is lord of the sabbath"* (Matt 12:8).
- *"All authority in heaven and on earth has been given to me"* (Matt 28:18).

He can give us his body as food in the Eucharist, in the form of the consecrated bread:

- *"I am the bread that came down from heaven"* (John 6:41).

For anyone else, these words would be a sign of outrageous pretension, unmitigated pride, or total blindness. But coming from the Son of God, the Messiah, they take on their full meaning, even as they transcend the usual categories of human reason.

2. Impossible teachings become possible

From a human standpoint, some of Christ's words seem too demanding for ordinary people. Even Christians are afraid of them. But with the grace of God, anything is possible, as the lives of many saints show.

3. Wise words for eternity

Being both man and God, Jesus naturally left behind many words of wisdom, some of which we have already considered above. He knew how to make his thoughts resonate with the concerns of his contemporaries, while at the same time storing up treasures of wisdom for future generations. We have already pointed out one example: his wisdom regarding what belongs to God and Caesar, which is the origin of the modern concept of the separation of religious and secular power.

4. The sudden courage of the apostles explained

At some point after Jesus' execution, the apostles underwent a profound metamorphosis. From cowards who denied Jesus and locked themselves in the upper room, they transformed into men full of boldness and conviction, proclaiming the message of Jesus to large and often hostile crowds. This was a miraculous change—their transformation came from their meeting with the risen Jesus. They saw him and touched him, and after that nothing could stop them. They came to understand that death was only a pathway to the resurrection, and that persecution was a necessary witness to faith as well as a human participation in Christ's sufferings. Seen in this light, their behavior is perfectly logical. Moved by an unshakable faith, each of them departed for the ends of the earth, emptied of possessions but filled with faith, rich in the supernatural powers that their Master left them as their inheritance—powers that allowed them to perform brilliant miracles.

5. The Apostles' success has no merely human explanation

It is true that they traveled a long way alone, without money, without weapons, without education, but they had the Holy Spirit, sent to them by Jesus after he ascended to heaven. They performed spectacular miracles, like when Peter resurrected the widow Tabitha in Jaffa. They lived what they preached: poverty, detachment, communion, forgiveness, and love. Their lifestyle, miracles, and sufferings had an extraordinary effect on those around them. The Apostles were not all talk. They lived their faith, and their faith was dearer to them than life itself. Blaise Pascal once wrote: *"I believe those histories only, whose witnesses let themselves be slaughtered,"*[585] and self-sacrifice was what ultimately convinced the masses in the early days of Christianity. The fate of the Christians persecuted under Nero did not stop the flood of conversions. On the contrary, the testimony of the martyrs was stronger than the persecution.

---

585. Blaise Pascal, *Pensées*, 1670, various VIII - Fragment No. 3/6, *The Thoughts of Blaise Pascal*, trans. C. Kegan Paul from the text of M. Auguste Molinier (London: George Bell and Sons, 1901), 117.

6. Jesus' lifetime was indeed the predicted time for the Messiah

His advent had been prophesied, and he came at the appointed time, as we discussed above. The humble shepherds of Bethlehem, and then later the inhabitants of Nazareth, did meet the Messiah, the anointed of Israel, who came not as a conquering warrior, but as an ordinary man, the opposite of what the chosen people had expected.

Our last hypothesis is the only plausible answer to the question of who Jesus really is, and it leads us to another irresistible conclusion: if Jesus is indeed the Messiah and the Son of God, then God must exist.

Final deliberation and verdict

We have heard all the witnesses, evaluated all their testimonies in detail, and noted their inconsistencies. You, the readers, are the jurors in our courtroom. Not one of you can close this chapter with an offhand comment or a vague feeling—you are the jury, and you must render a verdict!

Jesus: Who Could He Be? The reader has the resources at his disposal to reach a decision, since the possible answers are very few and well documented. Earlier chapters required technical skill, but here, one does not have to be an expert to have an informed opinion. Courage, common sense, and intellectual honesty are enough.

Do not go on the next page until you have made your choice, whatever it is.

20.

The Jewish People: A Destiny Beyond the Improbable

A thorn in the side of the materialist worldview

Proponents of a purely material Universe want world history and the history of humanity to be coherent, rational, and miracle-free. However, the extraordinary destiny of the Jewish[586] people proves to be a major obstacle to this desire for rationality, because it is a major anomaly, a spanner in the works of the materialist cosmos, and a thorn in the side of historians. A tiny nation with scant resources, the Jews have accomplished things all out of proportion to their size. Their destiny constitutes a serious challenge to any attempt to write a rational account of history.

---

586. Strictly speaking, the word "Jew" only applies to the members of the tribe of Judah who returned from exile; but in practice, it refers to all the descendants of Jacob and all those who practice Judaism.

And that's why we find them so interesting.

Let's take a look:

The Jews are likely the only people from antiquity who have preserved their original country, language, and religion.

- They have endured and survived multiple exiles, deportations, and even a systematic attempt at extermination.
- They are the only people to have reclaimed their native land 1,800 years after being expelled from it.
- They revived their ancient language 2,500 years after it ceased to be the vernacular of the people.
- Israel is one of the few countries, small and lacking natural resources, that has survived intense hostility from all neighboring nations, some of which still call for its complete destruction.
- Jerusalem, their capital city, holds little economic or strategic value but is a focal point of global geopolitical tensions and could potentially ignite a future world war.
- The Jewish people produced the most sold book in history—the Bible.
- They are the only people to have experienced reverse racism.
- Jewish contributions to the history of ideas and science are disproportionately significant compared to their population size.
- Israel is the only country where half of its citizens still view themselves as a "chosen people of God" and the cradle of the Savior, while also being one of the most advanced high-tech societies in the world.
- The modern state of Israel has achieved victories in battles as surprising and spectacular as those recorded in the Bible.

We are aware that this is a sensitive subject that could displease both Jews, who would prefer to live quietly, and those who don't like them and are disturbed by these extraordinary facts, either because they are motivated by reverse racism or because such anomalous historical realities challenge their materialist worldview.

Leaving out this chapter altogether would perhaps have been more prudent. Was that an option? Would that have been acceptable? The question of God's existence is of utmost importance. Should the extraordinary history of the Jewish people, which carries such weight in the debate, be left out for purely diplomatic reasons? We say no, and we are not the only ones who think it a worthwhile line of inquiry. The special destiny of the Jews has astonished intellectuals throughout history:

- French intellectual Blaise Pascal: "*This people is not peculiar only by their antiquity, but also remarkable by their duration, which has been unbroken from their origin till now. For while the nations of Greece and Italy, of Lacedæmon, Athens and Rome, and others who came after, have long been extinct, these still remain, and in spite of the endeavours of many powerful princes who have a hundred times striven to destroy them, as their historians testify, and as we can easily understand by the natural order of things during so long a space of years, they have nevertheless been preserved, and extending from the earliest times to the latest, their history comprehends in its duration all our histories.*"[587]

- American writer Mark Twain: "*The Egyptian, the Babylonian, and the Persian rose, filled the planet with sound and splendor, then faded to dream-stuff and passed away; the Greek and the Roman followed, and made a vast noise, and they are gone; other peoples have sprung up and held their torch high for a time, but it burned out, and they sit in twilight now, or have vanished. The Jew saw them all, beat them all, and is now what he always was, exhibiting no decadence, no infirmities of age, no weakening of his parts, no slowing of his energies, no dulling of his alert and aggressive mind. All things are mortal but the Jew; all other forces pass, but he remains. What is the secret of his immortality?*"[588]

---

587. Blaise Pascal, "Of the Jewish People," 119–124 in Blaise Pascal, *The Thoughts of Blaise Pascal*, trans. C. Kegan Paul from the text of M. Auguste Molinier (London: George Bell and Sons, 1901), 120. Written between 1658 and 1662.

588. Mark Twain, "Concerning the Jews," *Harper's Magazine*, March 1898. Available in Fordham University's *Internet Modern History Sourcebook*, https://sourcebooks.fordham.edu/mod/1898twain-jews.asp.

- British historian Arnold Toynbee: *"As for long life, the Jews live on—the same peculiar people—to-day, long ages after the Phoenicians and the Philistines have lost their identity like all the nations. The ancient Syriac neighbours of Israel have fallen into the melting-pot and have been reminted, in the fullness of time, with new images and superscriptions, while Israel has proved impervious to this alchemy—performed by History in the crucibles of universal states and universal churches and wanderings of the nations—to which we, Gentiles, all in turn succumb."*[589]

The extraordinary features of Jewish history deserve to be examined one at a time.

I. Probably the only people to have survived from antiquity to the present day, over 3,500 years

What has happened to the Goths, the Visigoths, the Ostrogoths, the Vandals, the Picts, the Saxons, the Huns, the Gauls, the Franks? And in the East, where are the Persians, Medes, Assyrians, Phoenicians, Philistines, Canaanites, Hittites, Jebusites, and all the rest? Gone! They have all disappeared, dissolved by the great machine of history that mixes and obliterates the identities of peoples through war, migration, and intermarriage.

Sociologists and historians define an ancient people as having "survived" if they meet three conditions:

- They live on the same land;
- They speak the same language;
- They've kept the same religion.

By these criteria, the French people today cannot be identified with the Gauls or Franks because their culture, language, and religion have all changed. For the same reason, the Italians are not the same as the Romans of two thousand years ago, nor are present day Egyptians

---

589. Arnold Toynbee, *A Study of History* (Oxford: Oxford University Press, 1934–1954), 2:55.

the same as the people who built the pyramids. The Babylonians and Persians lasted for centuries, then disappeared: there is nothing left of them, neither their language nor their religion.

The Jews seem to be the only ancient people to have survived—a real museum piece, a prehistoric specimen, a dinosaur of ancient history, more than 3,500 years old. If we were dealing with a people originating in a country as large as China, from a region protected by high mountains like Tibet, or from isolated islands like Japan, we could understand how they might survive despite the melting pot of history. But this is not the case with the Jews. On the contrary, the coastal plain of Israel was a thoroughfare that joined the great kingdoms of Egypt on one side and those of Mesopotamia and the Middle East on the other.

Many famous writers have expressed their astonishment concerning the longevity of the Jewish people throughout the centuries:

- Jean-Jacques Rousseau: *"But it is an amazing and truly unique spectacle to see an expatriate people, without either location or land for nearly two thousand years [. . .] dispersed over the earth, subjected, persecuted, scorned by all nations, and yet preserving its customs, its laws, its morals, its patriotic love. [. . .] The laws of Solon, of Numa, of Lycurgus are dead; those of Moses, far more ancient, are still alive. Athens, Sparta, Rome have perished and have no longer left any children on the earth. Zion, destroyed, did not lose hers; they are preserved, they multiply, spread throughout the world, and always recognize each other. They mingle among all peoples and never become confounded with them. They no longer have leaders, and are still a people; they no longer have a fatherland, and are still citizens."*[590]

---

590. Jean-Jacques Rousseau, *Social Contract; Discourse on the Virtue Most Necessary for a Hero; Political Fragments; and, Geneva Manuscript*, trans. Roger D. Masters, Christopher Kelly, and Judith R. Bush (Hanover, NH: University Press of New England [for] Dartmouth College, 1994), 33–34.

- Leo Tolstoy: "*What is a Jew? This question is not as strange as it may seem at first glance. Let's examine this free creature that was insulated and oppressed, trampled on and pursued, burned and drowned by all the rulers and the nations, but is nevertheless living and thriving in spite of the whole world. What is a Jew that did not succumb to any worldly temptations offered by his oppressors and persecutors so that he would renounce his religion and abandon the faith of his fathers? A Jew is a sacred being who procured an eternal fire from the heavens and with it illuminated the Earth and those who live on it. He is the spring and the source from which the rest of the nations drew their religion and beliefs. [. . .] A Jew is a symbol of eternity. The nation which neither slaughter nor torture could exterminate, which neither fire nor sword of civilizations were able to erase from the face of the earth, the nation which first proclaimed the word of the Lord, the nation which preserved the prophecy for so long and passed it on to the rest of humanity, such a nation cannot vanish. A Jew is eternal; he is an embodiment of eternity.*"[591]

- Nikolai Berdyaev, a sociologist historian of Russian background, wrote in 1936 before the Holocaust: "*How mystifying is the historic destiny of the Jews! The very preservation of this people is rationally inconceivable and inexplicable. From the point of view of ordinary historical estimates it should have vanished long ago. No other people in the world would have survived the fate which has befallen it. By a strange paradox, the Jewish people, an historic people par excellence who introduced the very concept of the historic into human thought, have seen history treat them mercilessly, for their annals present an almost uninterrupted series of persecutions and denials of the most elementary human rights. Yet, after centuries of tribulation which have strained its powers to the full, this people has preserved its unique form, known to all and often cursed. No*

---

591. Leo Tolstoy, "What is a Jew?" (1891). See Harold K. Schefski, "Tolstoi and the Jews," *The Russian Review* 41, no. 1 (Jan. 1982), 1–10, https://doi.org/10.2307/129561. Translation from https://blogs.timesofisrael.com/what-is-a-jew-written-by-count-leo-tolstoy-1891/.

other nature would have resisted a dispensation lasting so long without in the end dissolving and disappearing. But, according to God's impenetrable ways, this people must apparently be preserved until the end of time. As for trying to explain its historic destiny from the materialist standpoint, this is to court certain defeat. Here we touch upon one of the mysteries of history."[592]

II. Surviving extraordinary hardships, from the exile of the biblical era to the Nazi genocide

The Jewish people suffered a first exile in Egypt. From Jacob to Moses, the Hebrews were slaves of the pharaohs. They escaped in the Exodus and entered Palestine, probably between 1300 and 1200 BC. This exile is only known to us through the Bible.[593]

Their second exile was partial. The tribes living in the northern territory of Israel were deported to Nineveh around 722 BC. Here again, the Bible is nearly the only record of this exile.

The third exile followed the Kingdom of Judah's several defeats at the hands of the Babylonian king Nebuchadnezzar. In the wake of these defeats, a large part of the Kingdom of Judah's population, especially the educated portion, was deported to Babylon. This exile, lasting from 597 to 538 BC, is better documented. Extrabiblical testimony has survived in the form of contemporary cuneiform tablets.[594] The Jewish people's longing to return to Jerusalem from Babylon, and the eventual accomplishment of this return, is one of the main centers of gravity for Israel's hope, illustrated in the celebrated psalm *"If I forget you, O Jerusalem..."* (Ps 136:5).

The most recent exile has also been the longest, most significant, and harshest. It has ample testimony and historical documentation.

---

592. Nicolas Berdyaev, *Christianity and Anti Semitism* (New York: Philosophical Library, 1954), 5.

593. A notable exception, however, is the Mérenptah stele (1210 BC).

594. See the work of Wayne Horowitz, professor at the Hebrew University of Jerusalem.

Following two disastrous wars against the Romans in 70 and 135 AD—which ended in the complete destruction of Jerusalem and the Temple, the banishment of Jews from their native land, and the deportation of a large number of the survivors to countries all over the world—this exile lasted for more than seventeen centuries. The cruelty of their punishment is explained by the fear and humiliation that the Romans suffered during a brutal war in which the Jews held out for several years against a world superpower. In retribution for their audacity, and to give a memorable example to other subject peoples, the Romans set out on a campaign of unprecedented destruction. Flavius Josephus reports that nearly all the trees in the region about Jerusalem had to be cut down to sustain the siege of the city and, later, the crucifixion of the population. After 135, the Jews were forbidden to live in Jerusalem. Since the place held little interest, it was rapidly depopulated and remained a desert for several centuries. To blot out the memory of the city that had once stood there, the Romans renamed Jerusalem *Aelia Capitolina*, and the land *Palestine*. The Jews spread around the whole world. Some estimate that one million died in these two wars.

In addition to the unpardonable humiliation the Romans had suffered in these wars and the risk of sparking similar uprisings elsewhere in the Empire, rumors spread that the Jewish people would one day rule the world. The superstitious Romans were disturbed despite their total military and political supremacy, as two of the greatest Roman historians report:

- Tacitus, at the beginning of the second century, wrote: "*The majority firmly believed that their ancient priestly writings contained the prophecy that this was the very time when the East should grow strong and that men starting from Judea should possess the world.*"[595]

- And his contemporary Suetonius, in his "Life of Vespasian," wrote: "*There had spread over all the Orient an old and established belief,*

---

595. Tacitus, *Histories V.13*, in *Tacitus, Histories: Books 4–5. Annals: Books 1–3*, trans. Clifford H. Moore and John Jackson, *Loeb Classical Library* 249 (Cambridge, MA: Harvard University Press, 1931), 5.13.2.

that it was fated at that time for men coming from Judaea to rule the world."[596]

This review of the exiles suffered by the Jewish people helps us realize how unique their story really is. No other people have ever endured so much suffering and survived, maintaining their own identity throughout.

Some have argued that their story is not all that unique, citing possible comparisons with the Armenians, Lebanese, Polish, or some Africans in America who returned later to Liberia. Though these people certainly suffered great misfortunes, everything that happened to them differs greatly from the history of the Jewish people.

The six million dead in the Holocaust are a terrible witness to that fact. Never before had such a project of total extermination through industrial means been carried out. The untold suffering of the Jewish people blights the history of the twentieth century.

III. The only people to return to their native land eighteen centuries after total expulsion

Now let us turn to the enigma posed by the return of the Jews to their ancestral land. We will leave aside the three earliest returns, about which the historical record is too sparse, and concentrate on the most recent and well-known one.

In the history of mankind, this return is perfectly unique. Nothing else comes close to it. The sheer impossibility of it defies critical historians. The Jews' return to their homeland after so many centuries required a host of accidents and historical coincidences.

These included the following:

---

596. Suetonius, *Life of Vespasian IV*, in *Suetonius, Lives of the Caesars, Volume II*, trans. J. C. Rolfe, *Loeb Classical Library* 38 (Cambridge, MA: Harvard University Press, 1914), 4-5.

- For nearly two thousand years, a sizable portion of the Jews scattered throughout the world were not assimilated by the peoples among whom they lived.
- In the same period, Palestine remained unsettled and poor, with few exceptions. If this region had boasted the same natural resources as Provence in France, it would have been densely populated, cultivated, and prosperous, and a complete return would have been unimaginable.
- The whole time, many of the Jews continued to believe that this wholly chimerical return was possible.
- When persecutions around the world, especially beginning in the mid-nineteenth century, forced them to search for a homeland outside their countries of adoption where they had already put down roots, all other attempts to find them a new land—for example, a virgin territory in Uganda—failed.
- This project of a return to Palestine, which seems like such a biblical notion, was initiated, remarkably, by militant atheist socialists and communists. The religious Jews of Europe wouldn't hear of it: they were convinced that they had to wait first for the Messiah to return.
- Several prominent Jewish figures, thanks to exceptional service rendered to their adopted country or by pure chance, were able to work out the Balfour declaration of 1917 and the massive purchase of Palestinian lands then owned, for the most part, by local Muslim peasants.
- The Ottoman empire, which had forbidden the sale of Palestinian territory to Jews since 1901, conveniently collapsed in 1917.
- A friendly nation, the United Kingdom, obtained a mandate over Palestine.
- Wrestling with guilt for Europe's role in the Holocaust, major world powers pushed the UN to create a new state on May 14, 1948, situated in predominantly Muslim Palestine. Thousands of poor Palestinian families were arbitrarily expelled from their lands, setting the stage for ongoing conflict in the region.
- Despite their inferior numbers, Israelis defeated the previous occupants of the land and their new neighbors in every conflict (1948, 1967, 1973) that resulted from the UN's perilous and contentious decision.
- Later, the Iron Curtain fell (1990).

- Many of their neighbors were chronically weakened and paralyzed by division and internal conflicts, many of which continue today.

Around the middle of the nineteenth century, thousands of Jews from all over the world began to pack their bags and set off for Palestine. Arriving from Africa in 1840, Yemen in 1850, the Crimea in 1853, driven out by war, from Algeria and especially Russia, the Jews converged upon Palestine and settled down there. Gradually they began to outnumber the local population. In less than one hundred fifty years the number of Jews in Palestine had jumped from a few thousand to the six million we find there today.

Ottoman census records and various historical studies give us a good idea of Jerusalem's changing population. Even when we account for the unreliability of old figures, a clear trend emerges.[597]

Population of Jerusalem

Years	Jews	Muslims	Christians
1525	1,194	3,704	714
1849	1,800	6,100	3,700
1871	4,000	13,000	7,000
1905	13,300	11,000	8,100
1922	33,971	13,413	14,669
1944	97,000	30,600	29,400
1967	195,700	54,963	12,646
1990	378,200	131,800	14,400
2011	497,000	281,000	14,000
2016	536,600	319,800	15,800

---

597. The figures presented here have been extracted and summarized from the Wikipedia article "Demographic history of Jerusalem," https://en.wikipedia.org/wiki/Demographic_history_of_Jerusalem

1. An astonishing first prophecy

This is an absolutely unprecedented return. As we trace the multiple streams of migration that flowed in from Yemen in the south, the United States in the west, Africa in the southwest, Crimea in the north, Russia in the northeast, or Ethiopia's Falasha from the southwest—repatriated by a spectacular Israeli airlift in 1975—how can we fail to make the connection with this prophecy from Isaiah?

- *"Do not fear, for I am with you; I will bring your offspring from the east, and from the west I will gather you; I will say to the north, "Give them up," and to the south, "Do not withhold; bring my sons from far away and my daughters from the end of the earth—everyone who is called by my name, whom I created for my glory, whom I formed and made"* (Is 43:5–7).

To escape the unsettling intellectual consequences of this prophecy, some have claimed that it actually refers to Jews' return from Babylon and was written after that return, so what we have here is simply a historical observation dressed up as prophecy after the fact. But this doesn't quite square with the facts. Babylon is a single country located straight east of Jerusalem: neither north, south, nor west (much less at the ends of the Earth). The prophecy can't apply to the return from Babylon.

But if we compare the prophecy to the history of the modern return we have just seen, everything adds up: it perfectly describes events as they ultimately unfolded many centuries later.

2. Two astonishing Gospel prophecies give even sharper definition to this exile and return

Two of Christ's prophecies describe the time of this exile and return in even more spectacular terms. They are even more striking than Isaiah's prophecy, because they touch on our present age and the as-yet-unrealized future.

- First, it is written: *"Jesus came out of the temple and was going away, his disciples came to point out to him the buildings of the temple. Then he asked them, 'You see all these, do you not? Truly I tell you, not one stone will be left here upon another; all will be thrown down'"* (Mt 24:1–2).

- Then, we read: *"Jerusalem will be trampled on by the Gentiles, until the times of the Gentiles are fulfilled"* (Luke 21:24).

If we muse on the first prophecy and ponder the usual fate of ancient buildings, it becomes clear that the Temple's fate is unique. In almost all known cases of the destruction of temples or monuments, some ruins or remains are left standing—at least a few stones standing on one another or scattered about, testifying to the former presence of buildings on the site. We still have traces of practically all the contemporaneous temples of the Romans, Greeks, and Persians. We even find ruins of Carthage, despite the Romans' attempt to erase that civilization competely. The Athenian Acropolis still stands to this day, as does the Roman Colosseum. What remains of the Temple in Jerusalem? Nothing, not a single "stone upon a stone." The whole esplanade on which the Temple was built was razed to the ground. Not a trace of it exists, fulfilling Christ's word to the letter. Only the famous "wailing wall" remains, which was not part of the Temple proper, but only a retaining wall for the esplanade.

Ferdinand Prat explains the first part of the prophecy:

- *"The prophecy was fulfilled to the letter; anyone who walks today through the plaza where the mosque of Omar replaced the temple of Herod can see it with their own eyes. Imposing ruins remain of the most famous temples of Egypt, Greece and Rome; here the very ruins have perished. Every possible step was taken to annihilate them. When a soldier of Titus, guided by an invisible hand, threw a lighted torch under the sacred paneling, the fire caught with such speed and violence that it was impossible to extinguish it. Hadrian, erecting a sanctuary dedicated to Jupiter Capitoline in the place of the altar of the true God, continued the destruction, which was completed under Julian the Apostate. Keen to give the lie to the prophecies of Christ, Julian allowed the Jews to rebuild their Temple and covered the costs; but just when all the ruins had been cleared and they were ready to be rebuilt, swirls of flame issuing from the foundations forced out the workers, many of whom were burned alive. They had to stop the work, and it was never resumed. This is reported not only by the Fathers of the Church and contemporary Christian historians but by the unimpeachable witness Ammianus Marcellinus, who remained faithful to paganism and lived at the court of the emperor in Antioch, the same year (363) the events he reports took place."*[598]

The prophecy's skeptics sidestep this difficulty by arguing against all reason that this text, along with the rest of the Gospels, was written after the year 135. This would make it not a prophecy at all but, as in the case of the text from Isaiah, just a historical note dressed up as prophecy after the fact. While there are many reasons that argue to the contrary, we'll move on for now to consider Jesus' even more striking second prophecy. This one, as it turns out, is indisputable:

- *"Jerusalem will be trampled on by the Gentiles, until the times of the Gentiles are fulfilled"* (Luke 21:24).

While the dating of the Gospel account is certainly debated, almost no one argues that it was composed later than the end of the first century. Consequently, no one denies that this text was written nearly 1,900 years ago. This makes the prophecy's accuracy all the more striking.

It is a fact that Jerusalem has been continually trampled underfoot by pagans—"nations," as some translations put it—from 135 until the present day. Israel did conquer Jerusalem in 1967, but not the Temple at the heart of the city. The site is occupied by an important Muslim mosque, and Jews are not allowed to go there even to pray. In the Temple period, pagans were forbidden to approach the sanctuary. Violating this prohibition, even by setting foot on the outer court, was

---

598. Ferdinand Prat, *Jésus Christ, sa vie, sa doctrine, son oeuvre* (Paris: G. Beauchesne, 1938), Vol. II, 239.

punishable by death.[599] In this sense, Jerusalem is still being trampled underfoot by pagans—or the nations.

Faced with such extraordinary prophecies, materialist historians will need a lot of imagination to write a history of the Jewish people that is at once purely rationalistic and convincing.

The final words of Christ's prophecy seem to foresee the end of the current situation: *"until the times of the Gentiles are fulfilled."* Did he mean the end of time, or an age when all nations will disappear and the world will be governed by a single authority? This remains to be seen. But many Christians have long considered the return of the Jews to Israel to be a sign of the approaching end of time. Let's cite a few of them, focusing on those who wrote before the year 1900:

- **Saint Paul (first century)** touches on the conversion of the Jews at the end of time: *"I do not want you to be unaware of this mystery, so that you may not claim to be wiser than you are, brothers and sisters, I want you to understand this mystery: a hardening has come upon part of Israel, until the full number of the Gentiles has come in, and thus all Israel will be saved, as it is written."* (Rom 11:25–26).

- **Saint Jerome (347–420)** writes of a restoration at the end of time, distinguishing it from the return from Babylon: *"Clearly he (i.e. the prophet) is predicting the future restoration of the people of Israel and (God's) compassion for Israel after the captivity, which was fulfilled in part* iuxta litteram *under Zerubbabel, Joshua the priest, and Ezra, and which is described as going to be fulfilled* iuxta intellegentiam spiritalem *more truly and more perfectly in Christ. 'A time will come,' he says, 'when it will no longer be said*

---

599. The court of Herod had a low dividing wall that kept the Gentiles from entering the sacred space. We have evidence of this in the form of an approximately 2 ft x 1 ft stone with a Greek inscription. Discovered in 1871 by French archaeologist Charles Simon Clermont-Ganneau, it is currently held in the Istanbul Archaeological Museum. The inscription is actually a harsh proscription: *"No stranger is to enter within the balustrade round the temple and enclosure. Whoever is caught will be himself responsible for his ensuing death."*

that the people were brought back from Egypt through Moses and Aaron, but that they were brought back from the land of the North by means of Cyrus, the king of the Persians, setting the captives free.' He (also) says, 'and from all lands,' which did not occur in the time of Cyrus, but will be fulfilled at the very end, as the Apostle says: 'After the fullness of the nations has entered in, then all Israel will be saved.'"[600]

- **Thomas More (1478–1534)** wrote his *Dialogue of Comfort against Tribulation* while imprisoned in the Tower of London before his execution. In a passage about the return of the Jews announced by the prophecies, he posed a question that would have occurred naturally to someone facing imminent death: Will the end of the world come in my lifetime? Our martyred saint answers in the negative: "*But, as I say, methinketh I miss yet in my mind some of those tokens that shall by the Scripture come a good while before that—and among others, the coming in of the Jews, and the dilating of Christendom again—before the world come to that strait.*"[601]

- In 1673, **John Owen (1616–1683)**, Puritan theologian at the University of Oxford, wrote a book whose first chapter was entitled: "*The Jews dispersed at present throughout the entire world will all go back to their country.*" The titles of the second and third chapters adopt the same prophetic tone: "*The land will be made eminently fertile*" and "*Jerusalem shall be rebuilt.*"[602]

- **John Gill (1697–1771)**, pastor, theologian, and English university professor, addresses the same theme: "*For, without that the Jews upon their call and conversion shall return to their own land, in*

---

600. Jerome, *In Hieremiam Prophetam* [CSEL], 3, 64, 2; 1. Translation from Michael Graves, "'Judaizing' Christian Interpretations of the Prophets As Seen by Saint Jerome," *Vigiliae Christianae* 61, 2 (2007): 142-156, 148.

601. Thomas More, in the preface to book three of his "Dialogue of Comfort." Gerard B. Wegemer and Stephen W. Smith, eds., *The Essential Works of Thomas More* (New Haven: Yale University Press, 2020), 1195.

602. John Owen, *An Exposition of the Epistle to the Hebrews*, 2nd ed., vol. 1 (Edinburgh, 1812).

a literal sense, I see not how we can understand this, and many other prophecies."[603]

- More recently, **Pope Benedict XVI** has said: "*It is not difficult, I believe, to see in the creation of the State of Israel the fidelity of God to Israel revealed in a mysterious way.*"[604] Also: "*In this sense, the Vatican has recognized the State of Israel as a modern constitutional state, and sees it as a legitimate home of the Jewish people, the rationale of which cannot be derived directly from Holy Scripture. Yet, in another sense, it expresses God's faithfulness to the people of Israel.*"[605]

Many Jews and Christians (particularly Evangelicals) regard the fulfillment of this prophecy as a forerunner to the return of the Messiah, and certain circumstances suggest that history may be moving in this direction.

3. Some thought it ludicrous that the Jews would go back to Palestine, but they were wrong

The following miscalculations show just how improbable the return to Palestine seemed, humanly considered:

- In May 1944, **Frederick C. Painton**, writer and journalist, wrote a very pessimistic report of the state of the country for *Reader's Digest*: "*The Palestine problem will die out by sheer lack of Jews who would give up their own homelands to plant themselves anew in the sterile mountains of Judea.*"[606]

---

603. John Gill, *An Exposition of the Old Testament,* commentary on Jeremiah 30:3, available at https://johngill.thekingsbible.com/CommentaryVerse/24/30/3.

604. Benedict XVI to Arie Folger, August 23, 2018, "The Pope and the Rabbi: Correspondence between Pope Emeritus Benedict XVI and Arie Folger, the Chief Rabbi of Vienna," *La Stampa*, September 12, 2018, https://www.lastampa.it/vatican-insider/en/2018/09/12/news/the-pope-and-the-rabbi-1.34044561.

605. Benedict XVI, "Grace and Vocation without Remorse: Comments on the Treatise *De Iudaeis*," *Communio* 45, no. 1 (2018), 179.

606. Jewish Telegraphic Agency, "'Reader's Digest' Says Lack of Jews in Palestine Will Solve Ar-

- **Claude Ezagouri**, teacher for the Messianic community of Tiberias in Israel, recalls how his belief in the return of the Jews from Russia made him seem like a mystic: *"I remember in 1985 I was talking with family here in Israel. I was already saying then that tens of thousands of Jews would come from the USSR to Israel, fulfilling the Biblical prophecy. They mocked me, laughed. I was taken at the time for a mystic. I was told, 'It's impossible. You know very well that there is the Iron Curtain. It's impossible for the Jews to come to Israel.'"*[607]

IV. Another strange tale: resuscitating an ancient language after 2,500 years of disuse

Hebrew, the language of the Bible and the ancient Hebrews, disappeared as a living language after the Babylonian exile in 597 BC. From then on, it was the written language of the Bible, used only in the liturgy and rabbinical disputations. Yet around 1800, it began to experience a literary rebirth.

At the end of the nineteenth century, Hebrew became a living language again after twenty-five centuries of dormancy. It is the only "dead" language in the modern world to have been fully resurrected and is now spoken fluently by millions of people.

Since ancient Hebrew had a small vocabulary, it lacked words for describing the many realities of the modern world. Eliezer Perlman (1858–1922), later called Ben-Yehuda, set out to fix that. He published the first modern Hebrew dictionary and set up the Academy of the Hebrew Language, intended to enrich the language by adding words for contemporary objects and concepts. In 1917, England made Hebrew the official language

---

ab-Jewish Problem," *The Daily News Bulletin*, April 28, 1944, https://www.jta.org/1944/04/28/archive/readers-digest-says-lack-of-jews-in-palestine-will-solve-arab-jewish-problem).

607. From Claude Ezagouri, author of, *Israël, la terre controversée* (Montmeyran, France: Emeth Éditions, 2016).

of the Jewish homeland in Palestine. When Israel was founded in 1948, Hebrew was declared its official language.

V. A precarious position: small and poor in natural resources, Israel is surrounded on all sides by hostile neighbors, some of whom still call for its disappearance

Although there have been some developments in recent years, thanks to the signing of peace agreements with a few neighbors, it remains quite extraordinary that a country so small and sparsely populated, devoid of natural resources, and of little economic interest is surrounded by hostile neighbors with thirty times its population,[608] many of whom dream of the dissolution of Israel and a fair number of which are willing to die for this cause.

Any other small country in the same conditions would have been conquered by now. But Israel has held out for almost a century, without the help of a single allied neighbor at its borders. Along the more than six hundred miles of its borders are Syria (with a population of 18 million), Iraq (38 million), Iran (81 million), Jordan (ten million), Egypt (97 million), Libya (six million), Lebanon (six million), the Palestinian territories (two million), and we might add Saudi Arabia (with a population of 32 million), the Emirates (nine million), Qatar (three million), and Turkey (80 million), bringing us to a total of nearly 400 million potentially hostile neighbors. Although some of them have signed peace treaties, these agreements remain fragile and at the mercy of the revolutions that frequently occur in these countries. Israel is in a state of open armed hostility with its neighbors; Iran and other armed groups have proclaimed their intention to destroy the country entirely. In 2005, Iranian president Mahmoud Ahmadinejad stated his belief that the Islamic revolution could only be achieved through the complete annihilation of the Israeli people.

---

608. Israel has 8.7 million inhabitants, its neighbors more than 382 million.

And why have they been fighting over this little strip of land in Palestine for more than one hundred years? Of course, there is the comprehensible struggle of thousands of Palestinians seeking to regain their land and homes. Besides that, what else? For petroleum-rich soil such as we find in abundance in neighboring states? No, not a single drop! For a strategically essential route, like the Strait of Hormuz or the Suez Canal? Not at all. For mines full of rare metals? Not a single trace. For rich farmland? No longer. Modern Israel is primarily made up of the Negev, a great desert without oil; a Dead Sea that's drying up; the hills and mountains of Judea without a single gold, silver, or iron mine; and a coastal plain that, though more fertile, has no natural harbor. It is, moreover, one of the world's few countries to have no river of its own: the Jordan, now reduced practically to a trickle, is divided between four countries.

How is the nation's continued existence explicable under these conditions? Some have suggested that Israel is only propped up by the support of the Jewish diaspora plus American aid and military power. But is that a satisfactory explanation?

VI. Israel is the only country whose capital, despite being a city of little economic or strategic interest, remains a center of geopolitical tension and a flashpoint for a potential world war

Jerusalem is one of the main points of geopolitical tension in our times and could even one day spark a major world war. The city is objectively devoid of interest. It is perched atop nearly bare hills 2,300 feet above sea level, watered by no river, a stop on no important trade route. It has no mines and no particular agricultural richness. It is one of the few world capitals without direct access to a sea or river (except for one small stream, the Kidron), and possesses no significant water source. The water sources are situated outside the city; for thousands of years, Jerusalemites have relied on subterranean passages, wells, cisterns, and artificial irrigation for all their water.

VII. The people who produced the world's top bestseller

Both the Old and New Testaments of the Christian Bible, written almost entirely by Jews, have been translated into nearly two thousand languages. The Bible is by far the best-selling book in the world. An estimated two to six billion copies have been printed to date, putting it far ahead of Mao's *Little Red Book* or the Koran.

VIII. The Jews are the only people to maintain a disproportionately important role in the history of ideas and of science relative to their importance

The Jews have produced a surprising number of intellectuals, inventors, and artists. Some have suggested that persecution makes people stronger, but history in no way confirms this hypothesis: we find nothing similar among the Armenians, Palestinians, Lebanese, and the like.

Twenty-two percent of Nobel prize winners are Jews, while they represent only 0.2% of the world population:[609] 194 Nobel prizes, out of a total of 871, have been awarded to laureates of Jewish origin.[610] The following list gives the figures for each category:

- Psychology and Medicine: 55 out of 204, or 26.5% of laureates are Jews.
- Economic Sciences: 29 out of 69, or 41% of prizes awarded.
- Physics: 52 out of 193, or 26% of prizes awarded in this category.
- Chemistry: 36 out of 160, or 22% of prizes in this category.
- Literature: 12 out of 108, or 11% of the total.
- Peace: 9 out of 101, or 9% of laureates in this category.

The history of ideas has been indelibly shaped by great Jewish figures like Marx, Freud, and Einstein.[611]

---

609. See https://en.wikipedia.org/wiki/List_of_Jewish_Nobel_laureates.

610. From the US, besides Nobel Peace Prize winners, who aren't scientists, there have been sixty-seven Jewish laureates, 32% of the American prize-winners (Jews represented about 2.5% of the US population in the last century).

611. See for example the "informal list of Jewish inventions, innovations, and radical ideas" avail-

IX. The only people to be the victim of reverse racism

Racism is an ideology that considers some groups of people to be racially superior to others and ostracizes those considered inferior, with whom the racist would not want to mix at any cost—let alone see mix with his children. The racist rejects those whom he considers genetically inferior.

When applied to the history of the Jewish people, racism has a specific name: anti-Semitism. The Nuremberg Laws and the Holocaust later on were extreme manifestations of this idea that Jews belong to an inferior and degenerate race.

But there is another equally insidious side to anti-Semitism: the opposite belief, that Jews are superior to other races. This reverse-racism leads to the same outcome: hatred and rejection of the Jewish people.

In a long entry on "anti-Semitism" in the eleventh edition of the *Encyclopædia Britannica*, published in 1910, Lucien Wolf, the president of the Jewish Historical Society of England, claimed that anti-Semitism was *"exclusively a question of European politics,"* completely detached from any *"atavistic revival of the Jew-hatred of the middle ages."* He argued that the renewed persecutions against the Jews were entirely due to their civil emancipation and the astonishing social, economic, and cultural success of their *Gemeinde* (community) in Western Europe at the end of the eighteenth century.[612]

These questions have been the object of countless debates that we do not wish to enter upon. It is sufficient to note here that reverse-racism is a reality based partly on the supposed superiority of the Jews in the domains of finance, commerce, and intellectual speculation. It leads to practices of exclusion in a way opposite to common racism. Here's a representative example of this belief: in November 1967, a few months

---

able at https://boulderjewishnews.org/2009/an-informal-list-of-jewish-inventions-innovations-and-radical-ideas/.

612. Lucien Wolf, "Anti-Semitism," *The Encyclopædia Britannica* (Cambridge: The University Press, 1910), 2:134.

after the Six-Day War, General Charles de Gaulle remarked that the Jewish people is *"elite, sure of itself, and domineering."*[613]

X. The only country where half the population believes themselves God's elect and their lands the cradle of the Savior of the World, and which, despite such absurd daydreams, nevertheless ranks among the most technologically advanced countries in the world

Contrary to the trend in most developed nations, the Jews of Israel are growing more and more religious. This is in marked contrast with their European ancestors of the twentieth century, who were very republican, secular, and socialist, and who launched the economy of their fledgling nation with communitarian kibbutzim and small orange groves.

The demographics of Orthodox Jews today, who currently make up 12.5% of the population as opposed to the 1% they represented at the beginning of the 20th century, goes a long way towards explaining this development of the religious situation: they have an average of 6.5 children per couple, compared to two for secular Jews. But this revival of religiosity is also the consequence of major events such as the Six Day War, which we will discuss a little further on.

This demographic shift means that by 2060 religious people will represent 50% of the country's population, if the trend continues.

How can a materialist who holds that all religious faith is obscurantism and the particular beliefs of Judaism—like their conviction that they are God's chosen people and their expectation of the Messiah—are ridiculous ravings, explain how such a small desert country, resourceless

---

613. Press conference on November 27, 1967.

and exposed to danger on every side, has become one of the richest and most high-tech nations in the world within only a few generations?

Though the land of Israel was first devastated long ago by the Romans in 135, it was only reduced to its current, mostly desert condition over the last three centuries, due in large part to the massive clear-cutting of Palestinian wood for the Turkish railway system and to a tree tax instituted by an eighteenth-century Ottoman sultan. Records show that in order to reduce their tax burden, property owners in Palestine cut down the majority of their trees over a period of just a few years. This had dramatic consequences for the climate and environment, which rapidly became more desertlike and rocky.

Eighty years ago, jackals trotted across sand dunes in the region of Jerusalem. The land of Israel, between dangerously insect-ridden swamps and deserts stretching southward from Tel-Aviv, was scarcely hospitable to human life.

During a trip to Palestine in 1869,[614] Mark Twain could not find a single village within a thirty-mile radius—only a few Bedouin tents scattered here and there. Even as late as 1927, Floyd Hamilton wrote: *"In no other country are there as many ruins of cities and villages as in the Palestine of today."*[615]

Today, Israel's GDP per capita is higher than that of France, even though the country has few natural resources, and it has the economic handicap of a considerable military budget: no less than 6% of its GDP.

---

614. *"Stirring scenes like these occur in this valley no more. There is not a solitary village throughout its whole extent—not for thirty miles in either direction. There are two or three small clusters of Bedouin tents, but not a single permanent habitation. One may ride ten miles, hereabouts, and not see ten human beings. To this region one of the prophecies is applied: 'I will bring the land into desolation; and your enemies which dwell therein shall be astonished at it. And I will scatter you among the heathen, and I will draw out a sword after you; and your land shall be desolate and your cities waste.'"* Mark Twain, *The Innocents Abroad* (Newark: Bliss and Co., 1869), 485.

615. Floyd E. Hamilton, *The Basis of Millennial Faith* (Grand Rapids, MI: Wm. B. Eerdmans Publishing Company, 1942), 38

XI. A people who have stunned the world with unexpected and spectacular military victories

Several wars—in 1948, 1956, 1967, and 1973—followed the creation of the state of Israel. We will focus solely on the 1967 conflict, known as the "Six-Day War." Let's put this blitzkrieg in context.

In the spring of 1967, Egyptian president Nasser entered into a military alliance that brought together Iraq and Israel's three closest neighbors, Syria, Jordan, and Egypt, with the subsequent addition of Iraq. Together, these four countries completely surround Israel and significantly outnumber the Israelis, representing a total population of 48 million people versus 2.7 million, a difference of eighteen to one. Nasser sent one hundred thousand men to Sinai, blocked the Straits of Tiran at their issue into the Gulf of Aqaba—shutting off a vital sea route for Israel, whose passage had been guaranteed by the major world powers—and demanded the withdrawal of the 3,800 UN peacekeeping troops then stationed at the borders. None of the belligerent states would have dared to attack this neutral peacekeeping force that represented the major world powers and international order. But in a reckless and inexplicable move, the UN withdrew their soldiers without delay or prior consultation. War was then inevitable.

These were the forces involved:

	Israel	Egypt	Jordan	Syria	Iraq	Coalition Total
Aircraft	350	450	40	120	200	810
Tanks	800	1,400	300	550	630	2,880
Troops	264 000	270,000	55,000	65,000	75,000	465,000
Population	2.745 M	32.53 M	1.377 M	5.74 M	8.947 M	48.595 M

The Israeli force was enormously outnumbered. Surrounded on all three sides, it had to fight on its northern, eastern, and southern borders simultaneously. Syria and Egypt were armed, supplied, trained, and advised by the Soviet Union. The Jordanian army, known for its skill and bravery, had been trained and equipped by the British. Faced with these numbers, no one expected Israel to win the war. Many thought that in the aftermath of its defeat, the nation would cease to exist.

Here is how a leading French paper, *Le Monde*, analyzed the situation in the days before the conflict broke out:

- *Le Monde*, May 20, 1967: "*The foreign chancelleries are unanimous in their judgment that an Israeli-Arab war is not imminent. This relative optimism [...] flows from a logical analysis of the situation. It is highly improbable that Israel would desire to do battle on several fronts and fight simultaneously against Syrian and Egyptian forces.*"

- *NYT,* May 27—Cairo, May 26 (United Press International): "*President Gamal Abdel Nasser said tonight that any Israeli military action against the United Arab Republic or Syria would lead to all-out war. If war starts, he said, 'our main objective will be the destruction of Israel.'*"

- *Le Monde* (Paris), May 30: "*Not able to count on the support of the United States, Israel should hesitate to undertake such a dangerous enterprise alone. The Egyptian army, even according to western military experts, has been significantly strengthened since 1956 by Russian arms and by the experience it has acquired in Yemen. Further, the opening of a second front by the Syrians and the vulnerability of Israeli territory to massive bombardments should encourage Zionist leaders to proceed with caution.*"

- *NYT*, May 31—Cairo: "*King Hussein and President Gamal Abdel Nasser, until now avowed enemies, conferred unexpectedly here and signed the mutual defense pact binding their Governments to "use all means at their disposal, including the use of armed forces," to repel an attack on either nation. [. . .] Ahmed Shukairy, who has called for Hussein's overthrow in fiery broadcasts to Palestinians in Jordan, conferred quietly with President Nasser and Hussein this morning. Then, once the pact had been signed Mr. Shukairy accompanied the King on his return to Amman. [. . .] If Israel were to attack the United Arab Republic and Cairo's defense pacts with Syria and Jordan were invoked, General Fawzy would be able to mount an attack force drawing from the 300,000-man Egyptian army and the 60,000-man Syrian force as well as Jordan's army. He could strike Israel from the southwest, the east and the northeast.*"

- The same day, André Scémama, a *Le Monde* correspondent in Jerusalem, wrote: "*Israel now finds itself pinched in a vise between the united leadership of all the Arab armies.*"

- Also on May 31: "*Every day for months, Arab leaders have broadcast their desire to annihilate Israel. [. . .] Nasser declares that his goal is the total destruction of the Hebrew state.*"

- In *Le Nouvel Observateur* of May 31, Jean Daniel wrote: "*Is Israel on the verge of death? No doubt! Can we allow that? Under no circumstances.*"

- *Le Monde*'s headline on June 1: "*Israel Surrounded.*" Nasser brags: "*The Arabs are coming together at the critical hour.*" Indeed, "*the *alliance with Jordan is a formidable trump for the anti-Israeli forces. The royal army, the ancient Arab Legion forged by Glubb Pasha,*[616] *is one of the best-trained in the Middle East. Stationed in the region of Qalqilya, twenty kilometers from the Mediterranean coast, it finds itself positioned for an attempt to cut the Israeli territory in half if hostilities arise."*

- Also on June 1: *"The situation looks bleak for the Israelis. Colonel Nasser scores another point, though the commitments and promises of the maritime powers are still an unknown [. . .]. The reconciliation of President Nasser and King Hussein is a genuine coupe de theatre and for Cairo a diplomatic success of the highest order."*

- Cairo, June 1: *"Egypt holds its breath for the conflagration expected to come in the next few days or even hours."*

- *Le Monde*, June 2: *"Upon retaking possession of the premises of the Palestinian Liberation Organization in Jerusalem, closed since last January, Mr. Shukairi declared: 'We will accept nothing short of full liberation.' He concluded that 'when the fight comes, there will be few Jewish survivors.' [. . .] Claude Lanzmann stated: 'If Israel is destroyed, it will be worse than the Nazi Holocaust. Because Israel is my freedom.'"*

- French newspapers *France Soir*, *Le Figaro*, and *Combat*, June 3, 1967, report that the Israeli government has ordered 20,000 gas masks from Germany, hoping to deter Nasser from using chemical weapons as he allegedly did in Yemen.

- *Le Monde*, June 5, published a letter from Pierre Mendès France to Mapam (an Israeli political party with Marxist leanings): *"I share your worries in these grave hours. How could I not be moved when the right to exist of a country, a member state of the UN, is being*

---

616. John Bagot Glubb (1897–1986), called Glubb Pasha, was a British general with a unique life story. After seeing combat in France during WWI, he continued his career in the Middle East, commanding the Arab Legion from 1939 to 1956.

contested? When a people's very life is threatened by a coalition of all the powers that surround them?"

- June 5 *"The Voice of the Arabs,"* broadcast from Radio Cairo: *"Destroy them and lay them waste and liberate Palestine. Your hour has come. Woe to you Israel. The Arab nation has come to wipe out your people and to settle the account. This is your end, Israel. All the Arabs must take revenge for 1948. This is a moment of historic importance to our Arab people and to the holy war. Conquer the land."*

The Israelis struck the first blow on the morning of June 5. At the close of the first day of hostilities, 75% of the Egyptian air force, chiefly modern MIG fighter jets bought from the Russians, had been destroyed. In six days of fighting, Israel stunned the world by defeating all its adversaries simultaneously, taking the Golan Heights, Mount Hermon, the West Bank, Jerusalem, and all of Sinai.

No analyst or commentator could have imagined such a complete victory. But forced to give their readers some kind of explanation, they singled out the element of surprise Israel had capitalized upon and their enemies' long list of military blunders. The first argument has no merit at all, considering that the Arab coalition led by Egypt took the initiative in this whole war. They had crossed every line in the sand and brought Israel to the brink of war by closing the Straits of Tiran, invading Sinai by force, and expelling the peacekeepers. How could they have been taken by surprise? As for the errors and supposed disarray of the Arab forces, the newspapers and experts had piped a very different tune just the day before, as we saw above.

In the final analysis, these two hypotheses fail to explain such a staggering victory, and are nothing more than a feeble attempt to explain the inexplicable. No one has put forward a more satisfactory explanation.

As a result of the conflict, Israel expanded its borders significantly, taking the Golan Heights, parts of the West Bank, and (most importantly) all of Jerusalem. Many in Israel attributed their victory to a miracle, some jesting that the war took six days because God wasn't working on the seventh. Whatever we choose to think of this coincidence of the war's length with the six days of Genesis—one of several such coincidences—it's understandable why some people took the idea seriously. The excerpts below give some idea of the breathless astonishment soon expressed all over the world:

- *"Our generation has had the privilege of witnessing an immense revelation of the Divine Presence with the great miracle of the return of the people of Israel to its land. [. . .] When viewed in light of the biblical prophecies, the miracle of the Six Day War is particularly striking."*[617]

- *"The great Six Day War of 1967 lasted the same amount of time it took God to create the world in the Bible."*[618]

- *"In those six days, Israel defeated three Arab armies, gained territory four times its original size, and became the preeminent military power in the region. The war transformed Israel from a nation that perceived itself as fighting for survival into an occupier and regional powerhouse."*[619]

- *"The risk [the Israelis] ran was incommensurable. The Egyptians had an extensive and advanced anti-air defense system consisting of dozens of advanced missiles and hundreds of canons, generously provided by Russia. [. . .] For their part, most of the Israeli aircraft were old French planes with very limited capability to take on a large-scale operation. If they had been spotted before the attack while on route to their objectives, many would have been shot down, and Israel would have entered the war without an air force. [. . .] The whole system of [Egyptian] anti-air detection failed. The*

---

617. Hagi Ben-Artzi, *The Six-Day War Scroll: The Story of Yom Yerushalayim and the Six Days of Deliverance* (Jerusalem: Sifriyat Bet-El, 2017).

618. https://blogs.mediapart.fr/fxavier/blog/181109/israel-la-guerre-des-six-joursla-victoire-de-david-contre-goliath.

619. Avner Cohen, "The 1967 Six-Day War," *Nuclear Proliferation International History Project*, https://www.wilsoncenter.org/publication/the-1967-six-day-war.

> hand of Providence aided the determined pilots of the air force. [. . .] The author reports that the whole Israeli leadership was shocked by the extraordinarily positive results. He cites the air force commander, general Moti Hod: 'Even in my wildest dreams, I would never have dared imagine such an impressive outcome.'"[620]

- *Le Monde*, June 9: "Rabbi Goren, general chaplain of the armed forces cried out before the Wailing Wall: 'We have waited for this moment for 2,000 years. Today, a people reclaims its capital and a capital reclaims its people. They will never again be separated.'"

- *Le Monde*, June 15: "The Israelis wonder if they are dreaming, and many are seriously asking themselves whether the Messianic age has arrived. Overnight, ancient Israel returns as the Jewish state of today…We have heard Israelis far removed from religion evoke the god of armies in talking about the war that they have just won."

- *Le Monde*, June 20, about Egypt: "Many hundreds of superior officers have been forced into retirement or imprisoned. Many of them are accused of incompetence or dereliction of duty. But for the first time, there is a serious question of high treason. Investigations are underway to determine the causes for the total paralysis that seized the Egyptian air force from the commencement to the conclusion of military operations."

This last citation from the June 20 issue of *Le Monde* confirms the mysterious blackout that paralyzed the entire Egyptian air force the morning of June 5, as reported in Hagi Ben-Artzi's *The Six-Day War Scroll*.

Conclusion

The purpose of this chapter was not to explain away—if such a thing were even possible—the many twists and turns in the strange three-thousand-

---

620. Hagi Ben-Artzi, *The Six-Day War Scroll: The Story of Yom Yerushalayim and the Six Days of Deliverance*, trans. Danny Verbov, Kurt and Edith Rothschild, Mizrachi World Movement ed. (Jerusalem: Sifriat Beit El, 2016).

year history of the Jewish people,[621] much less to judge, praise, criticize, or condemn any peoples or nations. Our goal here has been to offer a real historical enigma to readers seeking signs of God's existence. The story of the Jews' return to Israel, with all its documented improbabilities, allows us to pass judgment on a dilemma that may be formulated with the following question: *Can the fate of the Jewish people be explained with a materialist historical narrative?*

> Is it possible that the history of the Jewish people, its endurance over millennia, its return to Palestine, the prophecies that foresaw it, its staggering victories in war, the number and fame of its intellectuals, the reverse racism to which it has been subjected, and all the other circumstances recounted in this chapter are all attributable to historical accident, the product of human forces and chance?

> The reader who joins us in thinking that this history falls "outside the realm of probability" will find here an argument in favor of the existence of a God who intervenes in history. And this is why the improbable destiny of the Jewish people belongs in our overview.

> To sustain his position, the materialist has to argue that this history does not exceed the bounds of probability for events in world history.

> We admit that this story may unsettle readers steeped in Western secular rationality and strict egalitarianism. But how should we approach this case? Take the easy way out and avoid talking about it? Certainly not. In science, as in many other fields, it's often the stubborn facts resistant to analysis that lead to great advances in thought.

---

621. We reproduce below text from Pope Emeritus Benedict XVI, who is considered a great specialist on the question: *"The formula of the 'never-revoked covenant' may have been helpful in a first phase of the new dialogue between Jews and Christians. But it is not suited in the long run to express in an adequate way the magnitude of reality. If brief formulas are considered necessary, I would refer above all to two words of Holy Scripture in which the essentials find valid expression. With regard to the Jews, Paul says: 'the gifts and the calling of God are irrevocable' (Rom 11:29). To all, Scripture says, 'if we endure, we shall also reign with him; if we deny him, he also will deny us. If we are faithless, he remains faithful—for he cannot deny himself' (2 Tm 2:12f)."* Benedict XVI, "Grace and Vocation without Remorse: Comments on the Treatise *De Iudaeis*," Communio 45 (2018): 163–184.

21.

Fátima: Illusion, Deception, or Miracle?

A rendezvous with the Sun

It is around noon on October 13, 1917. Cova da Iria is usually deserted, but today this impoverished and out-of-the-way meadowland not far from the small village of Fátima, about one hundred miles north of Lisbon, is crowded. Seventy thousand people have trudged here through rain-soaked mud. They scan the sky for any sign of the prodigious event that three illiterate children—Lucia, Jacinta, and Francisco—have predicted for three months running.[622]

For weeks, rumors of apparitions and messages from the Blessed Virgin Mary have been causing a stir in this officially anticlerical country. Newspapers buzz with the story of the three children who claim that the Blessed Virgin has told them that a miracle will take place at midday on October 13, for everyone to see. The mayor and police have attempted to hush up the affair by arresting and jailing the three little shepherds who are causing the new republican order such a headache. Seized, confined, and threatened over a period of more than two days, they are pressured to say they have been lying. But despite the scare tactics, they don't change their story. They have to be let go—they are only children, after all.

Their arrest does not have the desired effect. On the contrary, it increases the public's curiosity tenfold, and a dense crowd gathers to witness the promised event on October 13. It's a mix of fervent believers and curious

---

622. The first reference appears in *Il Memoria* 1, Doc. 3, July 14, 1917.

onlookers, some reporters,[623] a professional photographer,[624] and a few local politicians. Staunch anticlericalists and Freemasons have come as well, already relishing this opportunity to see the deception fall apart and witness the unmasking of a hopelessly outmoded superstition.

Almost all the observers had traveled a good part of the way on foot, since the roads in this poor region are not very passable and means of transport scarce. The photographer sets up his bulky camera amidst the downpour. The anticlerical reporters are hoping to put an end to these old obscurantist superstitions that to their minds serve only to exploit the naïveté of backwards peasants, who still cling to a religion that has, fortunately, been stamped out nearly everywhere else.

Many of the simple people are already on their knees praying. The curious stand around doubtfully. The rest begin to mock the supposed miracle of the sun, announced for noon. It is almost one o'clock, and nothing has happened. The sky is completely clouded over, it's pouring rain, and there's not the faintest sign of a miracle. Tired of waiting, some are preparing to leave, but Lucia asks them to stay and to close their umbrellas.

At one in the afternoon, the sky begins to clear, and the sun comes out. Suddenly around one thirty, an hour and a half "late," the impossible becomes reality. In fact, the promised miracle takes place right on time, as these extraordinary events begin punctually at solar noon, which happens, today in this location, at 1:21 p.m. Before the stunned crowd, there unfolds the most spectacular, grand, and astonishing miracle that has taken place since biblical times. It appears to them that the sun begins a frantic and frightening dance that goes on for more than ten minutes. A very long time.

---

623. Two well-known newspapers had reporters on the scene: *O Século* (Avelino de Almeida) and *Diário de Notícias* (journalist not identified).

624. Judah Bento Ruah, a trained engineer, is an atheist of Jewish background, from a well-known family that fled to Portugal to escape persecution in Spain. He replaced his uncle, a photographer, at *O Século*, the great anti-clerical and Masonic daily founded by the Grand Master of the Masonic Grand Orient, Sebastião de Magalhães Lima, Minister of Education in the Republican government since 1915 (see https://www.wikiwand.com/pt/%20Sebasti%C3%A3o_de_Magalh%C3%A3es_Lima). It is to him that we owe these eight photos of the crowds during the miracle, a real feat considering the technology available at the time.

There is ample photographic evidence (several photos are included below) of the crowd experiencing this event, which they described as uncanny and terrifying.

At the epicenter of a powerful shockwave

The effects of this phenomenon reached far beyond the small village of Fátima. It had repercussions at the national level: religious persecutions in Portugal ceased, and the faith experienced a renewal. On the international level, the shockwave of Fátima rippled miles away across Europe, even many years later. The eventual collapse of the USSR followed the long-delayed fulfillment of a request the Blessed Virgin made of the little shepherds. She had insisted that the pope and the bishops together consecrate Russia to her. John Paul II finally granted her request in the manner that she had specified, but more than forty years after Mary had expressed her desire.[625]

In 1986, scarcely two years after he had done so, a spectacular transformation began with the policy of glasnost in Russia and the Solidarność trade union in Poland, culminating in 1990 in the peaceful and complete collapse of the Soviet bloc, to the amazement of the entire world.

What really happened in Fátima on October 13, 1917?

It is certainly tempting to dodge the question: "How should I know? I wasn't there!" or, "It's impossible to say. It happened too long ago, far away from here," or "Miracle stories? How can they be included in a book that

---

625. Pius XII consecrated the world to the Immaculate Heart of Mary on October 31, 1942. On July 7, 1952, he repeated the consecration in his apostolic letter *Sacro vergente anno*, which states: *"Just as a few years ago we consecrated the entire human race to the Immaculate Heart of the Virgin Mary, Mother of God, so today we consecrate and in a most special manner we entrust all the peoples of Russia to this Immaculate Heart."* Russia would be consecrated again to the Immaculate Heart of Mary by Paul VI and three times by John Paul II. Only the last consecration in 1984, however, was considered valid because it was the only one made in union with all the bishops of the world, as Lucia pointed out. Frère Michel de la Sainte Trinité, *The Whole Truth About Fátima: Vol. 3, The Third Secret* (Buffalo: Immaculate Heart Publications, 1990), chap. 7.

wants to be taken seriously?" We hope our reader has the courage to set aside such simplistic and facile prejudices. Actually, as with the chapter on Jesus: who could he be? is, the answer to the question of what actually happened at Fátima is within the reach of anyone with the patience to read this chapter through to the end. In fact, there are very few possible answers, and conditions for eliminating almost all of them are optimal.

There are six potential answers:[626]

1. Nothing happened at all; it's a tall tale.
2. It was a natural phenomenon, a disturbance in the solar system resulting from cosmic events.
3. It was an unusual meteorological phenomenon.
4. The crowds experienced a collective hallucination.
5. It was a hoax.
6. It was a miracle.

And the conditions for evaluating the possibilities are ideal since:

- This prodigious event took place during the twentieth century in Europe, not so long ago, and not so far away.
- It took place in the presence of a huge crowd that included reporters and photographers.
- It was announced in advance.
- Poor and illiterate children foretold the event, and never received any benefit from it.
- Many resolute sceptics were present.
- The anticlerical milieu adds value to the testimony of the witnesses, especially those who were hostile to the idea of miracles.

---

626. We have excluded from the list of possible explanations some that are too unrealistic, such as the theory that the event was a UFO sighting, though several books have been written on the subject (see, for example, Gilles Pinon, *Fatima, un ovni pas comme les autres ?* (Paris: Osmondes, 2002), as well as the theory that it was a manifestation of the devil; as the existence of the devil presumes that of God, the question at the heart of the present book would be resolved by the simple acceptance of that possibility.

A miracle can be evidence

Is a chapter on miracles out of place in a book that takes a rational and scientific approach to evidence for the existence of God? Admittedly, the majority of intellectuals and scientists are put off by the idea of miracles. But supposed miracles deserve to be examined rationally. We all believe that the Universe is a rational place and that all events that occur in it are governed by universal and immutable laws. Precisely because of this, if these laws are violated in a well-documented way and there is no possible alternative explanation, reason demands that we accept the simplest account: the existence of a supernatural being capable of producing such a miracle. Denying a priori that a miracle can happen, on the basis that God cannot exist, is not a rational option.

The point of the miracles recounted in the Bible, the Gospels, and the history of the Church is often to demonstrate God's existence to those who witness them. And don't skeptics often demand clear proof that God exists? Well, that is just what the miracle of Fátima is! The Blessed Virgin herself affirmed this when she appeared on July 13, 1917: "*In October, I will perform a miracle that all may see and believe.*"[627] One object of the miracle was to induce the witnesses to believe the Virgin's message, but more importantly it was meant to give them, and us, evidence for God's existence. That is why the miracle of Fátima has a place in this work. As a miracle, it constitutes very strong evidence for the existence of God.

I. The political context of Portugal in 1917

At the beginning of the twentieth century, Portugal was dominated by anticlericalism

At the start of the twentieth century, militant anticlericalism was on the rise both in Europe—France, Spain, Italy—in Russia, and in new world countries like Mexico. In Portugal, anticlericalism was particu-

---

627. Translated from *Mémoires de sœur Lucie*, 4ᵉ mémoire, 7ᵗʰ ed. (Torres Novas, September 2008), II, Histoire des apparitions, chap. 5, July 13, 184.

larly virulent, part of a long and bitter struggle between the Masonic lodges and the Catholic Church.

In 1908, King Carlos I was assassinated along with the heir to the throne. His successor, Manuel II, was expelled in 1910, and Portugal was proclaimed a republic.

While the people of the countryside remained Catholic, the revolutionary bourgeoisie of the big cities were Freemasons, who at that time wielded considerable clout. The new regime immediately began to implement a plan of forced secularization more rigorous than even the laws adopted by the French Republic in 1905:

- In 1910, religious congregations dedicated to education were expelled from Portugal.
- All religious education was forbidden.
- The Church's property was confiscated.
- Civil marriage was established and divorce was legalized.
- Separation of church and state was proclaimed in 1911.
- Diplomatic ties were cut with the Vatican.

Backlash against the regime's violent anticlericalism caused political instability, and Portugal went through eight presidents and almost fifty governments in sixteen years.

In 1916, Germany declares war on Portugal

Though war had raged in Europe since the summer of 1914, Portugal did not enter the conflict alongside the Triple Entente (France, the UK, and Russia) until March 1916. Despite the formation of the Sacred Union government, the war effort aggravated the economic crisis and social disorder even as the country sent more than fifty thousand soldiers to France.

II. The local context

Fátima at the time of the miracle was a small village of two hundred inhabitants, located in central Portugal about 100 miles north of Lisbon. The local children spent more time herding sheep than attending school. Lucia de Jesus dos Santos (born March 28, 1907), her cousin Francisco Marto (born June 11, 1908), and his sister Jacinta Marto (born March 5, 1910), could neither read nor write. When the Virgin first appeared to them, the children were ten, eight, and seven years old.

III. The days leading up to October 13

May 13, 1917 at midday: the first apparition

On May 13, 1917, the three little shepherds were watching over their flock at Cova da Iria. At midday, the Blessed Virgin appeared and spoke to them, instructing them to return to the same place at the same hour on the thirteenth of each month, for the next five months.

Once home, Jacinta reported these facts to her parents, and Francisco corroborated her story. The report soon reached the ears of Lucia's sister Maria. When questioned, Lucia told her parents everything, and the news spread throughout the village.

June 13, 1917: the second apparition

The following month found the children back again, accompanied by nearly sixty villagers who came along out of curiosity. The children saw and spoke to someone who remained invisible to the onlookers. Nevertheless, they did perceive a great brightness, the murmur of a conversation, and some other luminous phenomena at the beginning and end of the apparition. This was enough to make a strong impression and stir up local interest.

July 13, 1917: the third apparition

News of the previous two apparitions had spread, and between two and five thousand people accompanied the children. As previously, the children were the only ones to see the "*Lady in white*." For a single instant, a few witnesses[628] saw an expression of terror come over their

---

628. Maria da Capelinha, for example.

faces. It was later revealed that the children had experienced a momentary feeling of panic when the Virgin showed them a vision of Hell.

This time, Lucia asked Our Lady if she would work a miracle. The apparition responded, "*In October [. . .] I will perform a miracle that all will see and believe.*" The news spread rapidly throughout the country.

August 13, 1917: the fourth apparition

The provincial administrator—a formidable, influential republican and Freemason named Artur de Oliveira Santos—was enraged at the importance attached to this matter and disturbed by the number of people who had come to witness the previous apparition. In an attempt to put an end to the affair, he arrived at the childrens' home at nine in the morning on the thirteenth and escorted them to his car. They were interrogated first at his home and then at the public prison of Vila Nova de Ourem. They were released on August 15.

Meanwhile, not knowing of the arrest of the little shepherds, a crowd of five to eight thousand people had made its way to Cova da Iria. As the children were not present, no apparition was seen, but the crowd did witness many extraordinary sound and light phenomena.

September 13, 1917: the fifth apparition

On September 13, twenty-five to thirty thousand people gathered at Cova da Iria. The majority of the witnesses experienced similar phenomena to those that occurred on August 13.

IV. October 13, 1917: the story of an incredible day

News of the impending miracle spread throughout Portugal. At noon on October 13, 1917, a large crowd—estimated at thirty to seventy thousand people[629]—gathered in that isolated place about two hours

---

629. Seventy thousand, according to Dr. Joseph Garrett, former professor of mathematics at the

by foot from Ourem, the closest town. The city bus of Torres Novas was pressed into service for the occasion. Someone counted the vehicles strung along the path: there were more than one hundred automobiles, one hundred thirty-five bicycles, and two hundred forty animal-drawn vehicles.

It hadn't stopped raining since 8:30 a.m., and the crowd was soaked to the skin.

The narrative of the prodigious event

Canon Casimir Barthas was present that morning and described the extraordinary event in his book *Il était trois petits enfants*: *"Suddenly the rain stops and the clouds, impenetrable since morning, dissipate. The sun appears at its zenith, like a silver disc that can be directly stared at without dazzling the eyes, and all at once it begins to turn like a wheel of fire, shooting sprays of light in all directions, changing color several times. The sky, the earth, the trees, the rocks, the group of visionaries, and the great crowd are successively colored yellow, red, blue, purple...*

The sun stops for a few moments. Then it takes up its dance of light again in an even more dazzling manner.

It pauses again to set off a fireworks display so fantastic that no pyrotechnician could imagine its like.

How to describe the multitude's reaction? The crowd of seventy thousand visionaries looks on ecstatic, motionless, holding their breath...

All of a sudden, the entire crowd, without exception, has the sensation that the sun has come loose from the sky and is falling upon them in zigzagging bounds.

University of Coimbra, who was present that day.

A great cry bursts from the crowd. "Miracle! Miracle!" some cry…"We are all going to die!" is heard from the other side… Others say, "How beautiful!"

Who can describe the emotional state of all this crowd? An old man, till then an unbeliever, waves his arms in the air, crying: "Virgin of the Rosary, save Portugal!" And similar scenes play out all over the plain.

The spinning of the sun, with the interval of rest, had lasted ten minutes. It was observed, let us repeat, by every single person present without exception: believers, unbelievers, peasants, townspeople, men of science, and even freethinkers. All, without any kind of preparation, without any other suggestion than the call of a girl inviting them to look towards the sun, perceived the same phenomena, passing through the same phases, at the day and at the hour announced a few months before as a great wonder.

The subsequent canonical investigation into the miracle revealed that the sun's movements had been seen by people who were five kilometers or more away and therefore unaware of what was happening at the Cova da Iria. They could not in any way have been influenced by the power of suggestion or fallen victim to collective hallucination.

The investigation also brought to light a very curious fact attested to by all those questioned on this subject. After the crowd recovered from their amazement enough to realize what was happening on earth, everyone observed with amazement that their clothes, all soaked by rain a few minutes ago, were now absolutely dry. No one minded at all that they'd been so wet."[630]

The professional photographer who had lugged his equipment all the way to Cova da Iria captured the scene on that famous October 13.

---

630. Translated from Casimir Barthas, *Il était trois petits enfants* (Montsûrs, France: Éditions Résiac, 1940).

Furthermore, testimonies were collected from witnesses for thirty-four miles all around.

John De Marchi reports that the phenomena were observed in the village of Alburitel (eleven miles to the east of Fátima) by the schoolteacher and her pupils, and by "*a godless man who had spent the morning making fun of the simpletons who had gone to Fátima just to see a girl.*" When the miracle happened, his arrogance collapsed: "*He was numbed, his eyes riveted on the sun. I saw him tremble from head to foot. Then he raised his hands towards Heaven, as he was kneeling there in the mud.*"[631] Again at Alburitel, Fr. Inácio Lourenço Pereira recounted: "*I was barely nine years old then. I attended primary school in my hometown, a small village perched on a lonely hill just in front of the mountain of Fátima, ten or eleven kilometers away. Around noon, suddenly we were alarmed by the cries and shouts of the men and women passing on the public road in front of the school. [. . .] Outside, in the square, the people who had gathered were crying and shouting, pointing to the sun, without even hearing the questions our distressed teacher put to them. [. . .] I stared fixedly at the sun; it seemed to me pale and deprived of its dazzling light; it looked like a snow globe spinning around. Then, suddenly, it seemed to descend in a zigzag way, threatening to fall to the ground. Distraught, absolutely distraught, I ran to stand among the people. Everyone was crying, waiting for the end of the world to come at any moment. [. . .] During the long minutes of the solar phenomenon, the objects close to us reflected all the colors of the rainbow... Our faces were sometimes red, sometimes blue, sometimes yellow, etc. These strange phenomena increased our terror. After ten minutes, the sun returned to its place just as it had descended, still pale and dull.*"[632]

Afonso Lopes Vieira (1878-1946), a jurist by training, was one of Portugal's most famous poets and a convinced atheist. He worked for

---

631. John de Marchi, *The True Story of Fatima* (1947; repr., Buffalo, NY: Fatima Center, 2008), and *Fatima: The Full Story* (Washington, NJ: AMI Press, 1986), 57.

632. Translated from Casimir Barthas, *Fátima, merveille du XXe siècle* (Toulouse: Fatima-Éditions, 1952), 136.

the anticlerical government until 1916 as a secretary in the Chamber of Deputies. He watched the marvel himself from his property of Sao Pedro de Moël, located near Leira at 18 miles from Fatima. He stated: *"On that day, the 13th of October, while I hadn't remembered the little seers' predictions, I was amazed by a dazzling sight in the sky, something entirely new for me, which I watched from this balcony."*[633]

He converted and even had a chapel built on his property. His story was recounted by the journalist Paula Sofia Luz.

The places where these witnesses reported seeing the event are located in four different directions and at various distances from Fátima. The miracle of the sun was therefore visible in an area of about twenty-five miles in diameter.[634]

And you will recall one important detail: the miracle of October 13 happened on the very day and at the exact time announced by the children months earlier.

These astonishing facts were reported over the following days by all the Portuguese press, Christian and anticlerical, as all the journalists present at the event saw exactly the same thing.

Anticlerical reporter Avelino de Almeida witnessed the event and wrote a sensational story for the secular daily *O Século*, where the miracle of the sun was front-page news; his report was entirely consistent with what the crowd saw.[635] The sheer number of newspaper articles is proof in itself: something incredible happened at Cova da Iria on October 13. Articles written by journalists hostile to the Church are especially helpful for our investigation: in setting themselves with all

---

633. Translated from Jean de Marchi, *Témoignages sur les apparitions de Fátima* (Fátima: Missões Consolata, 1994), 204.

634. We might also cite three contemporary private letters that also report that the Miracle of the Sun was visible in Torres Novas, a town neighboring Fátima (Documentation de Fátima, 3.1, Doc. 319, Carta de Adelaide Grego a uma amiga, Nov. 24, 1917; Cartas de Gonçalo Xavier de Almeida Garrett ao Padre Manuel Formigão, Doc. 334, Dec. 3, 1917, and Doc. 355, Jan 1, 1918).

635. The photos were published later in *Ilustração Portuguesa*, October 29, accompanying a new article by Avelino de Almeida.

their might against the possibility of a miracle, they allow us to sketch its contours in silhouette, as it were.

V. An informative press review

Below you will find a collection of the most salient reactions published in the secular, anticlerical, liberal republican press, plus the account from *A Aurora*, an anarcho-syndicalist weekly.

The accounts make for rather repetitive reading, but going through them all is essential because they demonstrate the irrational position of those who refuse to confront the reality of the event, confusedly combining alternative explanations without ever making a choice between illusion, hallucination, hoax, or meteorological phenomenon. We will focus on this collective incoherence.

All the newspaper articles cited below are available in the National Library of Portugal, the University of Coimbra General Library, the Public Library of Porto, or the Sanctuary of Fátima. We have translated them from the original Portuguese, highlighting the most important passages.[636]

Here are the most significant excerpts, presented in chronological order:

1. **July 21, 1917**, *O Século* entertains a *"financial scam,"* noting that the authorities have the case in hand: *"For a long time there has been a persistent rumor in this town that the mother of Jesus Christ was going to appear to two children. [. . .] This rumor, as might be supposed, aroused general curiosity, [. . .] drawing thousands of persons to the place—some, the unbelievers, to see something interesting; others, the religious, out of faith and devotion. [. . .] On the 13th, the day designated for Our Lady's apparition, we*

---

636. The translation of the Portuguese sources in this chapter was produced with help of Professor José Eduardo Franco (CIDH – Universidade Aberta/CLEPUL, Faculty of Letters of the University of Lisbon), with the assistance of Dr. João Diogo Loureiro (CLEPUL, Faculty of Letters of the University of Lisbon), Brother José Luis de Almeida (Bibliothèque du Saulchoir) and Helena Jesus (researcher at University of Paris IV).

went to the indicated spot. Thousands of people were already gathered, having dragged themselves there, driven by the desire to see what there was to see, some of them coming from far away villages. [. . .] At this was heard a sound like the rumble of thunder, and immediately the two children, who stood by a holm oak tree surrounded by many flowers, almost paradisiacal, broke into a distressing cry, making convulsive gestures and falling afterwards into ecstasy. Several people asked many questions of one of the children who had the privilege of seeing and hearing the saint, to which she responded that she had seen a kind of extremely beautiful doll, who spoke to her. She had, she said, a radiance about her head and called her close to her in a very fine and melodious voice. Among the many things that she said, the most important was to announce that she would appear on the 13th of the following month, in the same place. [. . .] The case seems highly ridiculous. Indeed, I would not have believed it were it not for the fact that the child deserved utmost confidence for being sincere and truthful, and that she was corroborated by others who told the story using the same words and citing the same facts. However, it is my opinion that this is a premeditated financial scam, with proposed revenue streams from the bowels of the mountains—a spring of mineral water recently discovered by some astute individual who, under the guise of religion, seeks to transform the Serra d'Aire into a miraculous pilgrimage site like the tedious Lourdes."*

2. **August 19, 1917,** an *O Mundo* headline reads "*Miracle Scam,*" with the subtitle, "*How to lead people astray—What has been happening in Fátima.*" They are already beginning to develop the hypothesis of hallucination or deception: "*Is it the hallucination of poor children who are regular churchgoers, or is it a clerical scheme? It is the duty of the authorities to find out, as we are sure they will, with as committed a republican at the head of the district as Mr. Artur de Oliveira Santos, who has always discharged his duty in an exemplary manner. The case certainly smacks of clerical exploitation.*" Later on in the same article, on "*how the scam started:*" "*The cler-*

ical scam started on May 13th. Three children [. . .] were walking on the vast plain of Cova de Santa Iria, grazing their docile sheep, when a lady dressed all in white appeared, according to Lucia. The lady told the little shepherds that they should learn to read and write and that on the thirteenth of each month she would appear to them again, beside an oak tree. On the thirteenth of October, she would descend for the last time from heaven to earth to bring peace to the world—to end the war."

A little further on: *"There is obviously more going on here than a child's fantasy. An illiterate child would not come up with these ideas all on his own or stick to them so confidently. They were coached ahead of time to play this role. What's more, Lucia is a frequent churchgoer. She went to confession four times in just one month. And she insists that she is keeping a secret that she cannot reveal until October 13. The priests swarmed about the place and—curious indeed!—found in the little one a resemblance of that chatty parakeet Bernadette of Lourdes."*

3. **August 23, 1917,** *O Debate* calls it *"a farce"* and *"a swindle"*: *"We refer to that ridiculous farce, whose success is already headline news, of the apparition of Our Lady in the outskirts of Vila Nova d'Ourem—played to an exclusive audience of just three children!"*

4. **September 16, 1917,** the *Semana Alcobacense* runs the headline: *"Once Upon a Time There Was a 'Miracle'"* and calls it a hoax: *"In the end, the so-called miracle of Fátima was a mirage. [. . .] Our Lady [. . .] promised to continue to meet her faithful there and appointed the thirteenth of each month for this purpose. News got around easily by word of mouth, and some newspapers eagerly spread the story, so that last Thursday when the first of the promised visits was to take place, thousands of people left their fields and houses to go to Fátima to see Our Lady, have a chat with her, and, who knows, perhaps give her a handshake [. . .] But one o'clock came and went, the time the spectacle was scheduled to begin, then two o'clock, three passes, and the entire afternoon goes by; you see the*

faces of all those people remembering the money they have spent and, bodies overcome by fatigue, [. . .] in the end they fall into the most outrageous swindler's story that has ever been spun."

5. **September 20, 1917,** *O Debate* speaks of a *"distasteful hoax"* and reassures its readership that the whole affair will *"end on its own:"* *"Beside a holm-oak appears Our Lady [. . .]whom nobody has seen yet, though thousands of people have gone there to see her. [. . .] The ridiculous and dishonorable farce that the priests have recently invented [. . .] is thereby mystifying those who due to their weakness of spirit are easily influenced and subdued through every cunning device [. . .] It goes without saying that we refer to the veritable swindler's tale to which the people in the outskirts of Vila d'Ourem have fallen victim, the story of the appearance of Our Lady on the thirteenth of each month at a certain place in Serra de Aires [. . .] Of course, no one has actually seen her! [. . .] The ridiculous farce has been staged so successfully that hundreds of deceived people have flocked to the place. [. . .] We do not call for the rigor of harsh repressive measures to address a scam like this. It will end on its own, when simple and sincere souls become convinced that it has been a scandalous deceit played on their good faith."*

6. **September 22, 1917,** *O Marinhense,* under the headline *"Miraculous!,"* also believe the *"marvelous"* case to be *"a farce:" "In a place near Fátima, around Vila Nova de Ourem in our district, a wonderful thing happened; a saint, covered in flowers, dressed in a tunic of many colors, of shining silk, and wrapped in a halo of light, appeared to some shepherds. She came down from the heavenly regions neither to predict the end of the war nor to reveal which number will hit the next lottery jackpot! She came only to rebuke disbelief, the sins committed by mortals who no longer compete to fill the church coffers, to feed the most holy Jesuits, or keep the Little Sisters of Charity alive, etc., etc. And so, each thirteenth of the month—fateful date—she returns to this place, to lay down these censures in the hearing of many people. And last Thursday it was estimated that the number of people who went to Fátima to see the*

holy virgin was twenty thousand!!! [. . .] Twenty thousand people who went to see the miracle of a scam-apparition!"

7. Furious at the enormous crowd of twenty thousand, the reporter feels obliged to issue this materialist call to order:
"*We must make people see that heaven or hell, or the land of good or bad actions is the world. That saints do not exist and that miracles are nursery tales. [. . .] There is nothing after death. It is urgent that we put an end to this infamous speculation and punish those who are not ashamed to deceive the ignorant. The apparition of Fátima is a lie, whose objective is to extort the meager savings these ignorant individuals have—at the expense of their well-being.*"

8. **September 27, 1917,** *O Debate* speaks of "*religious opportunism*": "*Heaven may very well be that delightful mansion of pleasure and happiness the vendors of religion around here preach. But it seems that "Our Lady" is bored there, for from time to time she walks this "Valley of Tears." [. . .] This year she chose to vacation in the neighboring municipality of Ourem, where she makes her appearance on the thirteenth of each month, according to the mountebanks' story. [. . .] What is happening in the municipality of Ourem is the same old ignoble and stupid religious scam. It's not worth fighting with.*"

9. **October 14, 1917,** after the miracle has already occurred, the *Jornal de Leiria* publishes an article too late that predicts failure for October 13th, "*The 'Lady' of Fátima:*" "*As we write these lines we have the presentiment that during the next six months the 'lady' will not return to make her apparitions, because she'd freeze in the cold, and the pilgrimages to Fátima could not be carried out with the same success that they have enjoyed until now.*

10. **October 13, 1917, the morning before the miracle, *O Século* published an essential article:** "*What do these apparitions consist of? The Virgin, in the form of a beautiful lady, has since last May on the thirteenth of each month descended to this valley of tears to show herself to three children, who in a voice of singular sweetness instructs*

them to pray and to give notice of her presence to all, warning believers and non-believers both that on the thirteenth of October—today—she would reveal the ultimate reason behind her visits and comfort by means of her heavenly vision all those who wish to be in grace. The fame of the miracle runs from north to south, and from all parts of the country countless persons have flocked to Fátima, thousands of individuals arriving to gather on the lucky moorland, many of whom claimed to have witnessed strange things."

Then: *"What will happen today in Fátima? We'll soon find out. Pious persons hope that the Virgin Mary will let them know when the war will end and that her kindness will go so far as to tell them when there will be peace. [. . .] And there are also those who imagine both a huge, sumptuously appointed church—constantly filled—and great hotels with all the modern conveniences, and stores filled with a thousand and one pious artifacts, souvenirs of the Lady of Fátima—and the construction of a railroad branch that takes us to the future miraculous sanctuary..."*

This last article and the next were penned by **Avelino de Almeida**,[637] a celebrated journalist of the first third of the twentieth century. Working with several important papers, including *O Século* (where he was editor-in-chief), *A Capital*, *O Primeiro de Janeiro*, and the *Jornal de Notícias*, he also distinguished himself as a theater and film critic, founding and directing the magazine *Cinéfilo*. He attended seminary but eventually distanced himself from Catholicism, becoming virulently anticlerical. He became a Freemason (Irradiação lodge) and founded *A Lanterna* (1909–1910), a radically anti-Catholic

Avelino de Almeida (1873-1932)

---

637. For more on Avelino de Almeida, see A. C. Vicente, "Almeida, Avelino de (1873–1932)," *Enciclopédia de Fátima*, C. Azevedo and L. Cristino (Parede, Portugal: Principia, 2008), 22–24; "Almeida, Avelino de," *O grande livro do espectáculo: personalidades artísticas: século XX*, L. Reis (Lisbon: Fonte da Palavra, 2010), 1:38–41.

paper. But the man was capable of honest reporting, as reading this first relatively balanced article shows, even before the next one, which will prove an earthquake.

11. **October 15, 1917, a bombshell**. The miracle of the sun is front page news, and *O Século* recognizes the reality of what happened: "*Amazing things! How the sun danced at noon in Fátima*": "*The dispassionate calculations of educated people completely alien to mystical influence puts the crowd at thirty or forty thousand persons. [. . .] And then we witness a unique spectacle, unbelievable for those who didn't see it themselves. [. . .] The sun resembles a matte silver plate, and it is possible to look on the disc without the slightest effort. It doesn't burn, it doesn't blind. You might say there was an eclipse taking place. And behold, a tremendous cry goes up, and the spectators closest by are heard to shout, 'Miracle, miracle! Wonderful, wonderful!' In the dazzled eyes of the people, whose attitude transports us to biblical times and who, white with astonishment, heads uncovered, face the sky, the sun trembles, the sun makes never-before-seen movements completely outside the bounds of cosmic laws—the sun "danced," as the peasants tended to put it. [. . .] The greatest number claims that they saw the trembling—the dance of the sun. Others, however, say that they saw the smiling face of the Virgin herself, swear that the sun spun like a wheel of fireworks, and that it dipped down, almost to the point of burning the earth with its rays. [. . .] It remains for the competent authorities to pass their judgement on the macabre dance of the sun that today in Fátima made hosannas burst from the breasts of the faithful and easily impressed—as trustworthy persons have assured me—the freethinkers and other persons without preoccupations of a religious nature who came to the now-celebrated moor.*"

12. **October 15, 1917** the *Diário de Notícias* felt compelled, despite itself, to run the headline "*The 'Miracle' of Fátima*," with the subtitle "*More than fifty thousand people come to the site of the apparition.*" The article acknowledges the "*extraordinary*" case and

testimonies of *"thousands and thousands of pilgrims,"* but speaks of *"suggestions"* without providing any explanation: *"Despite the light and irksome rain that began to fall early in the morning, an extraordinary number of people came to the parish of Fátima to witness the extraordinary event of the apparition that since the fifth week of Ascension has occupied so much of the people's attention [. . .] As a great number of people had their umbrellas open, the children ordered them to close them, and wonderful thing, according to the testimony of thousands and thousands of people, the sun appeared a matte silver color, with a circling agitation as if touched by electricity, according to the expression employed by illustrious people present at the event. And the thousands of people who experienced this, who knows, maybe they were blinded by the first light they had seen that day, fell to the ground weeping and raised their hands, which instinctively clasped together, into the air. [. . .] There were even people [. . .] who seemed to see the sun abandon its supposed orbit, break through the clouds, and descend to the horizon. This suggestion from these visionaries spread to others whom they told about this phenomenon. [. . .] The "miraculous hour" has passed."*

13. **October 15, 1917**, the review O *Portugal* ran the title, *"The sun in revelry,"* giving an ironic take on recent events: *"Certain inspired and lucky individuals had the chance yesterday to see the sun 'dance' there for the bands of Vila Nova de Ourem. [. . .] What surprises us is not that a dense and noisy crowd flocked to the place of wonders to participate in the heavenly revelations, so accustomed as we are already to the manifestations of indigenous belief; what surprises us is that the sun, a respectable star with all its papers in order, should take a role and start dancing like a village party reveler, despite his considerable age of thousands of centuries, which, if it has not given him white hair, has at least left certain suspicious spots upon his face that astronomers interpret as a sure sign of his old age. For many years now the sun has been considered an unmoving star, relative to our planetary system, and the*

affirmation of this truth cost its discoverer some unpleasantness. Now three little rustics come and take science out with the garbage, and through their influence at the heavenly court make the sun dance over the chosen spot at Fátima."

14. **October 16, 1917,** *A Capital* didn't know how to spin the story, so concluded that it was unable to explain what had happened: *"The case of the apparition at Villa Nova d'Ourem is raising controversies. We know that three young sages see an image and talk to her; we know that this repeats itself on particular dates; we know, finally, that thousands of people saw the sun 'take some turns.' This is clear, logical, and sure. [. . .] What we don't know is the name of the jokester behind this tremendous spectacle."*

15. **On October 16, 1917,** *O Portugal*, with no attempt at providing the slightest explanation, takes pleasure in a mocking headline, *"Prudent Reservation:"* *"In transcribing certain passages from Seculo's account of the miraculous events at Fátima, the* Dia *[the other paper] declares itself impressed by the facts and excuses itself from making a comment. This attitude is at least prudent. The small, rustic visionaries announced the end of the war, but they make no comment on the restoration of the monarchy; hence the reluctance on the part of the* Dia, *which naturally awaits on the thirteenth of some future month a substantial revelation on the subject before making its comments. Moreover, it is not just for the* Dia *to be reserved, since the Seculo's editor, who went to Fátima, is also waiting for experts to comment on the disordered movements of the sun, which the mob claims to have seen."*

16. **On October 18, 1917,** *Democracia do Sul* identified a suspect: *"The Jesuits:"* *"Near Vila Nova de Ourem, in Fátima, [. . .] there has been an infamous Jesuit scam with the fantastic and ridiculous sighting of a purported saint who has the idea, according to what they say, of appearing to children and telling them things about Portugal. This scam has gone on for quite a while now, with the criminal approval of the local authorities, who have not put a stop to such an infamy,*

ending once and for all this farce, which the Seculo is currently giving most helpful press through the force of its enormous publicity."

17. **On October 18, 1917,** *A Lucta* runs the headline *"The Lady of Fátima"* and reluctantly and scornfully admits the facts of the case: *"And what happened on that bit of uncultivated ground, where the Virgin had promised to come down and speak to the people of Portugal? Around two o'clock in the afternoon, one of the three children, who was praying near the oak, made a gesture, and the crowd knelt for a time. Then the sun, breaking through the clouds, burned with a brighter glow, and, surrounded by a dark circle, immediately became as pale as the moon, before turning to a watery blue, a sickly yellow and other faint shades. Finally irregular, jerky movements came over the sun—it trembled, trembled, and stood still. With the sudden changes in color, the landscape changed and lost its sharpness, its harshest contours softened into a vague mist. The women clasped their hands, white with astonishment, their pleading eyes fixed upon the sky, their postures dissolved in fervor or ecstasy [. . .] The miracle limited itself to magical mutations of the environment, the strange gymnastics of the sun, the pyrotechnics made of the clouds, the wonderful variations in the light. [. . .] But all my informants swear that they saw wondrous things happen to the sun—something which disturbs my own critical view not at all. I know very well that our people are poetic souls much affected by consoling images. [. . .] A prominent doctor from my area, once the most upstanding example of local Jacobinism, told me that he screeched like a man possessed, hands crossed, eyes fixed on the mark: 'There goes the Virgin! There she goes.'"*

18. **On October 22, 1917,** *O Mundo* in Lisbon ran the headline *"A Clerical Scam"* about *"the supposed apparition of the 'Virgin' in Fátima."* No explanation or justification, only a perceived obligation to undermine the credibility of something that seemed intolerable: *"This case of perfidious cheating cannot be tolerated in any way, and the authorities of the Republic must take immediate and decisive action to eradicate this evil before it spreads."*

19. **On October 26, 1917,** *O Democrata* ran the headline "*The Miracle of Fátima,*" mentioning the prophecy about the swift return of the soldiers but calling the whole thing a "*farce*" that couldn't stand up to "*rational*" scrutiny: "*They saw the sun dance, turn, descend, rise up—but why? As a logical consequence of looking long and fixedly at its light. As for the rest? They heard, they saw that Lady who decided to show herself to three little simpletons—because there is no one in this world more worthy of hearing and seeing her? [. . .] The virgin comes to tell the laborers of Fátima that the war will end soon and our soldiers will come home! [. . .] What a sad spectacle we're offering of religious belief and faith itself, listing facts that melt away at the slightest breath, the gentlest rational examination!*"

20. **On October 26, 1917,** *Chronica Carta de Lisboa* ran the headline "*The 'Miracle' of Ourem,*" gave a detailed account, and tried to chalk the vision up to "*the power of suggestion*" and "*incomprehensible crowd psychology:*" "*All those people stood at attention, in most anxious expectation, and all their umbrellas closed when the three visionaries, kneeling under an arch of greenery, ordered it. And oh, power of suggestion! Oh, incomprehensible crowd psychology!—the overwhelming majority of all those people saw it. Which proves that religious belief is still, and will be, so entrenched in the spirit of the masses, which has an extremely strong power of suggestion.*"

21. **On October 29, 1917,** Avelino de Almeida published in the national journal *Ilustração Portuguesa* a long four-page article (pages 353–356) on the October 13 miracle of the sun, accompanied by ten photos taken by Judah Ruah, his on-site photographer. The images show the dense crowd watching the sun and falling to their knees after witnessing a "*colossal miracle.*" In the article, Avelino gives little space to description of the event (which he had already detailed in previous articles), but he called upon the Catholic Church, the scientific community, and freethinkers to study the phenomenon observed by the crowds.

22. **On November 1, 1917**, *Ecos de São Pedro d'Alva* criticized *"the insipidity of Fátima"* and suspected a conspiracy: *"What has happened on the moor of Fátima, municipality of Vila Nova de Ourem, seems inconceivable. According to the report of O Seculo, [. . .] on the thirteenth of this month between forty and fifty thousand people flowed in from various parts of the country to witness a great miracle!!!—or was it the apparition of the Virgin!!! [. . .] There is no reason for this spectacle that lowers us in the eyes of the civilized world to go on, and it is the authorities of Vila Novem de Ourem who, in our view, are responsible for putting a stop to it, beginning with the detention of the three children who saw the sun dance and other things, until they confess who instructed them to see Our Lady, then calling the relevant court where the author or authors of this comedy will answer for their actions."*

23. **On November 11, 1917**, *A Aurora*, based in Porto, spoke of *"the apparition of Fátima,"* attempting to make it out to be an atmospheric phenomenon: *"[The pilgrims] had arrived there by every means of transportation, already predisposed for the contagious electric shocks of collective suggestion. And then there was the vision and miracle of Fátima, the new Lourdes, that will perhaps be marked by another pious basilica. All those good people, poor people[. . .] positively saw the sun as a matte silver disc that broke through the clouds of a rainy and dreary day, danced, fell, shook in hysteric convulsions, while, according to the story of the eldest of the children on someone's lap, the Virgin announced that the war was over and that the soldiers were on the way home. [. . .] If it were possible to reason with believers and visionaries, we might object, with simple logic, too easy for the troubled soul of a mystic, to the pettiness of a fruitless meteorological marvel."*

24. **On November 11, 1917**, the journal *O Mundo* speculated that the miracle was the result of invention, suggestion, or bribery, and proclaimed it to be: *"A propoganda session promoted by the Civil Registry Association against the dreadful way that*

the unscrupulous are abusing popular naïveté, hoping to better enslave and exploit the people of Portugal, casting it once more into the dense darkness of fanaticism, superstition, and belief, on the pretext of ridiculous miracles and ghostly apparitions they invented and for which they chose as their theater the little village of Fátima and perhaps bribed the parents or caretakers of three poor children."

25. **On November 18, 1917,** *O Mundo* published a story on "*the miracle of Fátima,*" deploring the importance the affair had assumed: "*The miracle of Fátima is still a topic of debate, when it's certain that this insignificant incident of provincial life ought to have been taken care of a long time ago, with two swift kicks in the rear to the brats who started it.*"

26. **On December 2, 1917,** a pamphlet was circulated against Fátima: "*Against the filthy profiteering of the laughable farce of Fátima, the Civil Registry Association and the Portuguese Federation of Free Thought vigorously protest. Citizens! [. . .] Let us all free ourselves, therefore, by tearing from our minds not only the stupid belief in crude and hilarious hoaxes like that of Fátima, but more especially from belief in the supernatural, in an alleged Almighty God.*"

This extensive review of press coverage is a treasure trove of information concerning the events that unfolded around October 13, 1917. Since the papers that published these articles were all anticlerical, it is clear that the journalists had no motivation to accept these events or give a platform to claims concerning a miracle. This makes their writings all the more valuable. On the basis of this material, we can deduce several key insights that allow us to eliminate several of the hypotheses listed above. So, what can we be certain of?

- **1. That the Portuguese government of the time was very anticlerical, particularly the mayor of Vila Nova d'Ourem,** who was also the regional administrator (cf. art. 2). Consequently, there is no possibility of conspiracy or complicity on the part of the state or local government in mounting a clerical hoax.

- 2. **That the children were uneducated and illiterate** (cf. art. 2), which makes the hypothesis of fraud more difficult to sustain.
- 3. **That the facts and figures** reported in these articles, as well as **the concessions they make to the factual nature** of the events, should be viewed as minimums, given that they were all printed in anticlerical papers. This holds, for example, for the size of the crowd that gathered at Fátima.
- 4. **That the miracle was announced in August, two months in advance, for a precise date**, and this information was widely diffused. As a result, anyone interested could have traveled to watch what they expected to be a sign straight from God or, on the contrary, to see a hoax or nothing at all happen, and so go home satisfied.
- 5. **That tens of thousands of people were present at the event.** There were at least thirty to forty thousand in the immediate area as well as two hundred thousand witnesses in the surrounding region. The exact number is immaterial; what matters is that even thirty thousand people is already a significant crowd. The unanimous testimony of so many people carries considerable weight (cf. art. 10, 11, 12, 14, 22).
- 6. **That a major event did occur** and was seen by everyone, even if what exactly happened and what caused it remains to be decided.
- 7. **That to our knowledge, no one who witnessed the scene directly** claimed to have seen nothing at all, even though many doubters and skeptics were present at Fátima (cf. art. 11).
- 8. **That we must be surprised and puzzled by the total absence of rational reaction on the part of the miracle-deniers.** It was fully within their capability to launch an official investigation, explore the site to uncover any remains of a secret pyrotechnic display, or interrogate witnesses and their families. In such a small village, there would be enough division and conflict between the villagers' stories that the whole truth would have come out. In reality, the only firework show that day was the carnival of mockery and insults, peremptory and incoherent claims without the slightest investigative proof to sustain them.

- 9. That, against all expectations and when no one thought that it would, the miracle happened as foretold at solar noon which, in Fátima, took place that day at 1:21 p.m.

Farce, hallucination, meteorological event, hoax, scam mounted by priests and Jesuits: you can read through an entire stack of articles making these claims without coming across a single shred of rational proof.

The miracle-deniers did not investigate and expose the priests and Jesuits they accused of having run the scam. And this despite the fact that they had all the means at their disposal to bring the whole conspiracy to light: the power, the police, the justice system, and the press!

Finally, how could they be satisfied with such contradictory and incompatible explanations? It is evidently impossible to square the accusations of a premeditated hoax with that of a sudden and unpredictable meteorological phenomenon.

Here again is a list of the various claims made by commentators and journalists to help us realize just how many times they come up in contemporary coverage of the event:

- Hoax: 9 articles→2, 3, 4, 5, 6, 7, 8, 9, 10.
- Speculation: 7 articles→1, 2, 7, 8, 16, 18, 26.
- Farce: 3 articles→3, 5, 19.
- Hallucination: 5 articles→2, 12, 20, 23, 24.
- Meteorological event: 1 article→23.
- Nothing at all: not a single article.

Only one rogue article talks sense: *O Seculo*'s article wherein Avelino de Almeida remarks that it is up to the authorities to make a pronouncement on the dance of the sun, the occurrence of which he does not call into question (art. 13).

But the authorities did not take up his invitation to run an investigation... So we will continue our own, with photos to support it.

VI. Photographic evidence

1. Cova de Iria in the rain

2. The same place at the time of the apparition of the sun

FÁTIMA: ILLUSION, DECEPTION, OR MIRACLE? 475

3. A glimpse of the crowd at the moment of the miracle

4. Another view of the crowd at the moment of the miracle

5. General view of Cova de Iria

6. Jacinta

7. Jacinta, Lucia, and Francisco

8. Lucia and John Paul II, the 13th of May, 2000

What the photos prove

These photos are very valuable. They show more than the moment of surprise itself. In fact, they confirm the reality of the event and the significance of the crowd while adding several more bits of evidence to our investigation.

    1. Photos 3 and 4 show a crowd looking up toward the sky, where no human-made hoax was possible, especially in those days.

Let's take a minute to compare Fátima to Lourdes. In the small Pyrennese village, the Virgin appeared in a rocky grotto, a place where it would have been conceivable to set up a sophisticated illusion by means of a hidden door or lights. At Fátima, that was impossible. Photo 5 shows a gently undulating plain that does not permit any high-altitude obfuscation.

2. The photos show the crowd all looking in the same direction, at a point in the sky where something is taking place. This helps eliminate the idea of a collective hallucination, if such a phenomenon even exists.

3. Photos 6 and 7 give us a glimpse of the timid children, uncomfortable with their sudden great notoriety. It's difficult to see them as perpetrators of a fraud.

4. Photos 1 and 2 confirm the chronological progress of the rain, then the clearing of the sky. Photo 2 shows the sun coming out in a clear sky, and this is important because it disproves the hypothesis of a simple meteorological event.

5. Photo 8 shows Lucia with Pope John Paul II at the beatification of Francisco and Jacinta on May 13, 2000. It confirms the importance accorded to the miracle of Fátima by the Church more than eighty years after the event.[638]

6. Scientific studies conducted on the dimensions and angles of the shadows appearing in the photos taken during the miracle have determined that the light source that produced them was at an elevation of 30 degrees, whereas the astronomical sun should have been at an elevation of 42 degrees at that moment.[639]

---

638. The apparitions were declared worthy of belief by the bishop of Leiria in 1930. The consecration of the world to the Immaculate Heart of Mary by Pius XII was in large part a response to the promises of the Virgin of Fátima. Paul VI also met Lucia at Fátima.

639. See the research of Father Philip Dallcur, who holds doctorates in both Applied Science and Philosophy, "Fatima Pictures and Testimonials: In-Depth Analysis," *Scientia et Fides*, January 9, 2021, https://apcz.umk.pl/SetF/article/view/SetF.2021.001/28737.

VII. What happened to the children?

Two of the little shepherds died soon after, as the Holy Virgin had foretold to them. Francisco was carried away by the Spanish Flu on April 4, 1919, and Jacinta died of the same disease on February 20, 1920. Lucia, however, lived a very long time. Having become a Carmelite nun, she died at the age of ninety-eight on February 13, 2005. As long as she lived, she never ceased to work for the realization of the Virgin Mary's demand that the pope and bishops of the whole world should consecrate Russia to her Immaculate Heart. The consecration did not validly take place until John Paul II performed it on March 25, 1984, two and a half years after the May 13, 1981 assassination attempt that should have killed him.

None of the three children ever modified their version of the events. None of them earned money or fame for their story.

VIII. Examination of six possible hypotheses

There are only six possible hypotheses to explain what happened that October 13, 1917.

1. Nothing happened at Fátima. It's all a tall tale.

No one has defended this hypothesis, and it is invalidated by the abundance of witnesses and photographs.

2. It was a natural phenomenon—a disruption in the solar system as a result of cosmic events

This hypothesis is swept aside by scientific observations.

On Thursday, October 18, the daily *O Século* ran a second-page story with a statement from Augusto Frederico Oom, the director of the astronomical observatory in Lisbon, who testified that *"if it were a cosmic event, astronomical and meteorological observatories would not fail to register it."*

In other words, the observatory at Lisbon reports that it detected no abnormal phenomenon on October 13—only a slight disturbance coming from the west.

In his pastoral letter on the cult of Our Lady of Fátima written on October 13, 1930, the bishop of Leiria wrote: *"This phenomenon, which was not registered by any astronomical observatory and so cannot be considered natural, was seen by persons of every condition and every social class with their own eyes [. . .] even people kilometers away. It simply cannot be explained as a collective illusion."*[640]

Indeed, let us recall that this phenomenon was observed by many witnesses up to twenty-one miles away.

Neither did any international observatory notice the slightest alteration in the sun's movement.

3. It was an unusual meteorological phenomenon

This hypothesis cannot stand in the face of the indisputable fact that the event was announced three months in advance. Three illiterate children could not have calculated the precise day and hour of such a phenomenon. Even if some extraordinary meteorological event could account for such a spectacle, it would have been totally impossible to predict three months out. We now know that it is impossible to forecast the weather more than fifteen days in advance.[641]

The hypothesis of a meteorological phenomenon was nevertheless defended by Diogo Pacheco de Amorim, a Portuguese doctor in mathematics. He contended that the "miracle of the sun" might have been tied to clouds whose high-altitude ice crystals can decompose light into different colors, as in the case of rainbows. Amorim posits that "air

---

640. Extract from his pastoral letter on the cult of Our Lady of Fátima, October 13, 1930.

641. In 1972, meteorologist Edward Lorenz explained that it would be impossible to make a forecast more than fifteen days in advance due to the weather's sensitivity to initial conditions. This is known as the "butterfly effect."

lenses" (of varying composition or temperature) may have disrupted the normal diffusion of light, influencing perception of the sun's apparent diameter, and might even explain the change in colors. Amorim recognizes, however, that *"we are not able to provide a complete explanation of such a complex and mysterious phenomenon, only compare it or break it down into elements that are comparable to known phenomena."*[642] But even if Amorim's theory might explain the light phenomena in part, it offers no explanation for the disordered movements of the sun.

Physicist Stanley Jaki has similarly suggested that the event may have been a meteorological phenomenon: *"What seems to have happened is that the transparent clouds veiling the sun formed an enormous natural lens."*[643] But once again, a lens does not create the impression of the sun dancing and falling towards the earth, as the witnesses claim to have seen it do. Further, the phenomenon Jaki describes has never actually been observed.

4. Nothing happened: it was a collective hallucination

This hypothesis is excluded for the simple reason that *"collective hallucinations"* do not exist in such large crowds. They can only take place in very small groups and are never observed in groups of a thousand people. In the context of thirty to seventy thousand people, it is entirely out of the question.

Hallucination is a phenomenon that has received serious study—for example, by Henri Ey, who describes it in his *Manuel de psychiatrie* [Psychiatric Manual] or in his *Traité des hallucinations* [Treatise on Hallucinations].[644] It is the perception of a nonexistent object stem-

---

642. Translated from Diogo Pacheco de Amorim, "O fenómeno solar de 13 de Outubro de 1917," *O Instituto* 122 (1961): 145–204.

643. Stanley L. Jaki, *God and the Sun at Fátima* (Royal Oak, MI: Real View Books, 1999).

644. Henri Ey, *Manuel de psychiatrie, deuxième partie*, and *Traité des hallucinations* (Meulan, France: Masson, 1973).

ming from an internal psychological experience that leads a subject to act as if he were experiencing a real sensation or perception, even though there is no objective exterior element justifying this sensation or perception.[645]

There can be different levels of hallucination (visual, auditory, tactile, or even psychic), but these are always a symptom of a person's disordered brain function. They generally occur in a context of imbalance or breakdown of the personality, commonly accompanied by various behavioral problems (agitation, fluctuating anxiety, inappropriate interactions with others, etc.). For this reason, *"collective hallucinations"* are by definition medically and logically impossible. The idea assumes that the same disorders are produced identically and simultaneously in different brains, without any objective exterior cause.

We should not confuse these conjectures with other phenomena of collective visions that are based on objective exterior elements:

- They can't be compared with UFO phenomena, that is, sightings of unusual objects in the sky (Lubbock,[646] Washington[647]), which are often accompanied by photos or objective elements like radar echoes.

- They can't be compared to magic tricks, which can all be reduced to the manipulation of real things.

- Nor can they be compared to the mirages people see on the horizon, which are the result of specific atmospheric conditions.

In the first place, collective hallucination is not a valid category in psychiatry, neurology, psychoanalysis, or clinical psychology. There is no study, theory, or experimentation on the topic. Scientists have never observed a collective hallucination affecting crowds.

---

645. J. Sutter, *Manuel alphabétique de psychiatrie*, ed. Antoine Porot (Paris: PUF, 1996).

646. https://en.wikipedia.org/wiki/Lubbock_Lights.

647. https://en.wikipedia.org/wiki/1952_Washington,_D.C._UFO_incident.

In the second place, "collective psychoses" or "affective contagions," which are only observed in small groups, are often induced by fear in a tense or anxious situation. They generally occur at night, not midday in broad daylight, in a plain like Cova da Iria, where photos taken before and after the miracle do not reveal any evidence of an induced trance or particular excitement.

In the third place, the group suffering the same hallucination has to share the same emotions, and its members must have a sufficiently strong process of suggestion. This is absolutely impossible in the case of thousands of different people, who were in this case believers, agnostics, or frankly hostile to religion. The hypothesis is all the more difficult to maintain given that several people saw the phenomenon at a distance of several kilometers from Fátima, even without foreknowledge of the coming miracle.

There are several documented cases of "collective psychoses:"

- The "*cursed bread*" affair (*pain maudit*):[648] On the night of August 24–25, 1951, the small town of Pont-Saint-Esprit, in Gard, France, was the scene of a very violent collective madness caused by rye ergot poisoning (rye ergot is the active ingredient in LSD). But the cause of this event is well established, and the hallucinations of the various people affected differed substantially.

- The case of the "*cook's ghost*":[649] In 1897, Edmund Parish reported that some of his fellow sailors had seen the ghost of their cook, who had died a few days before. The sailors had not only seen the specter but watched him walk on the water with his characteristic limp. In the end, the ghost "*turned out to be a piece of wreck, rocked up and down by the waves.*" Once again, the case is easily solved.

In both cases, there was a motive and an explanation.

---

648. Sandrine Cabut, "Les étranges symptômes du « pain tueur » de Pont-Saint-Esprit," *Le Monde*, July 31, 2014, https://www.lemonde.fr/sciences/article/2014/07/31/le-pain-tueur-sevit-a-pont-saint-esprit_4465400_1650684.html.

649. Edmund Parish, *Hallucinations and Illusions: A Study of the Fallacies of Perception* (London: Walter Scott, 1897), 311.

The events at Fátima, foretold three months in advance, are of an entirely different order.[650] All this leads us to the conclusion that collective hallucination is not a possible explanation for the events that transpired at Fátima on October 13, 1917.

5. It was a fraud

This is the only hypothesis compatible with the fact that the miracle was announced in advance and that no unusual natural event was detected. Fraud is also the hypothesis most frequently cited by the anticlerical journalists.

But it doesn't stand up to scrutiny. It would have been impossible to pull off an operation of such magnitude in the middle of the sky. The photos and witnesses all agree: something happened in the middle of the sky at the sun's zenith, at a time when the sky had become clear and bright again. And this is the one place where it is impossible to orchestrate a deception of this kind.

We also know that the phenomenon was seen as far as twenty-one miles away by Afonso Lopes Vieira. However, at that distance, an object located at the elevation of a 30 degree angle would be 19 degrees above sea level. We cannot see what human machination would be able to create an illusion at that altitude in 1917.

Furthermore, if it were a hoax, the organizers would have started their show at noon and not at 1:21 p.m., in all likelihood ignoring the detail of solar noon.

Further, we know that at least thirty thousand people witnessed the miracle, many of whom were merely curious or even unbelievers. What sort of special effects would have been necessary to deceive such a tough crowd? A fraud is so obviously impossible that no one has ever

---

650. The fact that the miracle took place at 1:30 (solar noon)—to the surprise of the participants, who were expecting something to happen at noon—excludes every possibility of a hoax.

offered even the slightest explanation for how such a spectacle could actually have been devised.

All the circumstances also plead against fraud: illiterate children, an impoverished village, hostile authorities, and the fact that for decades afterwards, no one has brought forward a shred of evidence for fraud and the three main witnesses never changed their stories. Would Lucia have spent eighty years in a convent just to cover up a lie? Or worse, would she have continued to deceive the pope himself eighty years later?

6. It was a miracle

The range of hypotheses is now considerably reduced, and we only have one option remaining. The circumstances of this miracle perfectly correspond to what we would expect:

- An inexplicable prodigy.
- Purpose was to encourage religious faith.
- Announced in advance.
- By uneducated children.
- Seen by an immense crowd.
- In the face of hostile local authorities.

This final possibility, the only one left, requires nothing less than the existence of God. The reader who has accompanied us this far will agree that this one position is not an unreasonable hypothesis—quite the contrary.

IX. The meaning of Fátima today

The modern echoes of Fátima lend even more proof to its supernatural origin. This miracle, as extraordinary as it was, has not remained frozen in time. Rather, it has had major historical repercussions over the last century.

One of the Holy Virgin's requests was that Russia be consecrated to her by the pope and bishops of the world. Although this request was made known to the pope in 1942, it was not fully realized under the conditions required for several successive pontificates.

Here is what we know for certain: On May 13, 1981, a professional assassin (a Turk named Mehmet Ali Agça, almost certainly hired by the KGB) fired on Pope John Paul II at close range with a high-caliber pistol. Against all expectations, the sovereign pontiff survived his wounds. From his hospital bed, the pope, struck by the coincidence of the date of the attack with that of the first apparition at Fátima, called for the text of the Fátima secret, which had lain neglected for a long time in the Vatican. He noticed that the consecration of Russia called for by the Virgin at Fátima had never been performed validly, as Lucia would later confirm during an interview. John Paul II applied all his power and authority to persuade the world's bishops to carry out Mary's request. The consecration of Russia took place on March 25, 1984.

What came next is well known. A reform movement was born in Russia in 1986 (Mikhail Gorbatchev's glasnost and perestroika), and a peaceful rebellion rose in Poland (Solidarność in Lech Walesa). A few years later, the wall fell; in 1989, the USSR totally collapsed without a single shot fired.

Whether or not we believe that there is a connection between these facts, it is certain that John Paul II firmly believed in it, raising the importance of the extraordinary event at Fátima.

A foregone conclusion

In our catalogue of evidence for God's existence, this chapter on Fátima has been much more than a picturesque detour or exotic excursion into religious folklore. What happened in that small Portuguese village was unique, unprecedented, and unheard of: a miraculous and inexplicable phenomenon in the sky, announced in advance and seen by large crowds, taking place in the precise location and at the exact time predicted.

At Fátima, everything took place in broad daylight, in both the literal and metaphorical sense. The whole country was alerted to the miracle beforehand. The events happened in the middle of the day before thousands of witnesses, making fraud out of the question. Considering the political and ideological context of the event, reason must exclude every other interpretive hypothesis. That is why the prodigious event of Fátima has as much convincing power today, for us, as it did for those who witnessed it at the time.

The miracle took place so that people might believe. There is no room for doubt about this extraordinary thirteenth of October. Everything took place in broad daylight and plain sight. The sun did not blind anyone who saw it dance. In the same way, the light of Fátima can cast out all blindness and help every reader reach a reasonable and informed decision.

22.

Is Everything Permitted?

After five chapters covering major historical enigmas, we will now explore a completely different path. The scenery will be much less spectacular, perhaps, but at the end we may find another way to detect the existence of God. Call it a little experiment. No, we won't look for God under a microscope or at the bottom of a test tube. This time we propose a much more exciting approach: an exercise in listening to the interior voice of our conscience, whose echo also can be perceived, though in a manner totally different to the echo of the Big Bang. St. Augustine once gave this worthy piece of advice: *"Do not go outside yourself, but enter into yourself, for truth dwells in the interior self."*[651] But how do we do that? Simply by listening intently to that part of ourselves in which God speaks to us, often without our realizing that it is his voice. We wish to speak of the soul, where our moral conscience resides. To this end, we ask you two questions designed to prick this moral conscience inscribed in your soul, if this conscience really exists.

Some will interject that morality doesn't need God to exist or for us to know it. Others will say, "Since God doesn't exist, there is no universal moral law; everything is relative." Instead of launching into a tedious and abstract philosophical discussion on these topics, let's plunge straight into a real-life situation. Pay attention to your reaction to the two following questions. Do you have an immediate "gut reaction," or inexplicable interior prohibition? If so, where does it come from?

First question: Can a democratic majority in your country pass a resolution, say, to gas Jewish people, reinstitute slavery, euthanize old people,

---
651. Augustine, *De vera religione* 39.72.

grant the right of abortion up to nine months, or legalize pedophilia. Does it have the right? Yes, or no? Probably you will say: "No, not even if the parliamentary vote is unanimous!"

Second question: You are offered one hundred million dollars to press a button that will instantly kill a family on the other side of the world. You have never met them before, and you are absolutely guaranteed that there will be no consequences and no punishment for their deaths. Think about it: your financial worries will be solved forever! After all, aren't humans just highly evolved animals, just a little more complex than the insects we step on everyday without remorse? And isn't the earth overpopulated? Isn't it in our general interest to reduce the human population that is taking such a terrible toll on our planet? After all, what is one family on the other side of the world? Wars, natural disasters, and epidemics kill thousands more every day.

If despite all these considerations you respond "No, never!" to these two questions, you must then reflect upon the reasoning for this astonishing refusal and therefore upon the moral norms impressed upon your moral conscience. Where do they come from? What causes the voice in us to cry "No!"? For if God does not exist, evil does not exist either, and therefore anything goes.

If God does not exist, evil does not exist, and everything is permitted

Effectively, if God does not exist, anything goes. The cosmos is the only thing that exists absolutely, and there is ultimately no difference between a man and a mosquito. Both are just the transitory by-products of matter and energy. If we begin from this rigorously materialist position, there is no basis for the reality of good and evil, or of objective moral values. When we crush a baby or a mosquito, we are doing nothing more than reorganizing the matter that constitutes it.

Many thinkers have wrestled with the implications of the existence of evil. Let's begin with these questions from Dmitri in Dostoevsky's *The Brothers Karamazov*: *"What if [God] doesn't exist? What if Rakitin's right—that it's an idea made up by men? Then if He doesn't exist, man is the chief of the earth, of the universe. Magnificent! Only how is he going to be good without God? That's the question. [. . .] But after all, what is goodness? Answer me that, Alexey. Goodness is one thing with me and another with a Chinaman, so it's a relative thing. Or isn't it? Is it not relative? A treacherous question! [. . .] Then everything is lawful, if it is so?"*[652]

Sartre, in Existentialism is a Humanism, brought out the profound logic of this idea: "*Dostoevsky once wrote: 'If God did not exist, everything would be permitted;' and that, for existentialism, is the starting point. Everything is indeed permitted if God does not exist, and man is in consequence forlorn, for he cannot find anything to depend upon either within or outside himself. He discovers forthwith that he is without excuse.*

For if indeed existence precedes essence, one will never be able to explain one's action by reference to a given and specific human nature; in other words, there is no determinism—man is free, man is freedom. Nor, on the other hand, if God does not exist, are we provided with any values or commands that could legitimize our behaviour. Thus we have neither behind us, nor before us in a luminous realm of values, any means of justification or excuse. —We are left alone, without excuse.

That is what I mean when I say that man is condemned to be free. Condemned, because he did not create himself, yet is nevertheless at liberty, and from the moment that he is thrown into this world he is responsible for everything he does. The existentialist [. . .] thinks that the man himself interprets the sign as he chooses. He thinks that every man, without any support or help whatever, is condemned at every instant to invent man."[653]

---

652. *Brothers Karamazov*, trans. Constance Garnett (New York: The Lowell Press, 1912), 669.

653. Jean-Paul Sartre, "Existentialism Is a Humanism," lecture given in 1946. In *Existentialism from Dostoevsky to Sartre*, ed. Walter Kaufman, trans. Philip Mairet (New York: Meridian, 1989).

Closer to our time, some well-respected materialist philosophers like Richard Taylor have reached the same conclusion: *"In a world without God, no purpose can be right and wrong, only our culturally and personally relative, subjective judgment. This means that it is impossible to condemn war, persecution or crime as evil. Nor can anyone admire brotherhood, equality and love. Because in the universe without God, good and evil do not exist—there is only the bare valueless fact of existence, and there is no one to say that you are right and I am wrong.*[654]

Richard Dawkins, another biologist and specialist in evolution, writes: *"If the universe were just electrons and selfish genes, [...] blind physical forces and genetic replication, some people are going to get hurt, other people are going to get lucky, and you won't find any rhyme or reason in it, nor any justice. The universe we observe has precisely the properties we should expect if there is, at bottom, no design, no purpose, no evil and no good, nothing but blind, pitiless indifference."*[655]

If the totality of the real is, in fact, nothing but an immense heap of matter hurtling through the void, it is not clear where absolute norms, imperative rules, or a hierarchy of sacred values come from. In the materialist paradigm, there is no higher authority that makes demands upon us. If God does not exist, then everyone is his or her own god, implying that we each "choose our own values."

From this viewpoint, all lifestyles are equally valid because there is no objective moral criterion by which we can judge or rank them. The only limit to our freedom of choice seems to be the autonomy of other individuals. Public morality is founded upon a utilitarian "no harm principle" that flows from this materialist worldview and its indifference to moral absolutes. Each person is allowed to do as he pleases, provided that he does not harm anybody else. The interest of each individual has the same weight in the scale of general interest. *"Everybody to count for one, nobody for more than one,"* according to Jeremy Bentham's famous maxim.

---

654. Richard Taylor, *Ethics, Faith, and Reason* (Englewood Cliffs, NJ: Prentice-Hall, 1985), 2–3.

655. Richard Dawkins, *River Out of Eden: A Darwinian View of Life* (New York: Basic Books, 1995), 132–133.

But what determines the absolute superiority of this minimalist view of human morality? What is it based on? The consequences of a rigorously pure materialist and utilitarian public morality have been made tragically manifest in so many of the past century's atheist regimes. Think of the communist leaders of Asia with their "sacrificial generations." They justified their policies of doing extreme evil to others by the promise of achieving a better collective good in the future.

But here a question arises:

If materialism is true, then why do we feel moral prohibitions?

If we are nothing but a mass of particles lost in the Universe, why do we feel compelled to observe strict moral absolutes? Why, for example, do we feel we must treat fellow human beings with an almost sacred dignity?

If you agree with the materialist vision, why do you refuse to press the button? Life is just a senseless game, so why not be a winner? If materialism is true, there is no absolute moral good or evil, but only the strong and the weak, the victors and the conquered, the lucky and the unlucky.

At this point you would certainly say that you feel a instinctive disgust, a prohibition, an immediate revulsion against committing the acts that we have proposed to you and that there's nothing philosophical or debatable about this: this has to do with instinct. Very well. But what is the source of this instinct, and what is its nature?

Darwinian evolution provides an explanation

Darwin provided an answer that is compatible with materialism.

Viewed in the light of evolutionary biology, our moral sense—whether it is felt in a particular moment in time or takes the form of legal codes such as the Ten Commandments or the Law of the Twelve Tables—has been selected in the course of evolution because it contributes to the survival of our species. In early humans, the impulse to avoid harming fellow humans, to protect them from suffering, to take care of the weak, or to treat others with reciprocity led to social cohesion and increased the likelihood of survival for individuals and groups. Populations with these qualities were better at surviving, adapting, and reproducing, so they grew more numerous. Eventually, these instincts were so generalized that they became part of our basic genetic baggage. At a much later date, these universal instincts were finally intellectualized, formalized, and written down, and the same fundamental moral laws were expressed by nearly every major human civilization. Such, in summary, is the sociobiological theory of morality. Just like our teeth, eyebrows, and opposable thumbs, our moral sentiments have evolved over the long history of our species. Michael Ruse, a celebrated expert in Darwinian biology, explains this view: "*Morality, or more strictly our belief in morality, is merely an adaptation put in place to further our reproductive ends. Hence the basis of ethics does not lie in God's will.*"[656] In other words, we must firmly believe that our moral instincts reveal absolute norms, even though we know they are merely the random by-product of our evolution as higher animals. They certainly do not reveal anything absolutely true but are just historical records of the qualities that have proved the most useful for our group's survival. Darwinians insist that this illusion of objectivity is essential to the efficacy of biological morality: "*The Darwinian argues that morality simply does not work (from a biological perspective), unless we believe that it is objective. Darwinian theory shows that, in fact, morality is a function of (subjective) feelings; but it shows also that we have (and must have) the illusion of objectivity.*"[657]

In conclusion: "*In an important sense, ethics as we understand it is an illusion fobbed off on us by our genes to get us to cooperate.*"[658]

Is this explanation plausible? Do you accept it? Is it compatible with what you feel?

Perhaps it is easy to deny the absolute character of moral values in the abstract. But would you really dare to affirm in all sincerity—as more

---

656. Michael Ruse and Edward O. Wilson, "The Evolution of Ethics," *New Scientist* 108 (October 1985): 50.

657. Michael Ruse, *Taking Darwin Seriously* (Cranbury, NJ: Prometheus, 1998), 253.

658. Michael Ruse and Edward O. Wilson, "The Evolution of Ethics," 50.

than a flight of theoretical bravado—that torturing a baby for pleasure, as a cat does with a mouse in accordance with its nature, is not objectively evil? That the prohibition against it is only a contingent aspect of our evolution?

Do you think that certain acts are forbidden, that some acts are absolutely bad, unthinkable in any situation, and do you refuse with all your might the idea that your internal sense of obligation could be just a Darwinian illusion? Then we have to ask why this is the case. The categorical refusal you feel when we ask you to press the button to earn one hundred million dollars does not come from a sense of prudence, a simple calculation of the chances of being caught, or the fear of going to prison. Nor are you just a prisoner of your genes, since someone in full knowledge of his genetic inclination and its origin still has the freedom to choose to follow it or not. What we have discovered in ourselves is an absolute prohibition. Naturally, we have to ask where it comes from. Since it seems we have exhausted our options for sources within the physical world, the only alternative is a source outside of it—"*of another order*," as Pascal would have said. This source of another order is our soul, wherein we hear this voice. The voice is not our soul itself, since when the voice speaks to us of our moral obligations, it does not address us in subjective, individual terms. Rather, it speaks of universal obligations, binding for all people.

From the soul to God

If we follow this line of reasoning, then, in the final analysis, our soul bears the imprint of a design that surpasses us. Spiritual and absolute, its origin must be a cause that is itself spiritual and absolute, transcending the order of physical things. The unconditional imperatives that this spiritual cause has put into our soul are the sign of a finality that is itself unconditional. There is a purpose to life, not defined by Darwinian evolution or by society. Its only author is the spiritual cause that generated our soul. Thus we have our answer to the question: if certain acts are repugnant to you, if you recoil in horror from the idea of wounding an innocent soul, that is because the voice of God is speaking in your soul.

And if we must live to man's fullest dignity, this is because God has given us the nature to participate in a design that is greater than ourselves.

Final considerations

Before you stop to reflect, there are two final considerations to take into account. If the chapter "Jesus: Who Could He Be?" or the chapter on Fátima left us with several possible answers to the question of God, in this case there are only two possibilities. Either you believe that absolute good and evil exist, and so you are obliged to believe in God; or you refuse to believe in him and so, in order to be coherent, you have to accept Darwinian materialism and all that comes with it. Finally, if evil does not exist, then neither does good: love, friendship, sharing, forgiveness—none of these things exist. All that really matters is our pleasure and our instincts.

Taking an honest look inside

Without requiring much reflection, from the outset you decided not to engage in genocide, euthanize the elderly, or torture babies. You also refused a hundred million dollars in exchange for a simple act of murder from which you would get away scot-free.

Having read this chapter, it is now your time to make your own way and probe the depths of your most inner self. When you hear the voice of conscience, do you believe in good and evil, and so in God? Or do you believe in Darwinian materialism, with all its consequences? It's your decision, your soul, your conscience.

23.

Philosophical Proofs Strike Back

Let's be honest: the philosophical proofs for the existence of God have never interested many people. Apart from philosophers, of course...

They are too lifeless, abstract, difficult. But it is of course impossible for a book whose purpose is to provide evidence for the existence of God to skip over them without comment. Know that if you persevere, you may very well find some benefit in this bitter draft.

These philosophical truths have a sad history. Back when everyone believed in God, philosophers constructed a multitude of proofs for his existence, but they were not of much use. Now that fewer people believe in God, guess what? Philosophers say the proofs are no longer valid! We exaggerate: there are still philosophers who propose and defend these ideas. True, these philosophers are in the minority, but that does not mean they are mistaken. In fact, we will try to show you that these happy few are correct.

To begin, we will rapidly survey the history of philosophical proofs for the existence of God.

It all began in Athens, Greece, under the Attic sun

In the fifth century before our era, Plato (428–348 BC), disciple of Socrates (470–399 BC), was the first philosopher to record a certain intuition that would go on to have a long afterlife: beyond the multitude of minor divinities found in woods, mountains, and seas, there is a supreme divinity, a "world soul" superior to the tiny souls that still populated nature in the pagan imagination. But Plato didn't stop at an intuition. He made an argument. The fact that reason can use logic and mathematics to understand the order of the world led him to conclude that beyond all that is organized there must be a great organizer: the *Demiurge* (the "artisan," in Greek). To craft the Universe, this divine being drew upon an inexhaustible source of models that could be combined in an infinite number of ways: the "Ideas." Plato doesn't say much about the nature of this Demiurge.

His most famous student, Aristotle, went further

Aristotle constructed a much more complex argument, arriving at God through the idea of motion. In simplified form, his argument goes like this: Everything moving is moved by something else, which is moved by something else, and so on. But we can't proceed forever from moved mover to moved mover. We have to stop somewhere. To put it another way, if something undergoes a change, there has to be something that changed it. But if the cause of this change is also something that changes, we have to find another cause, and so on. Now, according to Aristotle, there cannot be an infinite series of movers. The series must stop somewhere. Otherwise, there would be no movement at all. It is a bit like the cars on a train. If you ask why the train car is moving and someone tells you it is because the car in front of it pulls it, then you will pose the same question about the second car. If the person tells you that there are an infinite number of cars, you will say that this is impossible; there must be an engine. This is the essence of Aristotle's argument. But the series of motion in the natural world ends with a "Prime Mover" rather than a train engine. Further, it is not just any mover: the Prime Mover must be entirely immaterial. Why? Well, because material motors are never completely self-propelled: they always depend on an energy source outside themselves, a fact that pushes the series of causes and effects a step further back. The first term in the series of all movements in the Universe, therefore, cannot be a material motor. It has to be absolutely immaterial—or, in other words, spiritual, for what we mean by "spirit" is something both immaterial and active. That is why Aristotle considered this primary cause to be a *thinking* being. Aristotle's God is thus an entirely active being, with no passivity, whose principal object is to think himself. Aristotle therefore calls him the "thought that thinks itself."

After Aristotle, the Greek city states went into decline and soon fell under Roman rule. Philosophy lived on, but it was less interested in high metaphysics: it narrowed its focus to the question of individual happiness on this Earth. This was the age of Stoicism, Epicureanism, and Skepticism. The Stoics identified God with the energy that runs through all things. The Epicureans admitted that the gods might exist, but that it is best not to worry about them. The Skeptics—well, as their name implies, they refrained from affirming anything at all.

Under the Roman empire, after the birth of Christ, philosophy took up metaphysics again

The context was now more favorable: Hebraic monotheism began to conquer the world in the form of Christianity, and Platonism saw a revival among the pagans, who didn't want to be outdone by the Christians. And so, the philosophers went back to work. In the Neoplatonists' camp, Plotinus (205–270), an Alexandrian, tried to prove the existence of a supreme principle totally distinct from the Universe, which he called "the One." Unlike the God of Aristotle, Plotinus' One is so pure that it does not even think. It is an infinitely transcendent spring from which flows everything that is. On the Christian side, the great North African bishop and convert St. Augustine (354–430) produced an original proof that is worth focusing on briefly. Take $2 + 2 = 4$, or $A = A$. You know that these two propositions are true, and not only when you think about them. They are truths that always have been and always will be true. They are eternal, necessary, unchangeable, and completely independent. Even if the whole Universe disappeared, it would still be true that $2 + 2 = 4$. So Augustine asks a simple question: Where are the "things" that make these statements true? When you say "The cat is in the room," the things that make the statement true are the cat, the room, and the fact that the first is actually located within the second. For $2 + 2 = 4$, though? No material thing makes it true. But they are not merely ideas in my head, either, because even if I didn't exist, the equation would still be true. They are thoughts. Fine, but whose thoughts? Thoughts don't exist on their own. They can't be human thoughts because two plus two would still be four even if humans didn't exist. There is only one solution: there is a supreme, eternal, necessary Thinker who holds all these eternal truths in himself. This is God! With this argument, St. Augustine brings Plato's "Eternal Ideas" into Aristotle's "thought that thinks itself."

After St. Augustine, the Roman Empire fell, and philosophy found refuge in the monasteries of the West and the cities of the Middle East

In the early Middle Ages, the writings of ancient philosophers were recopied and conserved. Thanks to this work of transmission, philosophical creativity was able to flourish later on. The project of ancient philosophy was taken up again in earnest in the tenth century by Christians in the West and Muslims in the East. New arguments for God's existence proliferated in the West, most notably the "ontological argument" of Saint Anselm (1033–1109), an Italian Benedictine who became archbishop of Canterbury in 1093. Anselm tried to prove God's existence solely through his definition. The argument is so difficult that few understand it, and many logicians have considered it invalid, but it is worth noting that two of the best logicians the world has ever seen considered it to be valid: Leibniz (1646–1716) and Gödel (1906–1978).

In the thirteenth century, the great theologian St. Thomas Aquinas (1225–1274) proposed "five ways" to arrive at the existence of God, namely:

- **The proof from motion**, basically taken from Aristotle.
- **The proof from efficient causes**, which has the same structure as the former, except that the subject is existence rather than motion. Everything that exists receives its existence from something else. However, this chain cannot go on to infinity; we must eventually arrive at uncaused Existence.

- **The proof from contingency**, which asserts that contingent beings (beings that do not exist by necessity) must receive their existence from a non-contingent being (one that exists by necessity).
- **The proof from degrees of perfection**, which argues that the existence of imperfect things presumes the existence of an absolute perfection.
- **The proof by final causes**, which claims that the existence of beings oriented toward a determinate end cannot be explained by chance.

Other medieval philosophers did little more than perfect and develop the proofs of St. Anselm and St. Thomas. One man managed to combine the two: John Duns Scotus (1266–1308), a Scotsman who was called the "Subtle Doctor" because of the refinement and complexity of his thought. On the Muslim side, we must mention the two great thinkers Avicenna (980–1037) and Al-Ghazali (1058–1111). Avicenna made very powerful arguments that pointed toward those of Duns Scotus. Though less accomplished in pure abstraction, Al-Ghazali knew how to appeal to popular audiences and deployed a totally original argument based on the fact that the Universe had to have a beginning, and therefore a first cause outside of time. We'll return to this later.

The Middle Ages go out with a whimper

Unfortunately, the philosophy of scholastic theologians became so convoluted that no one understood it anymore. People hungered for something new.

The Renaissance and early modern era

In the Renaissance, people were primarily interested in man, not in finding proofs for God's existence. In the early modern period, however, interest in God revived, and there was a great efflorescence of arguments: Descartes, Malebranche, Leibniz, and Newton's collaborator Samuel Clarke were all great minds who produced a plethora of new proofs. We say "new," but they were really modernized versions of ancient arguments, adapted to suit the tastes of the age. Proofs became so numerous and well-developed that people began to classify them: there were *teleolo-*

gical arguments deriving from Platonic philosophy, proofs based on *the Ideas* (inspired by St. Augustine), *cosmological* arguments (based on the existence of the world, following Aristotle and St. Thomas), and purely *a priori* arguments (like Anselm's). All the arguments were carefully arranged in philosophical manuals.

Soon catastrophe struck: this scrupulously-built edifice buckled and collapsed

There were three reasons for this.

First, the experimental sciences born in the seventeenth century were so successful that many philosophers began to think, rightly or wrongly, that science was the only credible source of knowledge. Metaphysics was called seriously into doubt, along with its pretensions to demonstrate the existence of immaterial realities beyond the physical world. Two great skeptics left their mark on history: Scotsman David Hume (1711–1776) and Prussian Emmanuel Kant (1724–1804). The latter published the most formidable weapon ever deployed against proofs for God's existence: *The Critique of Pure Reason* (1781), in which he purports to demonstrate that when it comes to metaphysical questions (about God, the origins of the world, the infinity of time, etc.), reason is able to validly conclude two contradictory answers to the same question.[659] The message was clear: proofs pertaining to the existence of God have no value. Let's not be hasty: Kant was no atheist. He only claimed that it is as impossible to demonstrate God's existence as it is to demonstrate his nonexistence. That being the case, Kant believed in God not as the result of a theoretical proof, but because he heard within himself the voice of duty. The existence of this voice within us presupposes, for Kant, the existence of a superhuman "Legislator." But he does not positively affirm this. It is only a subjective conviction.

---

659. Emmanuel Kant, *Critique of Pure Reason* [1781], trans. Paul Guyer and Allen W. Wood. (Cambridge: Cambridge University Press, 1998), 498ff.

More radical philosophers

Second, some, carried away by their enthusiasm for the experimental sciences, ended up believing that nothing exists beyond matter moving in the void. Diderot, d'Alembert, La Mettrie, and d'Holbach are the main representatives of this tendency. Here we can speak of true-blue atheism, explicit claims for the nonexistence of God as a certain conclusion. This attitude has never received good press among philosophers, but this was the heyday of the French Enlightenment (in Germany and England, philosophers were more moderate).

The third knock-down blow against the philosophical claim for the existence of God was dealt by the "masters of suspicion"—Nietzsche, Marx, and Freud—at the end of the nineteenth, and beginning of the twentieth, centuries. Their approach was not to criticize the proofs themselves, but to entirely discredit the enterprise of formulating such arguments, because all metaphysical investigation either betrays resentment towards the world (Nietzsche), the wish to flee an unjust society (Marx), or the neurotic search for a father figure (Freud). These philosophers bring no technical objection against the arguments of previous philosophers, but they have convinced many people that one has to be morally sick, socially alienated, or neurotically obsessed to be interested in the question. As a result, they have dissuaded many from looking for answers.

After such an onslaught, it might seem that there is nothing left of the proofs for God's existence. For a time—perhaps two centuries—this was true. But theists had not had their last word. Starting in the 1960s in the United States and spreading across the entire anglophone world before gradually reentering continental Europe, metaphysics made a comeback. Why?

Very simply, because the atheist objections of the eighteenth and nineteenth centuries were not decisive

First of all, the "war machine" Kant brought against metaphysics had lost much of its force and prestige: excellent philosophers had shown that a number of his arguments simply did not work. In particular, they demonstrated the total falsity of the idea that when it comes to grand metaphysical questions, reason can demonstrate everything and its contrary. Furthermore, science after Kant has shown—thanks to Einstein—that it is perfectly possible to reason about the Universe in its totality, and, even more fatally for the Kantian system, that the Universe very probably had a beginning. These new findings totally contradict Kant's claims that such questions have no definitive answers.

Materialism and scientism, which hold that nothing exists besides atoms floating in an eternal void, are now totally discredited and viewed as outdated nineteenth-century ideologies. It is not true that everything that exists can be explained as electrochemical interactions between elementary particles: the human conscience, for example, is irreducible to this kind of reality—it is of another nature. Its nature and action are based on properties beyond the reach of mathematical physical science.

Finally, the "masters of suspicion" may over time have discouraged many people from being interested in God, but they have not actually disproved anything at all. The fact that some people have weak reasons for believing in God (life-denial, the desire to flee the world, the search for consolation in the face of an unjust society, nostalgia for an absent father) does not mean that God does not exist. The bad reasons denounced by the masters of suspicion do not negate any of the good reasons there may be for belief in God. In short, the psychological, moral or sociological critiques of religion do not, in reality, have anything to do with the question of whether or not God exists.

By the 1960s, all these objections to the existence of God had been carefully sifted, weighed, and ultimately found unconvincing. There was nothing left but to go back to the drawing board. That is what happened in Great Britain and the United States, where a silent revolution began in the philosophy departments of the 1960s and '70s. Numerous university philosophers returned to metaphysics. Not all of them intended to prove the existence of God, of course, but all at least admitted that the question deserved to be asked. Never before have so many and such good books been written about proofs for God's existence than in our time. To list a few contemporary names: Peter Forrest, Alvin Plantinga, Richard Swinburne, William Craig, Alexander Pruss, Robert Koons, Edward Feser, Joshua Rasmussen, Robin Collins, David Oderberg, Emanuel Rutten...

Beyond this list, we must evoke the name of the one figure who alone symbolizes the kind of reversal that has occurred in the Anglo-Saxon philosophical environment: Anthony Flew (1923-2010). During his entire university career, this philosopher was atheism's most energetic proponent on the British and American philosophical scene. But in 2004, he admitted publicly that he had finally changed his mind: after many reservations, he acknowledged the validity of arguments he had resisted his whole life. The clinching factor was, according to his own testimony, the revelation by physicists in the 1970s, of the fine tuning of the Universe (see Chapter 9 of this book). The discovery led him to reexamine the philosophical arguments and experience a genuine reversal of his previous convictions. He declared: *"Science spotlights three dimensions of nature that point to God. The first is the fact that nature obeys laws. The second is the dimension of life, of intelligently organized and purpose-driven beings, which arose from matter. The third is the very existence of nature. But it is not science alone that guided me. I have also been helped by a renewed study of the classical philosophical arguments."*[660]

This progressive revolution among philosophers has not quite trickled down to the general public, probably because philosophy is a difficult discipline inaccessible to many, but also—and perhaps especially—because contemporary Western society has been completely secularized. The idea that the existence of God could be proved would undermine the agnostic basis of the modern social order. Some are well aware of this danger; hence there has been a new resurgence of virulent atheism flowing especially from the pens of nonphilosophers like Richard Dawkins and Christopher Hitchens.

---

660. Antony Flew, *There Is a God: How the World's Most Notorious Atheist Changed His Mind* (New York: HarperOne, 2007), 88.

But this history of philosophical proofs has been quite long enough. Now it is time to delve into the arguments themselves. We have chosen to focus on three of them.

I. The Supreme Intelligence

The fact that the Universe is intelligible has amazed thinkers from Plato to Einstein. Mathematics offers a stunning description of the physical universe: it seems almost "woven" out of numbers. Math is not only good for expressing the laws of mechanics or chemistry; pure mathematical theories elaborated without any connection to physical science, and according to purely formal requirements, turn out, decades later, to provide the precise tools needed for describing the world. Complex numbers, Hilbert spaces, group theory... all these mathematical abstractions were discovered without any relation to physical reality, and yet they ended up offering scientists tools perfectly adapted to the formalization of quantum physics and Relativity. Let's put it another way. As long as we stick to elementary arithmetic, we can view mathematics as a pure abstraction that represents physical reality. Then it is unsurprising when the real world corresponds with mathematical formulae: they are derived from it. But when it comes to complex mathematical structures, elaborated without any link to concrete manipulations of everyday life like census-taking or land surveying, the empiricist theory no longer works, and we are faced with a true enigma. The seemingly miraculous correspondence between this higher math and reality almost seems like a "*happy coincidence.*" (Eugene Wigner, Noble laureate in physics, speaks in a famous article of "*the unreasonable effectiveness of mathematics.*")[661] Whatever philosophy of mathematics we adopt, we still find ourselves at an impasse: if we follow the realist school and think that math refers to real objects—immaterial and eternal entities—then it is not clear why physical reality, which is material and changing, should conform to mathematical laws,

---

661. In his 1960 book *Symmetries and Reflections: Scientific Essays* (Bloomington: Indiana University Press, 1967), Eugene Wigner (1902–1995), recipient of the 1963 Nobel Prize in Physics, included an essay entitled "The Unreasonable Effectiveness of Mathematics in the Natural Sciences."

since immaterial and eternal things have no effective causality. They do not act in the world. They exist apart, and that's the end of the story. Now if we side with the conventionalist school and think that mathematics is merely a language invented by man, a coherent convention that does not, however, describe anything in the real world, we are no better off, because this view does not clear up—and perhaps muddies the waters further—how the real world agrees with mathematics. In both cases, the correspondence between the free speculation of mathematicians and the physical reality of the world remains inexplicable apart from a coincidence bordering on the miraculous. This seeming impasse actually points toward its solution: mathematical formulae preceded the formation of the world; or, in other words, the world was conceived by an intelligent being. The French mathematician René Thom reached the same conclusion: *"We find math at work not only in the inflexible and mysterious action of physical laws, but also, in a more hidden but likewise undeniable manner, in the infinite succession of forms in the inanimate world, in the emergence and destruction of their symmetries. That is why Plato's theory—that the universe is being informed by eternal static ideas—is a very natural and philosophically simple explanation."*[662]

We should also hear the testimony of Australo-British physicist Paul Davies: *"The temptation to believe that the Universe is the product of some sort of design, a manifestation of subtle aesthetic and mathematical judgment, is overwhelming. The belief that there is 'something behind it all' is one that I personally share with, I suspect, a majority of physicists."*[663]

The argument goes as follows:

1. If the world was not conceived by an intelligent being, then the applicability of mathematics is a pure coincidence.

---

662. René Thom, *Apologie du Logos* (Paris: Hachette, 1990), trans. from the French.

663. Paul C. W. Davies, "The Christian Perspective of a Scientist," review of The Way the World Is, by John Polkinghorne, *New Scientist* 98, no. 1354 (June 2, 1983): 638–639. http://www.abc.net.au/rn/scienceshow/stories/2006/1572643.htm.

2. But it is very unlikely that the applicability of mathematics is a coincidence.
3. Therefore, it is likely that the world was conceived by an intelligent being.

The truth of the first proposition is demonstrated by the explanatory impasse we have described: neither empiricism, nor realism, nor conventionalism work. The second proposition is a simple observation of common sense. The conclusion follows logically.

The resolution of this impasse also allows us to clear up a disagreement between the various theories of mathematics: realists often ask themselves "where" and "how" mathematical entities exist. Now it seems absurd to suppose that ideas exist in themselves, or that they float in some intellectual realm in the sky without a thinker to think them (an idea without a thinker is like a song without a singer). But if an intelligent being created the world, the solution is at hand: mathematical ideas exist in a divine intellect that thinks them. Furthermore, if the world was created according to these mathematical ideas, it is obvious that the conventionalist thesis is doomed: the mathematical ideas conceived by men are neither arbitrary conventions nor simple tools, because they exist in the divine understanding. Thus it is the realists who are right.

We conclude with a small clarification: one might object that the intelligence we have been led to postulate need not be "God" in the full sense of the word. It could be a sort of "demiurge" or divine craftsman who organized pre-existing matter, much as Plato envisions in his *Timaeus*. But this is not the case for the following reason: there is no such thing as matter that is absolutely unorganized, absolutely chaotic. Everything, even the smallest of the elementary particles, can be described by mathematics. Thus there is nothing that can exist outside the action of the supreme intelligence. There is no "remainder." The only thing absolutely lacking in form is not a thing, but *nothing*. Consequently, everything that exists comes forth from this formative and creative intelligent being. This being is not only an *artist* but also a *creator*.

Let's move on to the next argument.

II. The One Necessary Being

This argument starts from a simple but dizzying question: *"Why is there something rather than nothing?"* As maddening as this question might seem at first, it is perfectly legitimate. From the standpoint of pure logic, there is a solid principle which since Leibnitz has been called the "principle of sufficient reason." This principle, which is the starting-off point for every investigation, can be expressed as follows: Everything that exists has an explanation for its being, either in itself, or in something else. You will observe that if this principle were false, we would live in a chaotic and unpredictable world in which all sorts of things could arise at any moment without any explanation. But by all accounts, this is not the case. Our experience of an ordered world is a solid basis for applying this principle without hesitation, and, in our case, asking why the Universe exists.

At this point in our reflection, seeing that science has been able to offer such excellent explanations of phenomena, we might be tempted to entrust it with the resolution of this last question. Our thought process would go like this. Take the Universe at any given moment: according to the laws of nature, the existence of the Universe at this moment is explained by the moment immediately prior. The moment that has just passed, for its part, can be explained by a previous state, and so on to infinity. The Universe, at any point in its existence, can be defined as the sum total of these states, with each one explained by the state immediately previous to it. In this way, everything is explained without any need to refer to an exterior cause. Our grand metaphysical question is at once deflated and resolved.

But this explanation doesn't hold.

By tracing the series of causes back to infinity, we do not explain the existence of the Universe, only its transformations

We can certainly explain why, *given the fact that the Universe exists*, we find it in this or that state at any given point in history. But the more fundamental question—the reason for the existence of the whole series of moments—still remains to be answered. Physics is not equipped to answer this question, because physics deals with things within the Universe. It necessarily accepts the existence of the Universe as a given, and proceeds from there. A scientific explanation for the existence of the Universe would have to refer to a state preceding the Universe; in other words, to presuppose its existence. In doing so, it can only postpone the question one more step, not resolve it. So we have to face the facts: even if the Universe has existed for eternity, and even if it had no beginning, the long series of linked causes and effects needs an explanation.

The reasoning goes like this:

1. Everything that exists has its reason for existing either in itself (necessary being) or in something else (contingent being).
2. But the totality of contingent things cannot have its reason within itself.
3. Therefore, the totality of contingent things has a cause in something else.
4. But this other thing must be a necessary being.
5. Therefore, the totality of contingent things has its reason for existence in a necessary being.
6. But the Universe is part of the totality of contingent things.
7. Therefore, the Universe has its cause in a necessary being.
8. But a necessary being lacks all the qualities proper to contingent beings: spatiotemporality, quantitative limitation, composition.
9. Therefore, the totality of contingent things of which the Universe is a part has its cause in a nonspatial, atemporal, and simple being. We will call this being God.

Proposition number 1 is the principle of sufficient reason. It defines the difference between necessary being (which has its cause in itself) and contingent beings (which have their cause outside themselves). How do we draw the distinction between these two types of beings? We start by asking whether we can coherently conceive of a given thing being different than it is (if we found it in another world instead of here, for example). If this is impossible to imagine, then the thing is a *necessary* being, something that has in itself the reason for being. If it is possible for it to have been different, then it is a *contingent* being, whose reason for existence lies outside itself in what we call a *cause*. Let's take a look at an example. It is impossible to imagine the equation $1 + 1 = 2$ being false, even on some other planet. So it seems that the immaterial reality described by this proposition is absolutely necessary in itself. The same holds for the laws of logic. These and other realities accessible to us exist out of absolute necessity. On the other hand, if you take the phrase "I was born on May 7," or "the Eiffel Tower is 1,063 feet tall," or, "an electron has a mass of 9.109×10^{-31} kg," it is obvious that we are dealing with contingent truths. It is very easy to imagine that these things could be different than they are. These phrases describe facts, of course, but there is nothing that logically prevents them from having been different. I might have been born on the eighth, and the Eiffel Tower could be four feet shorter, and the electron could have a different mass—in some other Universe, anyway. These values are what they are because of a cause outside themselves, even if we aren't sure what that cause is.

Propositions number 2 to 5 are logical consequences flowing from the first premise. If we were to take all the contingent beings (namely, all those things that have an explanation outside themselves) and put them into an imaginary bag, then we would have a large contingent being whose explanation would be outside itself. Since the bag contains all contingent beings, it cannot be explained by means of another contingent being. Therefore, its explanation must be found in a necessary being. This is consistent with what we said earlier about the pseudo-explanation of the Universe: an infinite series of contingent beings cannot explain itself, because it is itself contingent. Therefore, we have to look for an external cause.

Proposition number 6 affirms that the Universe belongs to the category of contingent beings. At this point, someone might object that even if composite entities are in fact contingent and require causes, the matter out of which such beings are composed does not need any cause or explanation outside itself. They will say that matter-energy is the necessary being we are seeking. In this view, matter belongs to the same category as mathematical and logical entities; it is, as Spinoza put it, *"that eternal and infinite being we call God, or Nature."*[664] No need to look for anything else: we have our necessary cause. But this cannot be. Matter does not contain its own explanation: rather, it is marked by all the signs of contingency. Even the scientists have recognized this: the fundamental characteristics of matter show no trace of necessity. It is perfectly possible to conceive of it being different without any contradiction. Here is what Steven Weinberg wrote on the subject:

"I have to admit that, even when physicists will have gone as far as they can go, when we have a final theory, we will not have a completely satisfying picture of the world, because we will still be left with the question "why?" Why this theory, rather than some other theory? For example, why is the world described by quantum mechanics? Quantum mechanics is the one part of our present physics that is likely to survive intact in any future theory, but there is nothing logically inevitable about quantum mechanics; I can imagine a universe governed by Newtonian mechanics instead. So there seems to be an irreducible mystery that science will not eliminate."[665]

In other words, we must consider all the matter in the Universe a contingent being. For it to be otherwise, the theorems used to describe it would have to be as blindingly obvious as the propositions of logic and mathematics. But that is manifestly not the case. The fact that an electron has a mass of 9.109×10^{-31} kg has all the signs of an arbitrary proposition with an almost cruel lack of explanation, as opposed to the equation $1 + 1 = 2$, which requires no further explanation, as it holds the explanation in itself.

Proposition number 7 follows from the previous one: if the Universe's

---

664. Benedict de Spinoza, *Ethics*, in *The Chief Works of Benedict Spinoza*, trans. R. H. M. Elwes (London: George Bell and Sons, 1901), part IV, preface.

665. Steven Weinberg, "A Designer Universe?" *Skeptical Inquirer* 25.5 (2001): 66.

matter is contingent, then it must have its explanation in a necessary being. Some have sought to resist this conclusion by saying that the principle of sufficient reason applies to what is *inside* the Universe, but doesn't have to apply to the Universe as a whole, which needs no explanation. Now we are no longer in Spinoza's camp but that of Sartre: the Universe is contingent, but it is absurd—that is, it has no explanation at all. The weakness of this proposition is that it relies on the existence of an exception to the principle of reason. But how is this justified? The burden of proof falls on those who make exceptions. They have a simple answer: there can be no explanation of the Universe as a whole… because there is nothing outside of it. How do they know there is nothing outside of the physical world? That is begging the question. Common sense demands that we hold these two solid truths: the Universe is contingent, and the principle of sufficient reason is valid.

The conclusion follows easily: if the Universe exists, then it has a causal explanation outside itself.

Proposition number 8 draws our attention to the particular characteristics of the first cause. Because this first cause explains the totality of physical beings, it cannot itself be physical. In other words, it is immaterial. Notice that if the necessary cause had spatiotemporal properties, then all the questions that apply to matter would apply to it as well: "Why does it have this particular set of dimensions, rather than another?," etc. In other words, it would be contingent. This first cause has to be free of everything that makes contingent beings demand an external explanation: it has to lack extension and quantitative determination of any sort, depend on nothing, have no parts, and be infinite (because for any given magnitude, we could ask why it is this size rather than any other). To put it positively: it has no cause, it is absolutely simple and immaterial. It is not *this* or *that* thing but absolute being, without any limit. In fact, if we were able somehow to "intuit" this being's essence directly, we would immediately recognize that, like a logical truth, it is absolutely obvious. We would have no questions: nothing about it would require explanation. This is the difference between what is contingent and what is necessary.

Why should we call such a being "God"? Two reasons. First of all, the first cause is immaterial. There are two types of immaterial being: abstractions and spirits. However, the first cause cannot be an abstraction (like a number or a mathematical function) for the good and simple reason that abstractions do not have any causal power (it is not the number 11 that wins the match, it is the eleven players on the team). Since the first cause, as its name suggests, enjoys a special causal power, it has to be something analogous to a spirit. Note that this point is consistent with the conclusion of our first argument. Second, the necessary being is *unique*. Indeed, imagine for a moment that there were two necessary beings. Being immaterial, they could not be distinguished in a spatio-temporal way. They could only be distinguished in their definition. But the special property of this necessary being is that it is existence itself, without definition, without limit, without any particular determination, without a shape that requires an outside explanation; in other words, it is being in its fullness. But there is only one way to be "pure being." Therefore, two necessary beings would be completely indistinguishable from one another and would be... the same being. Therefore, the necessary cause is a unique, all-powerful spirit. We could not paint a more accurate portrait of the figure commonly called "God."

Now on to the third argument.

III. The Creator of Time

In the previous argument, we said that the Universe, even if it were eternal, would require an outside explanation to account for its origin. In the final analysis, God might very well have created an eternal world that depends eternally on him. But are we so sure that it is conceptually possible to imagine a Universe existing for all eternity? All the same, this question deserves to be considered. This is the goal of the third argument.

Upon reflection, we find that the Universe cannot be eternal, for the simple reason that there cannot be an infinite past. Consider: if the past really were infinite, then the present would never come. The Universe would continually be passing through the infinite series of states that precede the present. Just as there is no way to reach infinity by adding numbers to zero, neither can you reach zero if you start from negative infinity. How can you get to a place that is impossible to reach? Even worse, it is impossible to start from "*less than infinity,*" because this operation doesn't even have a starting value. To start at infinity, said philosopher J. P. Moreland, "*is like trying to jump out of a bottomless pit.*"[666] This operation can be expressed by the equation $-\infty + 1 = -\infty$. No matter how many times we add one, we never make any progress. Therefore, given that the Universe does exist in the present, we have to conclude that the past is finite. There must, of necessity, have been a radical beginning. Notice that this argument is not based on astrophysics, the Big Bang Theory, or anything of that sort. It is purely philosophical, essentially consisting in a demonstration that the existence of an infinite past is a logical and metaphysical impossibility. Besides the impossibility of crossing an infinite span of time, we can also show that an infinite past presupposes a chain of causes with no origin, which leads to insoluble contradictions. A brief thought experiment will suffice to make this clear. Let's imagine a Universe composed of two particles that have existed forever. We'll call them Alpha and Beta. Alpha vibrates once every second, and each one of its vibrations has the capacity to switch particle Beta from a negative (-) to a positive (+) state, definitively. Now, if we observe particle Beta at any time t, logically we should find that it is in a positive state (+), since no matter what time t we choose, there have already been an infinite number of periodic vibrations from Alpha. The question is, which vibration precisely is responsible for the positive charge of Beta? The vibration that took place at $t-1$? That can't be it, since before the vibration at $t-1$ there was a vibration at $t-2$, and indeed this holds true for any $t-n$. No matter how far back we look, we still have to go a step farther. So we are faced with a contradiction: Beta must be in its positive state (+) because of Alpha's vibrations, *but no identifiable one of Alpha's vibrations can have caused Beta's positive state.* This is absurd. But the absurdity of this situation is caused only

---

666. J. P. Moreland and Kai Nielsen, *Does God Exist? The Debate between Theists and Atheists* (Amherst: Prometheus, 2009), 37.

by our assumption of an infinite chain of causes. Therefore, we can conclude that such a chain is impossible. No phenomenon, no event is explicable in terms of an infinite causal history.

So our intermediate conclusion seems solid enough—time must have had a beginning. Even this simple fact has dizzying consequences...

We are not talking about the beginning of something inside time but rather the beginning of *time itself*. In other words, it is a radical beginning. Let's be clear: just as it is absurd to ask if there is anything "north of the North Pole," it is absurd to ask what there was "before time." There was no before. For time to have a beginning means that all conceivable physical, spatiotemporal reality had to have a radical beginning (and this includes as many "multiverses" as you like). This origin is not preceded by any time, nor, therefore, by any spatiotemporal reality. Are we saying that there is no explanation for how all this came to be? Not at all! The principle of causality applies: everything that begins to exist has a cause. To deny this is to posit a magical world and uphold the idea that something can come from nothing. Therefore, the Universe has a cause. This cause is far from ordinary: it is beyond time (because it is the cause of time), it is beyond space (because everything that has extension moves, and everything that moves is within time), it is all-powerful (because it produced the whole physical world *ex nihilo*). Finally, this cause did not act before the first instant of the Universe, but at the first instant of the Universe, and it still exists, because it is atemporal.

Putting all the pieces together, the argument goes like this:

1) Everything that begins to exist has a cause.
2) The totality of spatiotemporal reality began to exist at one point in time.
3) Therefore, the totality of spatiotemporal reality has a cause.
4) If the totality of spatiotemporal reality has a cause, this cause is atemporal, nonspatial, uncaused, and all-powerful.
5) Therefore, the totality of spatiotemporal reality was caused by a nonspatial, atemporal, uncaused, and all-powerful being.

The first proposition is self-evident. The second is based on the impossibility of traversing a real infinity. The third follows logically from the first two. The fourth describes the properties necessary for the cause in question: the cause of time cannot be preceded by anything, nor can it itself be within time. The conclusion necessarily follows.

Let's take stock. Since the word "God" carries so much emotional baggage, it is natural that philosophers will continue to disagree, and perhaps never come to a unanimous agreement regarding these three arguments. That said, a significant number of contemporary philosophers do support the arguments we have just explained; among others, we might cite David Oderberg, Joshua Rasmussen, Robert Koons, and Alexander Pruss. Moreover, these arguments have convinced even highly intellectual serious atheists. Among these is Edward Feser, a contemporary American philosopher, whose story deserves our attention:

"I don't know exactly when everything clicked. There was no single event, but a gradual transformation. As I taught and thought about the arguments for God's existence, and in particular the cosmological argument, I went from thinking, 'These arguments are no good,' to thinking, 'These arguments are a little better than they are given credit for,' and then to, 'These arguments are actually kind of interesting.' Eventually it hit me: 'Oh my goodness, these arguments are right after all!' By the summer of 2001, I would find myself trying to argue my wife's skeptical physicist brother-in-law into philosophical theism on the train the four of us were taking through eastern Europe."[667]

If the three arguments we have presented are valid—and we believe they are—then one is perfectly justified in concluding that the Universe, as well as any contingent reality known or unknown to us, was caused by a necessary, simple, unique, immaterial, atemporal, uncaused, all-powerful, and intelligent being. It doesn't seem too much of a stretch to call that being "God"!

---

667. https://edwardfeser.blogspot.com/2012/07/road-from-atheism.html.

24.

Materialist Arguments against the Existence of God

There are many arguments against the existence of God, and some pose legitimate questions. Any reader who is so inclined can find them easily on the Internet and see that countless books[668] and articles[669] defend the atheist worldview. Here we will examine twenty claims that represent the typical arguments put forth on this subject. We will try to classify these arguments by type and present them as they are commonly formulated and reproduced, in their rough-and-ready form.

I. Skeptical arguments

1. "There is no proof for the existence of God. Otherwise, we would know it already"

The idea that there is no proof of God's existence is widespread and often taken for granted. The reasoning follows: "if there is no proof of God's existence, it is just because he does not exist." We hope that reading this book has convinced you otherwise. While the proofs presented here are not absolute proofs, and are more accurately described as merely relative proofs, there are many of them, they are compelling and they come from many different fields of knowledge.

---

668. Some notable examples are Sébastien Faure, *12 preuves de l'inexistence de Dieu* (1908); Bertrand Russell, *Why I Am Not a Christian: And Other Essays on Religion and Related Subjects* (1927); and Richard Dawkins, *The God Delusion* (2006).

669. For example, Wikipedia has a page called "Existence of God," where section 3.2 is devoted to arguments against the existence of God. https://en.wikipedia.org/wiki/Existence_of_God.

2. "The burden of proof is on the person who makes the claim"

When someone asserts that there is no proof for the existence of God, they often add the following remark: "At any rate, the burden of proof is on the person who makes the claim, not on the person who is asked to believe it." In other words, the person who affirms God's existence has to prove his case, and the one who denies it is free to disbelieve in the meantime.

"The person who affirms God's existence has to prove his case..."—and that is precisely what this book aims to do. But however compelling the evidence may be, to simply "believe in nothing while waiting for proof" is not a great mark of wisdom. Believing in nothing is itself a belief and an active choice.

3. "The existence of God is impossible to prove one way or the other, so it is better not even to talk about it"

This argument is similar to the first but more ambitious. If the first argument appeals to ignorance, this second one makes a positive claim about the nature and limit of our knowledge. The argument was first articulated by Immanuel Kant in 1781, long before the scientific discoveries of the twentieth century.

Kant did not intend to deny the existence of God. Rather, he argued that it was categorically impossible to give an irrefutable answer to the question. His view still enjoys widespread support among intellectuals.

Whatever its value in its own time, the Kantian view is now outdated. The major scientific discoveries of the late twentieth century demonstrate that science and metaphysics can come together. The beginning and end of the Universe, as well as the fine-tuning that regulates it, now attest to the fact that science can offer serious arguments in support of the necessity of God's existence.

4. "We don't need God to explain the origin or functioning of the Universe, so we can leave him out of the equation"

The previous argument claimed that the existence of God is impossible to demonstrate. This one asserts that his existence is unnecessary. According to this argument, the affirmation of this non-necessity, which seems liable to result in the deterministic views developed in the nineteenth century, allows one to conclude the following: "*If the Universe does not need God to exist and function, the simplest conclusion—which is generally the best—is to think that he does not exist.*" This was the position of French mathematician Laplace, presented in his famous quip to Napoleon.

The argument is based on weak premises, for we can easily imagine that God created a Universe that is comprehensible to man and that functions like a sophisticated and well-designed watch running on its own without any intervention on his part. The assumption that no creator is needed to explain the functioning of the Universe is itself no proof against a creator's existence.

But this argument has been completely disproved by the great discoveries of recent times. Since the argument rests on false assumptions, it does not establish the conclusion that God does not exist.

5. "The insignificance of mankind and the Earth proves that man has no special destiny. Therefore, he cannot originate from a god"

This notion stems from the idea that man and the Earth are insignificant in the context of a broader Universe. Yet this insignificance is at odds with the fact that man pretends to be the masterpiece of creation. From this perspective, man cannot be seen as the beneficiary of an immortal destiny granted by an all-powerful God.

Those who defend this argument point out that the Earth is not at the center of the Universe, that it is only a small planet rotating around the Sun and not the reverse, that our solar system itself is insignificant, nothing but a speck of dust in a galaxy, itself lost in the midst of billions of other galaxies somewhere deep in the Universe. Most of all, they highlight that "man" is only the descendent of primates and that his appearance stems only from completely fortuitous physio-chemical processes. For all of these reasons, the insignificance of man and the Earth seems to them indisputable and therefore incompatible with any claim of a divine origin.

Therefore, the belief in a divine author who developed a privileged destiny for man is only a reflection of man's desire for immortality. This desire stems purely from man's vanity and naivete.

This argument may have had some merit a century ago, but it no longer does today. Indeed, we now know that its foundational premises are false.

The Earth is not a little planet lost in the depths of the Universe. The Universe is like the surface of a sphere in that it has neither a center nor a remote corner. As for the size of our planet, it just happens that it is exactly the size that it should be. In other words, its size is perfect. Finally, the appearance of life and humanity is absolutely not a fortuitous physio-chemical phenomenon, as the chapter on biology has amply demonstrated.

6. **"Polls show that today few scientists are believers, in any case far fewer than in the general population; this proves that science leads to the belief that God does not exist, and this is logical because God doesn't exist"**

This important question was addressed in detail in Chapter 14, and we saw that the claim that science necessarily leads to materialism is not justified.

II. Philosophical arguments

7. "Can God create a rock so heavy he can't lift it?"

This question seems to have only two possible responses. Either he cannot create it, in which case he is not omnipotent; or, he can create it, but cannot lift it, in which case again he is not omnipotent. In both cases, he is not omnipotent. But God has to be omnipotent; therefore, he does not exist.

This flawed logic starts from an erroneous premise. Though God is, in fact, all-powerful, he cannot make things that are illogical or contradictory. Since he is reason itself, he cannot do one thing and also its opposite. He cannot make one thing true and false simultaneously, any more than he can make two plus two equal five.

Thus, God cannot create a rock heavier than he can lift, with no injury to his omnipotence.

8. "Can God commit suicide?"

If God can commit suicide, then he is not eternal. If he cannot, he is not all-powerful. Either case proves that he is not God, and, therefore, that he does not exist.

This reasoning is based on the same error as the previous one.

While these two arguments are of course simplistic, we must name and refute both of them. The reader should be aware that there are many other similar claims that we cannot mention here.

Bear in mind that Thomas Aquinas himself addressed and refuted the first of these arguments.

9. "A perfect God could not have created an imperfect world. Since the world is obviously imperfect, God does not exist"

The mistake in this argument is thinking that God can only create perfect things. First of all, if that were true, then he wouldn't be all-powerful. God can create imperfect things in order to lead them over time to a state of perfection.

In this view, man (who is not a "thing" but a being) is not born complete. He can study, educate himself, grow, achieve things, and in so doing determine what he will be later on in eternity through what he freely decides to do with his life on Earth. Man is partially the author of his own destiny, worth, and future glory (in the theological sense of the term). In sum, he is to an extent his own maker, his own creator.

Only machines, which are tools, are expected to be perfect and complete at the moment of their creation. Man, by contrast, is a free being, and this means that he has the privilege of evolving and perfecting himself over time.

If, instead, the world had been created in a state of perfection from the beginning and never deviated from that state, then neither human freedom nor human merit would exist.

A perfect Universe is either deterministic or has already reached its end. But ours is neither, for the moment.

Thus, we and our world are imperfect for two fundamental reasons: one, so that we can develop over time; two, so that this development can take place as a result of our free action.

In the Judeo-Christian view, the temporary imperfection of the world—or, in other words, its incompleteness—is not at all a testament to God's nonexistence. Quite the contrary, it reveals his ability to create free and evolving beings.

10. "If God is omniscient and knows the future, man is not free; if, on the other hand, God does not know everything, he is not all-powerful and therefore does not exist"

God's powers, such as we are able to imagine them, far surpass human understanding, for, if God exists, he is not superior to humans by only a few degrees—he is completely different from us. He is transcendent. We are only capable of having general ideas and very vague representations of God.

It is the same for everything that has to do with God's infinity, his eternity, his omnipotence, and his omniscience. From this perspective, the apparent contradiction between God's omniscience and human freedom should be interpreted as stemming from God's transcendence rather than from his inexistence.

11. "God is incompatible with chance. If God is perfect and omnipotent, then chance should not exist. But chance does exist, so God does not"

God's omnipotence does not mean that he has to micromanage every detail and instant of the life of the Universe.

His omnipotence only implies that he may do so if he wishes.

If God managed everything down to the tiniest detail, chance would not exist, and the history of the world would be entirely determined in advance. God would then be the designer of a great machine or algorithm, and man would be a perfect but entirely slavish automaton.

But, since God wants his creation to be free, what happens in the world cannot be fixed in advance, and chance must exist.

The twentieth-century discovery of the quantum nature of our world—which shows that it is indeterminate and radically subject to chance—is metaphysically essential. It provides indirect confirmation of the thesis that God exists while undermining the determinist positions held by atheist scientists of the nineteenth century.

Hence the existence of chance is actually a proof for God's existence, whereas the lack of chance would be a sign of his nonexistence.

Thanks to his desire for chance, God limits both the exercise and visibility of his omnipotence.

Anthony Suarez, a Swiss physicist and specialist in quantum mechanics, has said: "*Chance and guidance have the same origin and the same cause [. . .] Might chance have been one of the most important things that God created in the beginning?*"[670]

The same author also states: "*You cannot have both freedom and scientific determinism. Either freedom is an illusion, or we cannot consider classical physics complete.*" Again: "*In quantum physics, the laws of freedom can be surreptitiously inserted into natural causality.*"

12. "If God is perfect, then he is self-sufficient, and there was no need for him to create the world. But the world exists, so God does not exist"

The first three elements of the argument are entirely correct: God is perfect and self-sufficient, so he had no need to create the world.

---

670. From a conference held in London, "Does God play dice?" January 2–7, 2007.

The error in this reasoning appears when we jump from the assertion that "*God had no need...*" (which is true enough), to the claim that "*therefore, he definitely did not.*" The latter claim is false, because God is perfectly free to make things he has no need of—incidentally, so is man.

Indeed, God did not need to create us, since he is perfectly happy and self-sufficient in himself. According to the Judeo-Christian conception—from his overflowing goodness he desired that others partake in his happiness. It is a characteristic of love to spread and give itself freely, without any necessity and without expecting any reward in return.

Thus man and the world exist as a result of God's gratuitous love.

13. "The Universe is self-sufficient. It does not need a cause"

One of the main philosophical proofs for God's existence is that there is no effect without a cause. As the Universe is an effect, it must have a cause, and because one cannot go back indefinitely in the chain of causes, there must have been a necessary being—a first cause of all that exists—what we call God.

This reasoning would be false if the Universe were self-sufficient: in that case, it would not be an effect. It would be autonomous and eternal, like a long chain closed on itself, in which the links are all interconnected. If the Universe were not an effect, there would be no reason for its cause to exist. Therefore, God would not be necessary at all. If he were not necessary, then we can suppose he would not exist.

This argument does not claim to be a direct proof against God's existence but rather a refutation of a proof for his existence; however, it makes an unwarranted jump to the conclusion of God's nonexistence.

If the great scientific discoveries of the twentieth century have shown anything, it is that the Universe is not a "*Great Whole*" that is eternal, self-sufficient, and closed in upon itself. The late twentieth-century discovery of the Universe's eventual thermal death and the demonstration that there will never be a Big Crunch have shown that the "*universal, eternal, and self-sufficient Great Whole*" is a false conception of the Universe.

Thus, the philosophical argument that claims that the Universe is not an effect because it is a "*Great Whole*" has lost all value today.

III. Moral arguments

14. "A benevolent God would not tolerate the evil and suffering we see in the world, much less create it. But suffering and evil exist in the world, so God does not exist"

Understandably, this is the philosophical argument that most strongly captures the public's attention. As much as it is an argument against God's existence, it is often formulated as a reproach against God. The existence of evil and suffering is a profound mystery. The Old Testament "Wisdom books" of Job and Tobit are dedicated almost exclusively to exploring it. As it is not possible to summarize them here in only a few lines, we direct the reader instead to the Bible, which is in part designed to help us understand the problem of evil.

This claim has several points in common with the argument about the imperfection of the world, and the arguments we used to refute that objection also apply here. However, there are several significant differences between "the imperfection of the world" and "evil" that we will now discuss briefly.

According to the Bible, God is not responsible for the evil that exists in the world today. Evil entered into the world through man tempted by the devil. In this view, God tolerates evil on a temporary basis for the same reasons mentioned above: to permit man to exercise his liberty, giving him room to develop, build his identity, and determine his ultimate fate and improve himself through positive actions over the course of his earthly life.

While the suffering of the innocent is intolerable, we must recognize that here on Earth we are not in a position to understand God's actions.

Thus, from the biblical perspective, evil and suffering do not prove that God does not exist; they are, rather, the consequence of the bad use that angels and men have made of their freedom.

15. "Believers do not provide an example of moral excellence. Their bad and even scandalous lives prove that their religion is just a facade with nothing behind it"

Certainly the testimony of a life lived according to one's principles is more valuable than any amount of empty talk. The first Christians converted the world through their exemplary lives, many miracles, and the martyrdom that many of them freely accepted—not by recourse to long sermons.

By contrast, the example of bad Christians has had the opposite effect over the centuries.

The evil practices of those believers who profess the virtues of sharing, forgiveness, and brotherly love while doing the exact opposite are not proof against the God they claim to believe in. Rather, their behaviour is a testament to the great difficulty many people face in trying to live according to their beliefs.

IV. Arguments drawn from religion

16. "The existence of multiple, contradictory religions on Earth is incompatible with the existence of a single, benevolent God"

It is undeniable: there are a multitude of different and contradictory religions on Earth. Therefore, most of these religions are necessarily false. We can conclude from this that many human beings practice false religions, and that a lot of people are born, live, and die in error, through no fault of their own.

But if God were, by definition, good, just, and all-powerful, he would have communicated the truth to mankind. If he did not do so, that is either because he could not (in which case he is not all-powerful), or because he did not wish to (in which case he is neither good nor just). In either case, it follows, once again, that God does not exist.

A refutation of this thesis has first to consider that God has always faced two contradictory accusations. First, he is accused of being a tyrant who never leaves us alone and spends all his time forbidding humans to do all sorts of things, even the most ordinary pleasures. Second, he is accused of being indifferent or absent, allowing evil and those who practice it to flourish. At least one (and maybe even both) of these two accusations is false. Let's have a look.

It was the serpent who suggested to Eve that God was a tyrant: "*Did God say, 'You shall not eat from any tree in the garden?'*" (Gen. 3:1). Despite its inaccuracy (God merely forbade them to eat from just one of the trees), this accusation has stuck over the centuries.

A moment's reflection serves to show that this accusation is baseless. People do what they want every minute of their lives, and experience has shown that God does not rain lightning down on our heads when we fail to respect his commandments. God has left us "instructions" about how to lead our lives. These "instructions" or "marching orders" are inscribed in our conscience, but we are free to choose not to follow them. There is, therefore, no basis to the claim that "*God is a tyrant.*"

The opposite accusation, that "*God is absent and allows evil to happen*," is the more serious one. It can only be understood through the conjecture that God places great value on human freedom. Because of this, God cannot prevent man from doing evil without calling his freedom into question.

In this way, the multiplicity of religions and error is more a testimony of the freedom that God grants to men than a proof of his non-existence.

17. "The fact that there are many gods proves that none of them exist"

This objection is basically identical to the previous one and can receive the same response.

18. "Religions have always been the cause of violence and wars"

Those who make this argument point out that religions, considered the representatives of benevolent gods on Earth, have often engendered war and violence. Consequently, as these religions have brought about evil, they cannot really be established by a good God. So we can conclude either that God is evil or that he does not exist. Since God cannot be evil, he must not exist.

The premises of this argument are also false, and we only have to look at the misfortunes of the twentieth century for ample evidence of this fact. The wars waged over the last century by dogmatic materialists were far more cruel, violent, prolonged, and lethal than all the religious wars of the past put together. The crimes and violence committed by Hitler, Stalin, Mao Zedong, and Pol Pot have no precedent in human history.[671]

We then come to the unfortunate conclusion that man himself is inclined to evil and violence, and that religions and ideologies are often merely instruments or pretexts to indulge this evil inclination.

Recent history reveals that this argument i.e., that religions are the cause of violence and wars, is groundless.

19. "The existence of hell proves that God is not good. But since God must be good by definition, neither he nor hell exists"

To put it another way: *"Hell is an unlimited punishment. Man, on the other hand, is limited, ignorant, and largely irresponsible. Everything he does, even what he does wrong, is of a limited nature, and so an infinite Hell is, therefore, a disproportionate and unjust punishment. If God exists, he is necessarily just; since Hell appears to be an unjust punishment, God does not exist."*

---

671. Phillips and Axelrod's *Encyclopedia of Wars* cataloged 1,763 wars that have taken place throughout the history of mankind. Of these 1,763 wars, 123 were based on religious motives, or less than 7% of the total. Not only are the wars waged for religious reasons less numerous than some people suppose, but, according to the same encyclopedia, the number of people killed in these 123 "religious" wars is equivalent to 2% of the total number of persons killed in all historical wars. That means that "religious" wars in general have had fewer victims than wars waged for other motives.

Since this argument refers to the Judeo-Christian concept of Hell, we will respond from that perspective.

The existence of Hell, an eternal place of suffering with no hope of escape, is one of the saddest but best-attested realities in Scripture and Church teaching. It is always spoken of in direct and raw language:

• *"Then he will say to those at his left hand, 'You that are accursed, depart from me into the eternal fire prepared for the devil and his angels.'"* (Matt 25:41)

• *"And if your eye causes you to stumble, tear it out and throw it away; it is better for you to enter life with one eye than to have two eyes and to be thrown into the hell of fire."* (Matt 18:9; Mark 9:47)

• *"Enter through the narrow gate; for the gate is wide and the road is easy that leads to destruction, and there are many who take it. For the gate is narrow and the road is hard that leads to life, and there are few who find it."* (Matt 7:13-14)

In the New Testament, Hell (or its equivalents, "Gehenna," "the darkness," "eternal fire," etc.) is mentioned forty-five times, with much more frequency than most of the other realities Christ taught about.

Today, skepticism about the existence of Hell has penetrated all strata of society.

Certainly, it is very hard to believe that an insignificant and ignorant human will face an eternal fate at the end of his life, whether in eternal bliss or eternal damnation. It seems so disproportionate to his actions.

Earlier we examined a weighty philosophical argument that denies God's existence on the grounds that he could not have created an imperfect world. We saw that this imperfection, which is better described as incompleteness, is permitted on a temporary basis and is even necessary to ensure that man and the world can develop. But at its root, this initial argument is perfectly well-founded. In the long term (namely, eternity), God cannot create imperfect things. Therefore, there must come a time when we all enter into an eternity where nothing can change further.

The provisory character of the present world and the eternal character of the world to come are a necessary result of God's own existence. One can certainly deny the existence of God, but if you believe in his existence, it would be incoherent to deny the existence of eternity.

Eternity is a reality that is intrinsically linked to God's existence.

In the final analysis, the reality of an everlasting Hell, or even the possibility of its existence, is no proof that a good God does not exist. Rather, it is evidence that, because we have been created by an eternal and transcendent God, our fate too must be eternal and sealed at a given moment.

20. "The Bible is just a collection of primitive legends riddled with errors"

This common accusation is not intended as a proof against the existence of just any creator God. It rejects the existence of the Judeo-Christian God in particular. Pointing to the supposed errors present in the Bible, it concludes that whoever inspired this faulty text could not have been a God. Christians and Jews insist that their Bible was inspired by God, but the God they worship can't really exist, since he was mistaken about so many things.

This argument suffers from the profound misreading of the Bible. In fact, the truth is entirely the contrary, as we hope the two chapters dedicated to this subject have sufficiently demonstrated.

The existence of the many "*humanly inaccessible truths of the Bible,*" as we've called them, are actually strong evidence for the existence of a higher spirit.

Conclusion

For those who genuinely desire to explore the question of God's existence at a deep level and reach a sound decision about whether or not to believe, it is insufficient only to know the evidence in favor of it. We also need good responses to the diverse arguments and objections that are commonly brought forward to prove the opposite. That has been the purpose of this important chapter.[672]

---

[672]. We have not addressed some of the more far-fetched arguments that have been made throughout history against God's existence, such as his invisibility. Cabanis, an eighteenth-century anatomical doctor, exclaimed: *"I cannot find the soul with my scalpel!"* In a similar vein, a highly professional Soviet cosmonaut, Yuri Gagarin, cried out from his space vessel Vostok 1: *"I don't see any God up here!"*

CONCLUSION

25.

Materialism: An Irrational Belief

Materialism has always been just a belief, but today it increasingly appears as an irrational belief. Although many people may still choose to embrace it, that choice seems devoid of any rational justification. Materialism in many cases has become an intellectual justification for both individualism and the rejection of moral standards.

The evidence for God's existence presented in this book is modern, clear, and rational. Drawn from a variety of scientific and humanistic disciplines, these findings can be tested against real-world observations. While much of this evidence was unavailable as recently as the nineteenth century, scientific advances have since made remarkable strides. Even our historical understanding of who Jesus could be has evolved significantly over the past century.

We are living in extraordinary times. Although this shift has gone largely unnoticed by the general public, we are in the midst of an intellectual paradigm shift that fundamentally redefines our approach to the question of God's existence. The introduction's title, 'The Dawn of a Revolution,' might have at first seemed overly bold—even pretentious—but by the conclusion of this book, it is clear that it is justified.

The pieces of evidence are clear

You don't need to be a scientist to grasp the ideas and stakes behind this evidence—whether we're examining the Universe's beginning, its precise fine-tuning, the unlikely emergence of life on Earth, or the historical anomalies discussed in this book.

They are intelligible

These ideas are clear and accessible. Our arguments—drawn from cosmology, biology, mathematics, philosophy, ethics, and history—are all grounded in reason, careful analysis, and sober judgment.

The theoretical implications of this evidence can be tested against the observable Universe

True, the proofs we present are not absolute proofs. But, just like the many other scientific theories everyone accepts, they belong to the category of evidence that can be validated by examining their implications against the real Universe.

The evidence is abundant

The 25 chapters of this book bear witness to this. There is much additional evidence, but if we had included it all, the book you hold in your hands would be the size of an encyclopedia!

The evidence comes from many fields of knowledge

There are books devoted to the scientific evidence for God's existence, others focus on philosophical proofs, and still more on theological arguments. Our book is unique in offering a comprehensive panorama—including cosmology, philosophy, ethics, and historical enigmas. This richness of perspectives is possible only because evidence for God's existence appears in every field of human study.

Many who defended this evidence were persecuted

The terrible persecution of Russian and German scientists—whose stories we recounted in Chapter 8, 'The Big Bang: A Noir Thriller'—should give us pause. Had the thermal death and expansion of the Universe not been interpreted as evidence of a beginning, and therefore of a creator, such persecution would never have occurred.

What is the takeaway?

For believers, this book demonstrates that their convictions are based on solid rational foundations—even if others disagree. Our comprehensive survey of evidence equips them with the arguments needed to challenge the prevailing narrative, often touted by political correctness, that belief in God is irrational and should be confined to the private realm. In fact, the opposite is true.

For seekers who sometimes question the existence of spiritual realities or ponder why there is something rather than nothing, this book reveals just how unfounded the materialist worldview is and, conversely, how well-supported the theistic perspective is.

Finally, for materialists, this book highlights the formidable challenge that modern science poses to their views. To refute the thesis that a creator God exists, they must hold that every piece of evidence presented in this book is false, and consequently, they must simultaneously accept all of the following statements:

- There is an infinite number of universes beyond our own. (This is the only hypothesis they can rely on to escape the fine-tuning problem—even though it resembles science fiction and is completely devoid of evidence).
- At least one of these universes must be eternal; otherwise, one would have to have arisen from nothing.
- The leap from inert matter to life falls within a reasonable range of probability.
- Jesus was nothing more than a failed opportunist, or something to that effect.
- The unlikely and indisputable truths contained in the Bible exist purely by chance.
- The story of the Jewish people is not extraordinary.
- The extraordinary event at Fátima was nothing more than a hoax.
- Good and evil do not exist; anything goes.

Many committed materialists may not have yet realized how many unattractive assumptions underlie their position, assumptions they must accept to remain rational and coherent. They might be surprised to discover that, in effect, they are great believers, though in ideas that starkly contrast with the unbelief they once claimed. Recognizing this inconsistency could be a crucial first step.

For God made man to seek him

As Paul of Tarsus writes in Acts 17:26-27, "God made them [men], so that they would search for Him, and if possible, grope for Him and find Him, who in fact is not far from each one of us." These words encourage us to persevere in our search. We hope this book will help in that endeavour.

Appendix 1

Timeline of the Universe[673]

- 13,800,000,000 years ago: Big Bang – T = 0
- Between 0 and 10^{-43}: The simultaneous appearance of time, space, and matter/energy, with the nascent Universe expanding to a diameter of 10^{-35} meters at a temperature of 10^{32} degrees kelvin
- Between 10^{-12} and 10^{-6} seconds: Quarks and electrons appear
- Between 10^{-6} and 10^{-4} seconds: Protons and neutrons appear
- Between 3 and 15 minutes: Hydrogen, helium, lithium, beryllium, and boron appear
- 13,799,620,000 years ago: First emission of light (Cosmic Microwave Background Radiation, or CMBR), by a 3,000 K Universe
- Around 10,000,000,000 years ago: Creation of heavy atoms as first generation stars become supernovas at the end of their life cycle
- 5,000,000,000 years ago: The Sun, a third-generation star, begins to form
- 4,540,000,000 years ago: Formation of the Earth
- 4,520,000,000 years ago: Formation of the Moon
- 3,800,000,000 years ago: First unicellular living organism with DNA emerges in the sea
- 2,100,000,000 years ago: Appearance of the first multicellular living organism in the sea
- 542,000,000 years ago: The "Cambrian explosion:" practically all known animal phyla appear in the sea
- 480,000,000 years ago: Appearance of terrestrial plants
- 445,000,000 years ago: Eighty-five percent of Earth's species die off in the first of five great extinction events

---

673. When the date is a range, we give a median date in the interest of simplicity.

- 400,000,000 years ago: Insects appear
- 230,000,000 years ago: Dinosaurs appear
- 200,000,000 years ago: Mammals appear
- 150,000,000 years ago: Birds appear
- 65,000,000 years ago: Extinction of the dinosaurs between the Cretaceous and Tertiary period
- 45,000,000 years ago: Monkeys appear
- 3,000,000 years ago: Appearance of the genus *Homo*
- 1,900,000 years ago: Appearance of *Homo erectus*, the first in the *homo* group anatomically very close to modern man
- 300,000 years ago: First *Homo sapiens*
- 60,000 to 15,000 BC: First cave paintings in Cáceres, Chauvet, and Lascaux
- 3,500 BC: Writing invented in Sumeria
- 2,000 BC: Abraham
- 1450 to 1250 BC: Moses
- 1,000 BC: David
- 586 BC: Israel's exile in Babylon
- 475 BC: Confucius
- 450 BC: Parmenides states that "*nothing comes from nothing,*" deducing from this logical principle the eternity of the Universe and of matter
- 428 BC: Anaxagoras of Clazomenae states that "*nothing is born or dies; on the contrary, everything is assembled out of existing things and then dissolved,*" a principle that will much later be taken up by Lavoisier
- 400 BC: Buddha
- 384 BC: Aristotle is born. His monumental works on the concepts of God, the soul, and knowledge still inspire philosophers today
- 333 BC: Alexander the Great
- 300 BC: Euclid lays the foundations of geometry
- 240 BC: Eratosthenes of Cyrene correctly calculates the perimeter of the Earth
- 50 BC: Julius Caesar
- 10 BC: Caesar Augustus

APPENDIX 1: TIMELINE OF THE UNIVERSE

- 5 BC: Birth of JESUS CHRIST
- 150: Ptolemy describes the movement of the stars
- 400: St. Augustine of Hippo
- 529: First hospital established by Emperor Justinian
- 622: Mohammed founds Islam
- 1000: Pope Sylvester II introduces the decimal system in Europe
- 1094: The clock is invented
- 1150: The University of Paris is founded
- 1150: First European industrial production of paper, in Xativa, Spain
- 1163: Construction of Notre Dame cathedral begins in Paris
- 1250: St. Thomas Aquinas
- 1270: Invention of eyeglasses
- 1347: Ockham states his "*law of economy*," better known as "*Ockham's razor*," which can be summed up as follows: "*The simplest explanation is the most likely*"
- 1450: Invention of the printing press
- 1492: Christopher Columbus discovers America
- 1517: Luther starts the Protestant Reformation
- 1543: Copernicus publishes *On the Revolutions of the Celestial Spheres*
- 1582: Pope Gregory XIII authorizes the transition to the Gregorian Calendar
- 1609: Kepler states the first two laws of planetary motion
- 1633: Condemnation of Galileo
- 1609–1663: Invention of the telescope by Dutch scientists, later improved by Galileo and James Gregory and employed by Isaac Newton three years later
- 1687: Newton lays the foundation for classical mechanics and states the law of universal gravitation
- 1777: Lavoisier gives magisterial form to Anaxagoras' principle: "*Nothing is lost, nothing is created, all is transformed*"
- 1787: Buffon estimates the age of Earth to be 350,000 years
- 1800: Laplace asserts the deterministic principle that the future could be known with certainty if we understood all the laws of physics and knew the position and velocity of all particles

- 1805: Laplace responds to Napoleon's question *"And where is God in all this"* with *"I have no need of that hypothesis"*
- 1809: Lamarck discovers evolution, calling the theory "transformational"
- 1824: Sadi Carnot defines entropy in the course of his work on thermal machines
- 1838: Discovery of the ancient village of Capernaum by Edward Robinson
- 1839: Birth of cellular theory. Theodor Schwann claims that the cell is the elemental structure of all living organisms
- 1841: Richard Owen gives the name "dinosaurs" to the ancient animals discovered in the fossil record
- 1848: Marx and Engels publish the *Communist Manifesto*
- 1853: Arthur de Gobineau publishes his *Essay on the Inequality of the Human Races*
- 1859: James Clerk Maxwell publishes the classical theory of electromagnetic radiation, proving the existence of electromagnetic waves, including light
- 1859: Charles Darwin publishes *On the Origin of Species by Means of Natural Selection*
- 1860: Gustav Kirchhoff studies the light spectrum and poses the "black-body radiation" problem
- 1861: Louis Pasteur disproves the theory of *"spontaneous generation"*
- 1865: Clausius confirms Carnot's work and spells out the second law of thermodynamics
- 1869: Mendeleev formulates the periodic table of elements
- 1869: Friedrich Miescher isolates "nuclein," a molecule essential to the life of all organisms
- 1870: Ernst Haeckel resists the concept of entropy in the name of his philosophical ideals
- 1878: Ludwig Boltzmann formulates the entropy equations
- 1884: Gregor Mendel, a Catholic monk, founds the field of genetics with *"Mendel's Laws,"* which explain the way that genes are transmitted from one generation to the next
- 1888: Filaments discovered in 1875 by E. Strasburger and obser-

ved by W. Flemming in 1879 are dubbed "chromosomes" (colored bodies) by H. W. Waldeyer
- 1896: Wien's Law is formulated, which states that the longest wavelength emitted by a black body is inversely proportional to its temperature
- 1896: Freud publishes his first psychoanalytic articles
- 1900: Max Planck explains black-body radiation as "quantum action"
- 1900: Henri Poincaré first publishes the formula "$E = mc^2$," later made famous by Einstein
- 1902: Henri Poincaré publishes *La Science et l'Hypothèse*, calling both absolute time and absolute space into question and taking the speed of light as absolute
- 1905: Albert Einstein publishes his theory of Special Relativity
- 1908: Experiments by Jean Perrin (Nobel Prize in Physics 1926) demonstrate the existence of the atom
- 1911: Ernest Rutherford discovers the atomic nucleus
- 1916: The fine-structure constant, which quantifies the strength of electromagnetic force, is discovered by Arnold Sommerfeld
- 1917: Albert Einstein publishes his theory of General Relativity
- 1917: Freud publishes his *Introduction to Psychoanalysis*
- 1919: Arthur Eddington verifies the distortion of space-time during a solar eclipse, according to the angle predicted by Einstein
- 1920: Hermann Staudinger discovers macromolecules
- 1922: Alexander Friedmann, building on the work of Einstein, publishes the first theory of the expansion of the Universe
- 1923: Discovery of the Compton scattering effect, which shows that light is also a particle
- 1924: Louis de Broglie introduces the pilot-wave model and postulates a direct and real physical interpretation of matter waves
- 1924: Wolfgang Pauli defines the principle of exclusion in quantum mechanics
- 1924: Edwin Hubble and Milton Humason demonstrate that the cosmos is much larger than previously imagined: the Universe is not made up of a single galaxy, but a multitude of galaxies
- 1925: Erwin Schrödinger formulates the equations needed to de-

termine Broglie's waves
- 1926: The word "photon" is coined by chemist Gilbert Lewis to describe a quantum or a light particle, which weighs nothing and is *"wave and corpuscle"* simultaneously
- 1927: Werner Heisenberg defines the *"uncertainty principle"* in quantum mechanics
- 1927: Georges Lemaître publishes an article on the origins of the Universe in the *Annales de la société scientifique* of Brussels, postulating a *"primordial atom"*
- 1929: Observations by Hubble prove that the Universe is homogeneous, isotropic, and expanding
- 1930: Progressive discovery of the strong nuclear force
- 1931: Gödel's incompleteness theorems demonstrate the limitations of mathematics and logic, affirming that any logical system contains true but undemonstrable propositions
- 1931: Albert Einstein visits Edwin Hubble on Mount Wilson and acknowledges the expansion of the Universe
- 1935: Presentation of the EPR paradox by Einstein-Podolsky-Rosen, later resolved by Alain Aspect
- 1936: Enrico Fermi discovers the weak interaction
- 1938: Lev Landau (Nobel Prize in Physics 1962), student of Alexander Friedmann, is sent to a Soviet concentration camp
- 1938: Matveï Bronstein, student of Alexander Friedmann, is shot in Russia at the age of thirty-six for spreading ideas about *"the supposed beginning of the Universe"*
- 1945: Gamow publishes *The Creation of the Universe*, which for the first time describes a scenario for the beginning of the world
- 1947: The *Dead Sea Scrolls* discovered in the caves of Qumran
- 1949: Fred Hoyle coins the term "Big Bang" to mock Lemaître's ideas in a BBC interview
- 1949: George Gamow predicts the fossil radiation of the Universe
- 1949: The telescope on Mt. Palomar is operational
- 1953: Sir Francis Crick and James Watson discover the double helix structure of DNA, revolutionizing our understanding of life (Nobel Prize in Chemistry 1962)

- 1956: Tjio and Levan discover the forty-six human chromosomes
- 1960: Oscillating Universe Theory stubbornly maintained, in an attempt to avoid the absolute beginning stipulated by the standard Big Bang model
- 1964: Discovery of the hyperweak force
- 1964: Arno Penzias and Robert Wilson, researchers at Bell Telephone Company, accidentally discover fossil radiation (CMBR) at 2.725 K, for which they earn the 1978 Nobel Prize in Physics
- 1969: Armstrong sets foot on the Moon
- 1973: Brandon Carter theorizes the anthropic principle
- 1977: Prigogine describes chaotic systems on the molecular level, opening the door to the idea that infinitesimal variations due to quantum indeterminacy can change the course of the Universe
- 1982: Alain Aspect at Orsay demonstrates quantum entanglement, establishing the existence of instantaneous interactions faster than the speed of light
- 1984: Formalization of string theory, which led to multiple nonvalid theses
- 1987: Discovery of the town of Bethsaida, in Galilee
- 1992: George Smoot and the COBE satellite map the oldest light in the Universe in almost perfect thermal equilibrium with a CMBR of 2.725 K and tiny irregularities that explain the structure of today's Universe; their findings confirm the Big Bang theory
- 1994: Arvind Borde and Alexander Vilenkin affirm the necessity of an absolute beginning to the Universe in their article "Eternal Inflation and Initial Singularity," in *Physical Review Letters*
- 1998: Saul Perlmutter, Brian P. Schmidt, and Adam Reiss unexpectedly show that the expansion of the Universe is accelerating, putting an end to the hypothesis of an eternally cyclical Universe (Big Bounce), passing from Big Bang to Big Crunch; they win the 2011 Nobel Prize in Physics
- 2000 (+/-): Numerous discoveries demonstrate the fine-tuning of the Universe
- 2003: The first map of the human genome and its twenty-five thousand genes is completed on April 14, after twelve years of work by

twenty-five thousand researchers
- 2003: Theorem by Arvin Borde, Alan Guth, and Alexander Vilenkin demonstrates that the past cannot be eternal and that there had to be a moment of initial singularity
- 2004: Simon Conway Morris, paleontologist and professor at Cambridge, speaks of "possible functional forms predetermined since the Big Bang" in his book *Life's Solutions*.
- 2004: Discovery of the pool of Siloam in Jerusalem
- 2006: George Smoot speaks of the *"face of God"* when he accepts the Nobel Prize for his map of the oldest light in the Universe
- 2009: The Nobel prize is awarded to three researchers who demonstrated how the ribosome translates DNA code into proteins
- 2010: Analysis of Neanderthal DNA (350,000–50,000 BC) by Svante Pääbo
- 2012: The Higgs boson particle discovered at CERN in Geneva, Switzerland

In the distant future:

- 4,500,000,000 years: Death of the Sun
- 1,000,000,000,000–100,000,000,000,000 years: The gas necessary for star formation is exhausted
- 10^{32} to 10^{34} years: Disintegration of protons into smaller particles and the disappearance of neutrons, which can only live fifteen minutes on their own
- 10^{100} years: Thermal death of the Universe and the end of all activity when the Universe has expanded fully

Appendix 2

Physical Measurements Great and Small

Physical measurements, from shortest to longest:

- 1.616×10^{-35} m: The Planck length (the smallest quantum possible)
- 10^{-15} m: The atomic nucleus
- 10^{-10} m: The atom—it takes 70 billion atoms to make up a pinhead, and the atom is 100,000 times larger than its nucleus
- 10^{-10} m: Water molecule
- 10^{-9} m: Nucleotide of DNA (A, C, G, or T)
- 10^{-9} m: Certain aminoacids essential to life synthesized in Stanley Miller's experiments
- 10^{-8} to 10^{-9} m: Proteins that are macromolecules of various sizes, ranging between a few dozen to hundreds of angstroms (10^{10} m)
- 20×10^{-6} m (20 microns): A single human cell—itself made up of 176,000 billion molecules (99% of which are water, 70% of the weight of the cell)
- 10^{-5} m (0.01 mm): Plant cell
- 10^{-4} m (0.1 mm): Mite
- 10^{-3} m (1 mm): Ant
- 10^{-1} m (10 cm): Smartphone = length of 1 billion water molecules
- 1 to 5 m: Man/car = 1 billion times the length of a protein
- 2×10^{4} m (20 km): Diameter of the Paris region = 1 billion times the length of a cell
- 4×10^{7} m (40,000 km): Circumference of Earth
- 3.8×10^{8} m (380,000 km): Distance from Earth to the Moon
- 1.5×10^{11} m (150,000,000 km): Distance from Earth to the Sun
- 6×10^{12} m (6,000,000,000 km): Radius of the solar system to Pluto

- 10^{16} m (4.3 light years): Distance to the closest star, Proxima Centauri
- 10^{21} m: Diameter of the Milky Way
- 10^{22} m: Distance from Earth to the Andromeda galaxy
- 10^{27} m (95 billion light years): Size of the Universe

Orders of magnitude for a few important figures:

- 10^{11}: Number of stars in our galaxy, the Milky Way
- 10^{12}: Number of galaxies in the Universe
- 10^{18}: Number of insects on the Earth
- 10^{21}: Number of H_2O molecules in a drop of water
- 10^{23}: Number of grains of sand in the Sahara Desert[674]
- 10^{23}: Number of stars in the Universe
- 10^{25}: Number of water droplets in all the seas and oceans put together[675]
- 10^{30}: Number of bacteria on the Earth
- 10^{40}: The ratio between the radius of the cosmos and that of the electron, which is also the ratio between the force of gravity and the electromagnetic force (something Paul Dirac called "*strange coincidences*")
- 10^{60}: Number of Planck times (10^{-43} seconds) since the Big Bang (13.8 billion years)
- 10^{80}: Number of atoms in the Universe

---

[674]. The Sahara has a surface area of about 1,000 km by 1,000 km and a depth of 50 m, which would hold approximately 10^{23} grains of sand, each of 0.1 m radius.

[675]. The volume of water in oceans on the Earth is on the order of 10^9 km², and the volume of a drop of water is 50 mm³.

Appendix 3

Benchmarks of Orders of Magnitude in Biology

Parts of the human body (ranked by abundance):

- 7,000,000,000,000,000,000,000,000,000 (7×10^{27}) atoms in the human body composed of forty-one different chemical elements
- 40,000,000,000,000 (4×10^{13}) bacteria of approximately five hundred different types
- 30,000,000,000,000 (3×10^{13}) cells of two hundred different types
- 25,000,000,000,000 (2.5×10^{13}) of these are red blood cells, which make up 84% of human cells
- Another 85,000,000,000 (8.5×10^{10}) of these are neurons
- 3,000,000,000 (3×10^{9}) base pairs in the human genome or that of corn: this is very few compared to that of the onion (16×10^{9}) or of *Paris japonica*, a herbaceous plant from Japan with one hundred fifty billion base pairs (150×10^{9})[676]
- 100,000 different kinds of molecules
- 50,000 different kinds of proteins
- 20,000 genes
- 2,000 different enzymes

Parts of a human cell

- 176,000,000,000,000 (1.76×10^{14}) molecules in each twenty-micron cell

---

676. This is the C-value paradox: the C-value representing the size of a given genome does not correlate to the complexity of the organism itself.

- 174,000,000,000,000 (1.74×10^{14}) of these are water molecules (98.73% of the cell's molecules are water molecules, making up 65% of the cell's mass)
- Another 1,310,000,000,000 (1.3×10^{12}) of these are other inorganic molecules, or 0.74% of the molecules of the cell
- 19,000,000,000 (1.9×10^{10}) proteins of 5,000 different types in each cell
- 280,000,000 (2.8×10^{8}) hemoglobin molecules per red blood cell
- 50,000,000 (5×10^{7}) DNA molecules in each cell
- 574 amino acids in each hemoglobin molecule
- 23 pairs of chromosomes

Parts of the simplest known bacteria (Candidatus Carsonella ruddii, discovered in 2006)

- 159,662 nucleotide bases in its smallest decoded DNA molecule
- At least 250 genes (estimated)
- 182 types of protein

Number of atoms that make up different proteins (always complex macromolecules)

- 551,739 atoms in the 30,000 amino acids that constitute titin, the largest protein known in humans
- 3,000 in the 150 amino acids that make up the smallest protein

Results of laboratory attempts to recreate the conditions necessary for life (Stanley Miller and other equivalent experiments)

- 500 atoms in the components obtained from laboratory experiments: a few dozen amino acids each containing 10 to 40 atoms
- 13 of the 22 amino acids were obtained under conditions at the time thought to roughly mimic conditions on the early earth
- 1 of 4 nucleotide bases was obtained from the experiment

Glossary

Absolute zero: Absolute zero is the lowest temperature that can exist. It corresponds to the lower limit of the thermodynamic temperature scale (-273.15 degrees Celsius) and is the state in which the entropy of a system reaches its minimum value, or zero. At this temperature, all molecules are motionless.

Anisotropy: An isotrope is something that is uniform in all directions. Primary anistropy, the directional dependency of the cosmic background radiation, refers to the tiny variations that conditioned the Universe's future development.

Anthropic principle: Expression introduced by astrophysicist Brandon Carter in 1974 to describe the work of Robert Dicke and many other scientists on the extremely fine adjustment of the Universe's fundamental parameters (initial conditions, constants, laws) that allows the possibility of life and our existence. The anthropic principle has been the object of multiple philosophical interpretations. We thus generally distinguish the "weak" anthropic principle (which, without seeking a deeper explanation, is satisfied to note that without these multiple optimizations, life would not have been able to develop and lead to humanity), from the "strong" anthropic principle (which claims that the fundamental parameters on which the Universe depends are set so that it allows the birth and development of observers within it at a certain stage of its development). We cannot put these two interpretations on the same level, because the "weak" anthropic principle is nothing more than a resignation of reason.

Big Bang: Term coined over the airwaves of the BBC by English astrophysicist Fred Hoyle in the 1950s as an ironic way of describing the model of the Universe that begins with an explosion.

Big Bounce: Cyclical model of a "phoenix" Universe that imagines an endless repetition of Big Bang and Big Crunch, in which each Big Crunch is immediately followed by a Big Bang, comparable to a rebound effect (bounce).

Big Crunch: a model of the evolution of the Universe that ends with a collapse, that is to say with a phase of contraction following a phase of expansion which, after having slowed down, is reversed, perhaps allowing another Big Bang to occur.

Big Freeze (or Big Chill): See "Thermal death of the Universe."

Big Rip: Cosmological model in which all the Universe's structures, from the smallest atom to the greatest galaxy cluster, are destroyed in the end by an increasingly strong force that distends, dislocates, tears, shreds, and finally annihilates them.

Concordism: An approach that interprets sacred texts so that they accord with science. To be labeled "concordist" today is an accusation and almost an insult. However, if a divine revelation is authentic, it must "concord" with science, for science also deals with the real world. It is in this sense that in the vision of Pope John Paul II reason and revelation appear as "*two wings that lead to truth,*" a single truth. But since divine revelation does not relate a priori to science, nor to history or philosophy, it is important to discern clearly what it says and what it does not say, in order to avoid any ignorant or unjustified concordism.

Cosmic microwave background radiation: Name given to the first electromagnetic radiation emitted by the universe 380,000 years after the Big Bang. It corresponds to radiation emitted by a black body in thermal equilibrium at a temperature of 2.725 K, coming from all directions in the sky. Suspected as early as 1948 and discovered by chance in 1964, it corresponds to the oldest possible image of the Universe and exhibits minute variations in temperature and intensity. These anisotropies, studied in detail since the early 1990s, provide a wealth of information on the structure, age, and evolution of the Universe.

Cosmogony: Theory, model, or mythological account that describes or explains the formation of the Universe, the Earth, celestial bodies, and mankind.

Cosmological constant: A parameter related to the repulsive force that stretches the Universe, added in 1917 by Albert Einstein to his 1915 equations of General Relativity, in order to make the theory compatible with his conviction that the Universe was static. Faced with the evidence of the Universe's expansion, Einstein retracted the constant, regretting it as his "*biggest mistake.*" In 1997, the discovery of the Universe's acceleration made it necessary to reintroduce this cosmological constant, which does exist and remains very mysterious.

Cosmology: The science of the physical laws of the Universe, its structure, and its formation.

Dark age: The term "dark age," often used to refer to the centuries following the collapse of the Roman Empire or any disastrous or negative historical period more generally, is used by extension to designate the period between the formation of atoms, when the cosmic background radiation was emitted, and the formation of stars, and a future period, after the death of the stars.

Entropy: Drawn from a Greek word meaning "transformation," the term was coined in 1865 by Rudolf Clausius to express the principle of irreversible energy dispersion, increase in disorder, and degradation of any given isolated system over time. As it correlates to disorder, we can affirm with Claude Shannon, the founder of information theory, that entropy is the opposite of information.

Epistemology: Field of philosophy that studies the theory of knowledge. It analyzes, studies, and critiques all sciences, their assumptions, basic logic, value, scope, methods, and findings.

Fine-structure constant: A dimensionless quantity that represents the strength of the electromagnetic force and the strong nuclear force. It has a value of approximately 1/137, commonly represented by the symbol α. The electromagnetic force is at work in numerous physical phenomena: in the interactions between light and matter, in quantum physics, and at the basis of the cohesion of atoms and molecules, by "holding" electrons together.

Flat Universe: The theory of Relativity posits that the local geometry of space is modified by gravity. In a flat, or "Euclidean," space, the sum of the angles of a triangle is precisely equal to 180°. A universe with positive curvature represents a spherical, or closed, universe in which the sum of the angles of a triangle is greater than 180°. A universe with negative curvature is hyperbolic, or open, and the sum of the angles of a triangle is less than 180°. What about the Universe itself on a very large scale? Is it "flat" or does it have a positive "curvature" (a spherical, closed universe) or negative "curvature" (an open, divergent universe)? The latest measurements from the Planck satellite strongly suggest a universe with a very slight positive curvature—essentially, a spherical and closed universe—making it difficult to conceive of an alternative.

General Relativity: Gravitational theory published by Albert Einstein in 1917. It includes and supersedes Isaac Newton's theory of universal gravitation, stating in particular that gravity is not a force but rather a manifestation of the curvature of space-time. This theory predicted effects such as the expansion of the Universe, gravitational waves, and black holes, which were later verified.

God: A being transcendent to our Universe, one who is eternal and all-powerful, non-spatial, non-temporal, non-material, the first cause of all that exists, according to the definition of classical philosophy and religion.

Hilbert space: Abstract space that makes up the framework of quantum mechanics. Proposed by the mathematician David Hilbert, these spaces extend the classical framework of Euclidean spaces (the usual three-dimensional space) to spaces of infinite dimensions. The quantum states form a Hilbert space.

Incompleteness theorem: The two famous incompleteness theorems proposed by Kurt Gödel in 1931 marked a turning point in the history of logic, delivering a negative answer to the question of mathematical coherence posed by Hilbert's Program more than twenty years earlier. Gödel's theorems demonstrate that in any sufficiently complex logical system, there necessarily exists at least one true but unprovable proposition.

Initial singularity: The word "singularity" describes the uniqueness of something or someone. Applied to the origin of the Universe, it denotes the unique point in space-time at which the quantities describing the mass-energy density and the curvature of space become infinite and the known physical laws break down.

Kelvin: Unit of the international absolute temperature measurement system (unit symbol K), where zero kelvin corresponds to -273.15 degrees Celsius, the lowest possible temperature.

Logos: An extremely rich concept used since the time of the Greek philosophers to signify at once language, speech, discourse, reason, rationality, meaning, intelligence, law, logic, logical argumentation, and the divine. Plato, Aristotle, Newton, Leibniz, and so many others did not conceive of barriers between science, philosophy, metaphysics, or theology. The distinctions amongst these genres have only existed since the sixteenth century. The majority of the early Greek philosophers were scientists. They worked from logos, and their opponents were not scientists but poets, those who favored pathos, that is to say feeling, emotion. In the Christian tradition, God himself is the Logos (cf. John 1: 1-15).

Multiverse: An a priori unverifiable hypothesis of the existence of multiple universes parallel to ours, generated by mechanisms like inflation and the membranes of superstring theory. The scientific character of this hypothesis is strongly contested because it is absolutely impossible to verify the theory directly, as these imaginary universes would be inaccessible to us.

Ontological proof: Argument that aims to prove the existence of God from the definition of a perfect being. First delineated by Boethius (6th century), developed by Saint Anselm of Canterbury (11th century), worked over by Descartes (17th century) and Leibniz (18th century), it was criticized by Kant (18th century) and many other philosophers before being revived in the mathematical language of modal logic by Gödel (20th century), and recently verified by the ultra-powerful artificial intelligence utilized by Christoph Benzmüller (21st century).

Ontology: Branch of philosophy also known as the "philosophy of being" or "first science," dealing with the nature of existence and the meaning of the word "being."

Primeval atom: The "hypothesis of the primeval atom" is the name given to the cosmological model developed by Fr. Georges Lemaître in the early 1930s to describe the earliest phase of the Universe. The word "atom" is used here allegorically and does not imply a connection with the world of atomic particles.

Quantum tunneling: A purely quantum effect by which certain quantum waveforms can have a nonzero probability of passing through a barrier, despite the fact that nothing comparable occurs in classical mechanics.

Ribosome: A ribonucleoprotein complex (composed of proteins and RNA) found in eukaryotic (with nucleus) and prokaryotic (without nucleus) cells. The ribosome's function is to synthesize proteins by decoding the information contained in messenger RNA. The origin of the ribosome is one of the great mysteries in cellular biology.

Second principle of thermodynamics: Formulated by Sadi Carnot in 1824, it postulates the irreversibility of physical phenomena, in particular during heat exchanges, and the notion that disorder can only increase during a real transformation in an isolated system. Since the nineteenth century it has continued to be restated and refined by Clapeyron (1834), Clausius (1850), Lord Kelvin, Ludwig Boltzmann (1873), and Max Planck from the nineteenth century to today.

Science: Science means knowledge. As there are different types of knowledge, there are different types of science. Their common trait of what we call today science is that they are approached through the "logos," that is, rationality. Thus, cosmology is the logos applied to the cosmos; biology is the logos applied to living things; archaeology is the logos applied to what is ancient. It is the same for geology, psychology, paleontology, ecology, oceanology, oncology, cardiology, dermatology, neurology, pharmacology, climatology, criminology, futurology, graphology, epistemology, ethnology, eschatology, theology, ontology, oenology, ophthalmology, etc. The meaning of the word "science" has sometimes fallen into more and more restrictive uses, up to Popper's criteria that claim to exclude everything that is not "falsifiable." But science is also and above all else that which scientists actually practice, and it is difficult to reduce its scope in too arbitrary a manner.

Space-time: Mathematical entity within the framework of Einstein's theory of General Relativity that combines the three dimensions of space and the dimension of time. It replaces the classical conception of absolute space and time.

Special Relativity: Theory developed by Albert Einstein in 1905 from the principle that the speed of light in vacuum has the same value according to all Galilean or inertial frames of reference. The resulting equations led to predictions of phenomena that defy common sense (though not one of these predictions has been proved wrong by experience), one of the most surprising being the slowing down of clocks in motion, which helped construct the thought experiment often referred to as the "twin" paradox. The principles of special Relativity, already glimpsed in the work of Henri Poincaré from 1902, have a strong philosophical impact: they force us to pose the question of time and space differently, as they eliminate any possibility of the existence of an absolute time and duration in the whole of the Universe, an idea held since Newton.

Strong nuclear force or strong interaction: Force responsible for the cohesion of the atomic nucleus, according to the standard model of particle physics. It acts at short range to bind quarks together into protons and neutrons. It also pulls protons and neutrons together to form an atomic nucleus.

Thermal death of the Universe: Also known as the Big Freeze or Big Chill, this is the apparently inevitable end towards which our Universe will evolve in the very distant future, by which any thermodynamic process allowing movement or life will eventually become impossible (absolute entropy).

Tunnel effect: A purely quantum effect by which certain quantum wave forms have a non-zero probability of passing through a barrier, though crossing this barrier is not possible according to classical mechanics.

Weak nuclear force or weak interaction: Force responsible for the radioactive disintegration of neurons (beta radioactivity), which acts on all known types of elementary fermions (electrons, quarks, neutrinos). It is one of the four fundamental interactions of nature, the other three being the strong nuclear force, the electromagnetic force, and the gravitational force.

Index of
Personal Names

Aba Shalev, Baruch: 296, 305-306

Abbott, Larry: 203

Abraham: 343, 394, 412, 540

Adam: 357-358, 360, 369

Adam, Jehan: 368

Adrian: 427

Ahmadinejad, Mahmoud: 433

Akiva, Rabbi: 407

Al-Ghazali: 69, 120, 499

Al-Kindi: 69, 120

Albert I, King of Belgium: 167

Albert the Great, Saint: 69, 120

Alexander lll: 130

Alexander the Great: 540

Alfvén, Hannes: 95-96, 112

Almeida, Avelino de: 448, 458, 464, 469, 473

Almeida, José Luís de (brother): 459

Alpher, Ralph: 92-93, 147-148, 175, 177-178, 201

Amorim, Diogo Pacheco de: 479-480

Anan the Younger: 384

Anaxagoras: 54, 540-541

Andrew, Saint: 403

Anfinsen, Christian: 264

Anselm of Canterbury, Saint: 330, 498-499, 500, 555

Arago, François: 75

Arber, Werner: 242, 283

Archimedes: 343

Aristotle: 120, 124, 231, 291, 307, 333, 347, 496-498, 500, 540, 555

Armstrong, Neil: 545

Arrhenius, Svante: 71, 96, 124

Artaxerxes: 407

Ashtekar, Abhay: 115

Aspect, Alain: 101, 544-545

Athronges, false Messiah: 407

Augustine, Saint: 369, 487, 497-498, 500, 541

Augustus, Caesar: 359, 540

Avicenna: 499

Bagot Glubb, John: see "Glubb Pasha"

Balanovsky, Innokenty: 129, 147

Bar Kokhba, Simon, false Messiah: 407

Bar Serapion, Mara: 384-385

Barnes, Luke: 186, 196

Barrow, John D.: 190, 275

Barthas, Casimir, Canon: 455-457

Bartholomew, Saint: 403

Barton, Derek: 273

Bauer, Bruno: 382, 391

Behe, Michael: 287

Bekenstein, Jacob: 80

Ben-Yehuda: 432

Ben-Artsi, Haguy: 445

Benford, Gregory: 226, 271

Benedict XVI, Pope: 431, 446

Bentham, Jeremy: 490

Benzmüller, Christoph: 233, 330-331, 555

Berdyaev, Nicolas: 420-421

Beria, Lavrentiy: 140

Bernadette of Lourdes, Saint: 461

Berthelot, Marcellin: 71-72, 124

Bethe, Hans: 92, 147-148

Bieberbach, Ludwig: 166

Binet-Sanglé, Charles: 383, 397

Boethius: 555

Bogdanov, Igor and Grichka: 109, 146, 172, 209, 566

Bohm, David: 176-177

Bohr, Niels: 35, 137, 140, 160, 279, 312

Boltzmann, Ludwig: 35, 56, 67, 70, 72, 78, 130, 183, 542, 556

Bonaventure, Saint: 69, 120-121

Bondi, Hermann: 112

Bonner, William: 177

Borde, Arvind: 109, 112, 115, 119, 173, 217, 219, 222, 224, 226, 303, 545

Bormann, Martin: 156, 173

Born, Max: 129, 160, 167, 169-172, 208, 274

Buddha: 51, 540

Bukharin, Nikolai: 40

Broglie, Louis de: 543

Bronstein, Matvei: 127, 129, 137, 141, 143-144, 224, 544

Buffon, Comte de: 25-26, 34, 541

Burbidge, Geoffrey: 114, 206

Cabanis, Pierre J. G.: 531

Calvin, Melvin: 236

Capelinha, Maria da: 453

Carlip, Steve: 209-210

Carnot, Sadi: 35, 55, 65, 67, 72, 542, 556

Cartan, Henri: 151

Carter, Brandon: 35, 190, 545, 551

Cassou-Noguès, Pierre: 322, 324-326

Celsus: 348, 362, 386

Caesar, Julius: 343, 359, 385, 412

Chain, Ernst: 284

Carlos I, King of Portugal: 452

Chew, Geoffrey: 281

Christ: 343, 344, 377, 381-390, 404, 427-429, 459, 497, 529, 541 see also "Jesus of Nazareth";

Christos: 387

Chuquet, Nicolas: 368

Church, George: 247, 285

Cicero: 376

Clapeyron, Émile: 556

Clarke, Samuel: 499

Claudius, Emperor: 385

Clausius, Rudolf: 35, 55, 65-67, 71-72, 542, 553, 556

Clermont-Ganneau, Charles Simon: 429

Clerk Maxwell, James: see "Maxwell, James Clerk"

Cohen, Daniel: 245, 286

Collins, Francis: 290

Collins, Robin: 503

Columbus, Christopher: 541

Compton, Arthur: 275, 543

Confucius: 540

Conway Morris, Simon: 284, 546

Copernicus, Nicholas: 25-26, 34, 42, 44, 91, 279, 302, 376, 541

Coppens, Yves: 288

Cornwell, Elisabeth: 298, 306, 572

Couchoud, Paul-Louis: 382, 391

Craig, William Lane: 97, 121, 503

Crick, Sir Francis: 35, 124, 243, 245-246, 251, 281-281, 544

Curie, Marie: 140

Cyrus II, Persian King: 343-344, 346, 430

D'Alembert, Jean Le Rond: 231, 501

D'Holbach, Paul Thiry: 501

Daniel, Prophet: 405-406

Daniel, Jean: 441

Darius: 343-344

Darwin, Charles Galton (grandson of the naturalist): 171

Darwin, Charles: 10, 25-27, 29, 34, 171, 232-233, 257, 285, 287, 290, 303, 491-492

Dauvillier, Alexandre: 95

David, King: 343, 540

Davies, Paul: 182, 187, 211, 267, 277, 328-329, 505

Dawkins, Richard: 48, 117, 309, 367

Dawson, John and Cheryl: 318

De Marchi, John (priest): 457-458

De Sitter, Willem: 75

Delbrück, Max: 169

Demaret, Jacques: 186, 273

Demirsoy, Ali: 289

Denton, Michael: 247, 252, 258, 286, 565

Descartes, René: 103, 330, 499, 555

DeWitt, Bryce: 113

Dicke, Robert: 35, 94, 188, 190-192, 195, 209, 270-271, 551

Diderot, Denis: 231, 501

Dirac, Paul: 35, 187, 190, 194, 291, 548

Doroshkevich, Andrei: 149

Dostoevsky, Fyodor: 153, 488-489

Drexler, Anton: 154

Duhem, Pierre: 71

Duns Scotus, John: 499

Duve, Christian de: 245, 257, 283

Dyson, Freeman: 189, 274

Dzerjinski, Felix: 137, 139

Eccles, John: 283

Ecklund, Elaine H.: 178

Eddington, Sir Arthur: 68, 73, 84, 88, 90-91, 154, 177, 543

Efstathiou, George: 102, 226, 275

Ehrenfest, Paul: 86, 135, 157-158, 161-162

Eichmann, Adolf:173

Einstein, Albert: 12, 24, 35, 44, 54, 67, 72, 74-75, 83-91, 100, 103, 114, 123-124, 129, 132-137, 143, 146, 154-168, 170, 173, 177, 190, 201, 207, 296, 307, 309-315, 319-322, 326, 328, 331, 333, 351, 373, 435, 502, 504, 543-544, 553-554, 557

Eliade, Mircea: 352, 529

Elijah, Prophet: 405

Engels, Friedrich: 56, 71, 124, 542

Eratosthenes: 540

Eropkin, Dmitri: 128, 144

Espagnat, Bernard (d'): 279

Euclid: 319, 343, 348, 540

Eusebius of Caesarea: 387-388

Eve: 358, 360, 527

Everett, Hugh: 112, 216

Ey, Henri: 480

Ezagouri, Claude: 432

Ezra: 429

Faraday, Michael: 333

Fauth, Philipp: 164

Feder, Gottfried: 156

Fermat, Pierre de: 39

Fermi, Enrico: 169, 201, 544

Feser, Edward: 292, 503, 515

Feynman, Richard: 187, 194, 208, 264

Flemming, Walther: 543

Fleury, Vincent: 278-279

Flew, Anthony: 257, 292, 503

Fock, Alexander (father of Vladimir): 141

Fock, Vladimir: 131, 135, 141, 142, 149

Forrest, Peter: 503

Fox, Sidney W: 235, 236

Franck, James: 172

Franco, José Eduardo: 459, 567

Frederiks, Vsevolod: 128, 136, 146-147

Freud, Sigmund: 25-28, 435, 501, 543

Friedmann, Alexander: 24, 35, 74, 85-89, 92-93, 96, 112, 132-139, 141-143

Gagarin, Yuri: 531

Galen: 386

Galileo: 25-27, 29, 34, 44, 224, 302-303, 349, 541

Gamaliel: 410

Gamow, George: 148-149, 175-178, 224, 544
Gaulle, Charles de: 437
Gaunilo of Marmoutiers (monk): 330
Gennes, Pierre-Gilles de: 261, 287
George VI, King: 175
Gibbs, Willard: 71
Gill, John: 430, 431
Gingras, Yves: 306
Gisin, Nicolas: 280
Gitt, Werner: 272
Glubb, Pasha: 442
Gobineau, Arthur de: 542
Godart, Odon: 95
Gödel, Kurt: 7, 32, 35, 37, 101, 233, 291, 307, 317-333, 498, 554, 555, 573
Goebbels, Joseph: 156, 167
Gold, Thomas: 112
Gorbachev, Mikhail: 153
Goren Rabbi: 445
Grassé, Pierre-Paul: 289
Greene, Brian: 202, 271
Gregory XIII, Pope: 541
Gregory, James: 541
Gross, David: 281
Guinzburg, Alexander: 151
Guth, Alan: 195, 197, 216-217, 219, 222, 224-225, 303, 546
Gutkind, Eric: 309, 313
Haeckel, Ernst: 56, 71, 124, 542
Hahn, Hans: 318
Halévy, Marc: 268
Hamilton, Floyd: 438
Harold, Franklin: 252
Harriot, Thomas: 349
Harrison, Edward: 227, 270
Hartle, James: 113, 117
Hausdorff, Felix: 127, 173, 174
Hawking, Stephen: 205, 206, 218, 222, 268, 272, 279, 280, 329
Hegel, Friedrich: 130
Heisenberg, Werner: 169, 207, 212, 242, 265

Helmholtz, Hermann von: 66-68
Hempel, Carl Gustav: 172
Herman, Robert: 92-93, 177-178, 201
Herod Agrippa: 409
Herod Antipas: 405
Herod the Great: 344, 403, 405, 407 427, 429
Hesiod: 355
Hess, Victor Francis: 172
Hewish, Anthony: 267
Heydrich, Reinhard: 156, 173
Higgs, Peter: 546
Hilbert, David: 172, 219, 280, 317, 319, 320, 333
Himmler, Heinrich: 153, 156, 165, 173
Hipparchus: 348
Hirsch, William: 383, 397
Hitchens, Christopher: 503
Hitler, Adolf: 163-164, 166-167, 173, 528
Homer: 344
Hooker, Joseph: 88, 139, 232
Hörbiger, Hans: 164
Hoyle, Fred: 257-260, 277, 307, 544, 551
Hubble, Edwin: 74, 87-90, 111, 139, 161, 543-544
Humason, Milton: 543
Hume, David: 124, 330-331, 500
Hussein, King of Jordan: 441-442
Isaiah, Prophet: 327, 375, 405, 426, 428
Isidor Isaac, Rabbi: 273
Isidore of Seville: 376
Ivanenko, Dmitri: 137, 141, 143
Jacob: 415, 421
James the Major, Saint: 403
James the Minor, Saint: 384, 403
Jaki, Stanley: 480
Jastrow, Robert: 266, 278
John, Saint: 555
John the Baptist, Saint: 405

John Paul II, Saint: 274, 449, 476-478, 485, 552
Jeans, Sir James Hopwood: 280
Jefferson, Thomas: 383, 393
Jeremiah, Prophet: 405, 431
Jerome, Saint: 376, 387, 429, 430
Jesus of Nazareth: 8, 32, 338, 343, 375, 380, 381-414, 426, 428, 450, 459, 494, 535, 537, 541, 574-575
Yeshua: 383 (also see "Christ")
Jesus, Helena: 459, 567
Job: 346, 354, 374, 525
Jordan, Pascual: 169, 276
Josephus, Flavius: 384, 386-388, 422
Judas the Galilean: 407, 410
Jude, Saint: 403
Julian the Apostate: 427
Justinian, Emperor: 541
Kafka, Franz: 153
Kaita, Robert: 14, 277, 278
Kajita, Takaaki: 198
Kant, Immanuel: 330, 500, 502, 518, 555
Kapitsa, Pyotr: 142, 149
Kastler, Alfred: 265
Kauffman, Stuart: 289
Kelvin, Lord: 56, 65-68, 556
Kenyon, Dean H.: 287, 307
Kepler, Johannes: 44, 302, 541
Kirchhoff, Gustav: 542
Klein, Felix: 170, 172
Kohn, Walter: 278, 279
Koonin, Eugene: 255
Koons, Robert: 503, 515
Kozyrev, Nikolai: 127, 128, 147, 224
Krauss, Lawrence: 117
Krutkov, Yuri: 86, 133, 135
La Mettrie, Julien de: 501
Labrot, Philippe: 237, 238, 288
Lactantius: 376
Lamarck, Jean-Baptiste de: 25, 26, 34, 542

Landau, Lev: 127-128, 137, 141-144, 149, 224, 544
Langevin, Paul: 140
Lanzmann, Claude: 442
Laplace, Pierre-Simon de: 24, 26, 34, 519, 541, 542
Laue, Max von: 165
Laughlin, Robert: 204, 220, 265
Lavoisier, Antoine de: 54, 540, 541
Leclerc, George-Louis(Comte de Buffon), see "Buffon"
Leibniz, Gottfried W.: 231, 312, 330, 498, 499, 555
Lemaître, Georges: 93, 95-96, 98-100, 111-112, 114, 140, 159, 175, 544, 556
Lenard, Philipp: 159, 160-161, 164-165
Lenin: 124, 130
Lennox, John: 12, 116, 188-189
Leuba, James H.: 290, 583
Levan, Albert: 545
Lewis, C. S.: 396, 399
Lewis, Geraint: 196
Lewis, Gilbert: 186, 544
Linde, Andrei: 115, 217
Lipson, Henry: 278
Lipton, Bruce: 290
Lopes Vieira, Afonso: 457, 483
Loureiro, João Diogo: 459, 567
Lov, V. E.: 148
Luke, Saint: 388, 394-395, 398, 403
Lucian of Samosata: 386
Lucretius: 124, 231
Luminet, Jean-Pierre: 87, 93, 95, 137
Luther, Martin: 541
Luz, Paula Sofia: 458
Lycurgus: 419
Lyell, Charles: 26
Mach, Ernst: 70, 124
Mohammed: 541
Malebranche, Nicolas: 499

Manson, Neil: 185, 226, 293
Manuel II, King of Portugal: 452
Mao Tse-tung: 39, 124, 528
Mark, Saint: 394-397, 399, 403
Marcellinus, Ammianus: 428
Margulis, Lynn: 252
Magdalen, Saint Mary: 393
Marto, Saint Francisco: 447, 453, 476-478
Marto, Saint Jacinta: 447, 453, 476-478
Marx, Karl: 25, 27, 34, 56, 124, 130, 435, 501, 542
Mather, John: 208, 209
Matthias, Saint: 403
Matthew, Saint: 388, 403
Maxwell, James Clerk: 73, 333, 542
Mayr, Ernst: 286
McDonald, Arthur B.: 198
McFadden, Johnjoe: 289
Meitner, Lise: 172
Menahem, false Messiah: 407
Mendel, Gregor: 91, 542
Mendeleev, Dmitri: 542
Mendès France, Pierre: 442
Menzhinsky, Vyacheslav: 139
Miescher, Friedrich: 542
Miller, Stanley: 238, 288, 550
Millikan, Robert: 266
Minkowski, Hermann: 170
Mints, Y. V.: 397
Moses: 109, 223, 267, 361, 386, 3998, 419, 421, 430, 540
Molotov, Vyacheslav: 139, 140
Monod, Jacques: 250, 284, 288
More, Saint Thomas: 430
Morgenstern, Oskar: 320
Morowitz, Harold: 259
Morris, Simon: see "Conway Morris, Simon"
Moti Hod, Aluf (General): 445
Musselius, Maximilian: 128, 146
Mussolini, Benito: 382

Nebuchadnezzar: 421
Napoleon: 519
Narlikar, Jayant: 114
Nasser, Gamal Abdel: 439-442
Nernst, Walther: 161
Nero: 385, 413
Neumann, John von: 307, 318, 319, 333
Newton, Isaac: 25, 26, 34, 43, 44, 185, 286, 302, 333, 541, 555, 557
Nietzsche, Friedrich: 124, 128, 383, 397, 501, 567
Novikov, Igor: 149
Numa: 419
Numerov, Boris: 128, 145, 146
O'Keefe, John: 278
Oberth, Hermann: 164
Ockham, William of: 225, 541
Oderberg, David: 503, 515
Odifreddi, Piergiorgio: 331
Olbers, Heinrich: 74
Oliveira Santos, Artur de: 454, 460
Onfray, Michel: 382, 391
Oom, Augusto Frederico: 478
Oppenheimer, Robert: 161, 169, 333
Origen: 348, 362, 369, 386
Oró, John: 236
Owen, John: 430
Owen, Richard: 542
Pääbo, Svante: 546
Page, Donald: 269
Painton, Frederick C.: 431
Parish, Edmund: 482
Parmenides: 54, 124, 343, 540
Pascal, Blaise: 17, 413, 417, 493
Pasteur, Louis: 284, 333, 542
Paul, Saint: 429, 538
Paul III, Pope: 376
Paul VI, Saint, Pope: 449, 477
Pauli, Wolfgang: 169, 171, 207, 208, 275, 543
Pecker, Jean-Claude: 111

INDEX

Peebles, James "Jim": 94, 191
Penfield, Wilder: 285
Penrose, Roger: 116, 119-120, 210-211, 218, 264, 322, 323, 324
Penzias, Arno A.: 108, 109, 142, 149, 178, 188, 191, 223, 267, 361, 545
Pereira, Inácio Lourenço: 457
Perepyolkin, Yevgeny: 128, 143
Perlman, Eliezer: 432
Perlmutter, Saul: 75, 78, 545
Perrier, Pierre: 279, 566
Perrin, Jean: 543
Philip, Saint: 403
Phillips, William D.: 266
Philoponus, John: 69, 120
Pius XII, Saint, Pope: 449, 477
Peter, Saint: 402, 413
Pilate, Pontius: 385, 386
Planck, Max: 276, 373, 313, 543, 556
Plantinga, Alvin: 503
Plato: 231, 232, 343, 344, 495, 496, 504, 506, 555
Plekhanov, Georgi: 131
Pliny the Younger: 385, 386
Plyushch, Leonid: 128, 150-152
Plotinus: 497
Podolsky, Boris: 544
Poe, Edgar Allan: 74
Poincaré, Henri: 70, 543, 557
Pol Pot: 528
Polkinghorne, John: 273, 505
Polo, Marco: 376
Polyakov, Alexander: 291
Ponnamperuma, Cyril: 236
Popper, Karl: 41-42, 46, 224
Pouchet, Félix-Archimède: 232
Prat, Ferdinand: 382, 427, 428
Prigogine, Ilya: 66, 73, 268, 282, 283, 545
Pruss, Alexander: 503, 515
Ptolemy: 541
Pythagoras: 343, 344, 348, 377
Ramakrishnan, Venkatraman: 249

Ramses: 343
Rasmussen, Joshua: 503, 515
Rathenau, Walther: 161
Reeves, Hubert: 281
Reeves, Perry: 290
Renan, Ernest: 383, 393, 396, 400, 403
Riemann, Bernhard: 75, 333
Riess, Adam: 75, 333
Robinson, Edward: 542
Röhm, Ernst: 163
Roll, Peter: 191
Romanus III Argyrus, Emperor: 376
Rosen, Joe: 280, 544
Rosenberg, Alfred: 156, 158, 165, 171-173
Rothman, Tony: 276
Rousseau, Jean-Jacques: 419
Ruah, Judah: 448, 469
Rubbia, Carlo: 272
Rumford, Count: 66
Ruse, Michael: 492
Rust, Bernhard: 172
Rutherford, Ernest: 73, 543
Rutten, Emanuel: 503
Sakharov, Andrei: 138, 144, 149, 152, 153, 198
Salaman, Esther: 154-156, 311
Salet, Georges: 288
Salisbury, Frank: 254
Solomon, King: 343, 394
Sandage, Allan (Dr.): 187, 276, 307
Santos dos Jesus, Lúcia, Sister: 453, 448-449, 454, 461, 476-477, 484-485
Sartre, Jean-Paul: 124, 489, 511
Sautoy, Marcus du: 321, 333
Scémama, André: 441
Schaefer, Henry F.: 272
Schäfer, Lothar: 277
Schäfer, Peter: 388, 389
Schawlow, Arthur: 266
Scheitle, Christopher P.: 178

Schins, Juleon: 332
Schleicher, Kurt von: 163
Schmidt, Brian P.: 75, 78, 545
Schrödinger, Erwin: 543
Schwann, Theodor: 542
Shannon, Claude: 244, 250, 553
Shapiro, Robert: 245, 258, 259
Shukeiri, Ahmed: 442
Simon (false Messiah): 407
Simon the Zealot, Saint: 403
Slipher, Vesto: 75, 88
Smalley, Richard: 267
Smolin, Lee: 178, 262
Smoot, George F.: 208-209, 211, 269, 545-546
Socrates: 343, 344, 384, 495
Solon: 419
Sommerfeld, Arnold: 207, 543
Sonigo, Pierre: 244
Speer, Albert: 156, 168
Sperry, Roger: 288
Spinoza, Baruch: 124, 312, 330, 331, 510, 511
Stalin, Joseph: 140, 142, 145-147, 149-150, 166, 528
Stark, Johannes: 157, 159, 160-161
Staudinger, Hermann: 543
Steitz, Thomas A.: 249
Stengers, Isabelle: 66, 73, 282, 283
Stern, Otto: 127, 129, 168-172
Stirrat, Michael: 298, 306
Strasburger, Eduard: 542
Strauss, David F.: 383, 397
Suarez, Antoine: 332, 523
Suetonius: 385, 387, 391, 422, 423
Susskind, Leonard: 203, 220
Swinburne, Richard: 503
Sylvester II, Pope: 541
Tabitha (widow): 413
Tacitus: 385, 387, 391, 422
Talin, Kristoff: 306
Tamarkin, Jacob: 129, 132-133, 138

Taylor, Richard: 490
Tegmark, Max: 276
Teller, Edward: 170
Tertullian: 404
Thales: 343, 344, 348
Theudas, false Messiah: 409, 410
Thom, René: 505
Thomas, Apostle: 403
Thomas Aquinas, Saint: 121, 376, 498-500, 521, 541
Thompson, Benjamin, Count Rumford. See "Rumford, Count"
Thomson, George Paget (sir): 223
Thomson, William: see "Kelvin (Lord)"
Tiberius: 385
Tipler, Frank: 190, 275
Titus: 427
Tjio, Joe Hin: 545
Tobit: 525
Tolman, Richard: 111
Tolstoy, Leo: 420
Townes, Charles: 268
Toynbee, Arnold: 418
Trajan, Emperor: 385
Trinh, Xuan Thuan: 196-197, 226, 269-270, 328
Trotsky, Leon: 143
Twain, Mark: 417, 438
Uhlig, Herbert: 274
Ulla: 388
Urey, Harold: 251
Venter, John Craig: 238
Vilenkin, Alexander: 109, 113, 115, 117, 119, 204, 217, 219, 222, 224-225, 263, 303, 545-546
Vinci, Leonardo da: 324
Voltaire: 182
Von Braun, Wernher: 278
Vorontsov-Velyaminov, Boris: 148
Wald, George: 282
Waldeyer, Heinrich W.: 543
Walesa, Lech: 485

Watson, James: 35, 243, 245, 544
Weinberg, Steven: 74, 109, 188, 205-207, 268, 510
Weisskopf, Victor Frederick: 169
Weyl, Hermann: 172, 319
Wheeler, John: 80, 89, 113
Wickramasinghe, Chandra: 257, 258, 260, 277
Wigner, Eugene: 170, 504
Wiles, Andrew: 39
Wilkinson, David: 191, 209
Wilson, Robert W.: 94, 106, 109, 142, 149, 178, 191, 213, 223, 265, 361, 545
Witten, Edward: 115
Wolf, Lucien: 436
Wolfenden, Richard: 248
Woltzenlogel Paleo, Bruno: 330
Woodson, Sarah: 285
Yockey, Hubert P.: 237, 244, 250, 257, 287, 288
Yonath, Ada E.: 249
Yoshikawa, Shoichi: 274
Zeldovich, Yakov: 149
Zerubbabel: 429
Zhdanov, Andrei: 140, 147
Zichichi, Antonino: 274
Zwicky, Fritz: 111

Acknowledgments

This book has been a collaborative effort over many years. Each chapter was written in consultation with specialists in each area covered. Their assistance has permitted us to offer an accurate and up-to-date compilation of the evidence for God's existence available today. We thank the following persons in particular for their time and expertise.

For her collaboration in writing this book.

Agnès Paulot, graduate of the prestigious École Normale.

For consultation on scientific subjects:

Jean-Robert Armogathe, graduate of the prestigious École Normale, former chaplain of ENS Ulm, Associate of Letters, PhD, Doctor in Political Science, and emeritus Director of Studies at the École Pratique des Hautes Études, member of the International Academy of the History of Science and the Accademia Ambrosiana of Milan, co-founder and international coordinator of the journal *Communio*, which has organized and participated in review groups with correspondents from the Academy of Sciences.

Vincent Berlizot, Polytechnique graduate specializing in the epistemology of science.

Michael Denton, biochemist, medical geneticist, former professor at the University of Otago in New Zealand and former Director of the Genomics Lab of Prince of Wales Hospital in Sydney, Australia, and world-class specialist in genetic eye diseases.

Yves Dupont, graduate of the prestigious École Normale, Associate Professor of theoretical physics, teacher of advanced mathematics in the second year of the preparatory class for the "grandes écoles" at Collège Stanislas de Paris.

Marc Godinot, emeritus Director of Studies at the École Pratique des Hautes Études, Earth and Life Sciences section.

Jean-François Lambert, neurophysiologist, researcher emeritus at Université Paris-VIII.

Jean-Michel Olivereau, former Professor of Neuroscience at Université Paris-V, who shared with us many of the scholarly notes he has collected over the last twenty years. He has been an invaluable resource.

Pierre Perrier, active contributor to international research, member of the Academies of Science and Technology of France and of the United States.

Fabien Revol, biologist, Doctor of Philosophy and Theology, researcher at the Université Catholique de Lyon.

Jean Staune, founder of the Université interdisciplinaire de Paris (UIP), philosopher of science and essayist, diploma in paleontology, mathematics, administration, political science, and economics. We are especially indebted to him for his generous gift of time to this project.

Rémi Sentis, graduate of the prestigious École Normale, Doctor of Science, director emeritus of research, president of the Association des Scientifiques Chrétiens, author of *Aux origines des sciences modernes* and *L'Église est-elle contre la science?*

Antoine Suarez, physicist and philosopher, specialist in quantum mechanics.

For their contributions to Chapter 8, "The Big Bang, a Noir Thriller":

Igor and Grichka Bogdanov, doctors of physics and mathematics, respectively; authors of numerous books including *Dieu et la science*, *Le visage de Dieu*, *La pensée de Dieu*, and *La fin du hasard*.

For their collaboration on Chapter 21, "Fátima":

Dr. José Eduardo Franco, professor at the University of Aberta and of the Faculty of Letters at the University of Lisbon.

Dr. João Diogo Loureiro, professor of the University of Lisbon's Faculty of Letters.

Helena Jesus, researcher at Université Paris-IV.

For their contributions to Chapter 22, "Is Everything Permitted?":

Frédéric Guillaud, graduate of the prestigious École Normale, Associate of philosophy, author of numerous books including *Dieu existe: arguments philosophiques*, *Catholix reloaded: essai sur la vérité du christianisme*, and *Par-delà le bien et le mal de Nietzsche*.

Richard Bastien, Canadian reporter, economist, and essayist, author of *Crépuscule du matérialisme* and *Cinq défenseurs de la foi et de la raison*.

For their assistance with Chapters 17 and 18 about the Bible and Chapter 20, "The Jewish People: A Destiny Beyond Improbability":

Christophe Rico, Director of Polis—The Jerusalem Institute of Languages and Humanities, Professor of Greek Philology at the École Biblique of Jerusalem.

Charles Meyer, member of the bar in Brussels and Israel, vice-president of the Association France-Israël and the European Alliance for Israel, author of numerous books.

For their help with research, citations, and references:

Peter Bannister, former collaborator at the Université Interdisciplinaire de Paris (UIP) and the chair of *Science and Religion* at the Université Catholique de Lyon. President of the *Science et Sens* association and force behind the website *www.sciencesetreligions.com*.

Alexis Congourdeau, history teacher and photographer specializing in the history of Jerusalem.

Patrick Sbalchiero, PhD in history, journalist and author of around twenty books.

For copy-editing and type-setting:

Madison Zahaykevitz, Maj-Britt Frenze and Emily Demary.

For type-setting:

Caroline Hardouin, illustrator and recipient of the *Meilleur ouvrier de France* in graphic design.

Detailed Table of Contents

Table of Contents 7

The Book Everyone Is Talking About 9

Preface 19

INTRODUCTION 21

1. The Dawn of a Revolution 23

2. What is Evidence? 37

3. Implications Arising from the Two Theories:
"A Creator God Exists" versus "Nothing Exists
Beyond the Material Universe" 51

   I. Study of the implications of the thesis "nothing exists beyond the material universe" 54

   1. The Universe cannot have had an absolute beginning 54

   2. The Universe cannot end in thermal death, since such an end implies an absolute beginning 55

   3. Deterministic laws apply universally, and things are distributed randomly 56

   4. Miracles cannot exist 57

   5. Prophecies and revelations cannot exist 57

   6. Good and evil are not absolute and are therefore open to unlimited democratic decision-making 57

   7. The spirit world—including devils, angels, evil spirits, possessions, exorcisms—does not exist 57

   II. Study of the implications of the thesis "a creator God exists" 57

   1. The Universe can be expected to have a purpose or end goal 57

2. The Universe can be expected to be ordered, beautiful, and
      intelligible 57

   3. The Universe can be expected to have had a beginning 57

   4. Miracles are possible 58

   5. Prophecies and revelations are possible 58

   III. What a coherent materialist must accept as true 58

EVIDENCE WITHIN THE SCIENCES 61

4. The Thermal Death of the Universe:
Story of an End, Proof of a Beginning 63

   Introduction 64

   I. History of the discovery of the thermal death of the Universe 65

   II. The scenario with the greatest consensus today 78

5. A Brief History of the Big Bang 83

   I. The Big Bang and the birth of cosmology
      in the twentieth century 83

   II. In the beginning was the Big Bang 98

6. Attempts at Alternatives to the Big Bang Model 111

7. Convergent Evidence for an Absolute
Beginning to the Universe 119

8. The Big Bang, A Noir Thriller 127

   I. Soviet suppression of the Big Bang Theory 128

   II. Nazi opposition to the Big Bang 153

   III. The Big Bang in the West after 1945 174

9. The Fine-Tuning of the Universe 181

   I. The story of the discovery of the anthropic principle 190

   II. What are these mysterious fine-tunings? 192

1. In the beginning, the coupling constants that determined the relationships between the four fundamental interactions were very precisely fixed 192

2. In the first instant after the Big Bang, the ratio between the amount of energy in the Universe and its rate of expansion had to be fixed with phenomenal precision 195

3. The fine-tuning of the weak force allows antimatter to disappear 197

4. The masses of the fundamental structures of the Universe (electrons, protons, neutrons) are also adjusted to perfection 199

5. Keeping the highly unstable neutron going: another essential fine-tuning 200

6. The cosmological constant: one of the finest of fine-tunings 201

7. The tuning of the strong and weak nuclear forces is equally astonishing and indispensable 204

8. The tremendous conservation of beryllium, which enables the generation of the carbon essential for life 205

9. The "magical" adjustment of the electromagnetic force has also astounded the greatest scholars 207

10. Cosmic microwave background anisotropy is also fine-tuned 208

11. The Planck Constant—which universally regulates the energy levels of all atoms—deserves the nickname "theological constant" because, without it, all chemistry would be impossible 212

Conclusion 212

10. The Multiverse: Theory or Loophole? 215

11. Preliminary Conclusions: One Small Chapter for Our Book, One Giant Leap for Our Argument 221

12. Biology: The Incredible Leap from Inert to Living Matter 229

I. A gap that ultimately turned out to be an abyss 229

II. The most complex inert matter obtained by experimental means 235

III. The simplest life-form 238

1. DNA: Double helix, double mystery 243

2. Proteins or the 3-D letters of the alphabet of life 246

3. The ribosome: a mysterious translator 247

4. Enzymes: incredible time-savers 248

5. Two languages and a translator 249

6. Metabolism: the whirlpool of life 250

13. One Hundred Essential Citations From Leading Scientists 263

I. Cosmology – Physics – Chemistry 263

II Biology and Life Sciences 281

III. Mathematics 291

IV. Philosophy of Science 292

14. What Do Scientists Believe In? 295

I. Let us begin by examining the major studies on religious belief among scientists 295

1. A 2009 study by the Pew Research Center 295

2. A 2003 study by geneticist Baruch Aba Shalev 296

3. In a 1989 survey of heads of research units in the exact sciences at the French National Centre for Scientific Research 297

4. James H. Leuba's studies 298

5. A 1998 Nature study 298

6. A survey of 1,074 members of the British Royal Society carried out by E. Cornwell and M. Stirrat 298

II. Let us now examine the fact that the proportion of believers among scientists is smaller than the proportion of believers among the general population, and let's see what factors might cause this difference 299

1. The influence of socio-economic status on belief 300

2. Correlation is not causation 300

3. Long-standing conflicts between science and religion 302

4. Many scientific discoveries favoring the existence of God are very recent 303

5. Another thing to consider when drawing a causal relationship between belief and scientific knowledge 305

Conclusion 305

Two more surveys of scientists on the question of religious belief 305

How can we account for the significant differences between the conclusions of these various studies? 306

15. What Did Einstein Believe In? 309

16. What Did Gödel Believe In? 317

Conclusion 333

EVIDENCE FROM OUTSIDE THE SCIENCES 335

Introduction 337

17. The Humanly Inaccessible Truths of the Bible 339

I. What is a humanly inaccessible truth? 340

II. What out-of-reach truths did the Hebrews know about the cosmos and humanity thanks to the Bible? 341

III. Is our Bible the same as the one the Hebrews had? 342

IV. Are we sure that they interpreted the Bible's claims in the same way we do today? 343

V. How advanced were the Hebrews? 343

VI. How advanced were their neighbors? 344

VII. Some of these truths "fall from the sky" 345

1. The Sun and Moon are nothing more than luminous bodies 345

2. The Universe had an absolute beginning and was created from nothing by a God who is exterior to it: a biblical notion contrary to all cosmogonies 349

3. The Universe is progressing towards its end, following the unidirectional arrow of time 351

4. The human body is made up of nothing but physical matter 353

5. Nature and the elements are not inhabited by divine beings 356

6. Humanity traces its origin back to one man and one woman, implying that there is no innate hierarchy among men 357

18. The Alleged Errors of the Bible, Which Are Not Errors.................363

    Constraints and challenges...364

  I. The errors most commonly ascribed to the Bible................................366

    1. Supposed "error" number one: The world was created in seven days 366

    2. A second "error" in the Bible: "In the beginning, God created the heavens and the earth"...370

    3. "Error" number three: "According to the Bible, light is created even before its source, the sun and the stars"................................. 373

    4. A fourth "error": "The Bible teaches that the earth is flat"................374

    5. Fifth Error: God made man out of earth in a single day (Genesis 2:7), while in reality man was not the product of a separate creation but descended from primates and hominids, following a process of evolution over millions of years 378

19. Jesus: Who Could He Be?.. 381

  I. Jesus never really existed. He is a myth that was invented later..........384

    1. Ancient authors and historians on the witness stand.......................384

    2. Jesus, a myth? A fairly new strategy with an obvious motive 390

    3. A myth? Since when?... 391

    4. Jesus an absurd and sacrilegious myth? That makes no sense!..........392

  II. Jesus was a great sage 393

    1. Crazy words, words of an insane man...394

    2. Infinitely pretentious words 394

    3. Words impossible to live by...394

    4. Sacrilegious words..395

    5. A great sage at the head of a gang of liars and crooks? 395

  III. He was a madman, a crazy mystic 396

    1. Disconcerting but wise words 397

    2. Why didn't the Talmud portray Jesus as a madman?.........................399

    3. Requiem for a madman 399

  IV. Jesus was a failed opportunist.. 400

    1. Why didn't the high priests look for Jesus' corpse? 401

    2. Why would the disciples take the charade so far? 402

    3. Are these the words of an opportunist? 403

    4. Why was the apostles' preaching so successful? 403

  V. He was a prophet 405

  VI. Jesus is the Messiah and an extraordinary man, but only a man 406

  VII. Jesus is the Messiah and God incarnate 411

    1. The madman's words start to make sense 411

    2. Impossible teachings become possible 412

    3. Wise words for eternity 412

    4. The sudden courage of the apostles explained 413

    5. The Apostles' success has no merely human explanation 413

    6. Jesus' lifetime was indeed the predicted time for the Messiah 414

  Final deliberation and verdict 414

20. The Jewish People: A Destiny Beyond the Improbable 415

  I. Probably the only people to have survived from antiquity to the present day, over 3,500 years 418

  II. Surviving extraordinary hardships, from the exile of the biblical era to the Nazi genocide 421

  III. The only people to return to their native land eighteen centuries after total expulsion 423

    1. An astonishing first prophecy 426

    2. Two astonishing Gospel prophecies give even sharper definition to this exile and return 426

    3. Some thought it ludicrous that the Jews would go back to Palestine, but they were wrong 431

  IV. Another strange tale: resuscitating an ancient language after 2,500 years of disuse 432

  V. A precarious position: small and poor in natural resources, Israel is surrounded on all sides by hostile neighbors, some of whom still call for its disappearance 433

VI. Israel is the only country whose capital, despite being a city of little economic or strategic interest, remains a center of geopolitical tension and a flashpoint for a potential world war.........434

VII. The people who produced the world's top bestseller......................435

VIII. The Jews are the only people to maintain a disproportionately important role in the history of ideas and of science relative to their importance 435

IX. The only people to be the victim of reverse racism 436

X. The only country where half the population believes themselves God's elect and their lands the cradle of the Savior of the World, and which, despite such absurd daydreams, nevertheless ranks among the most technologically advanced countries in the world 437

XI. A people who have stunned the world with unexpected and spectacular military victories 439

21. Fátima: Illusion, Deception, or Miracle? 447

I. The political context of Portugal in 1917 451

II. The local context 452

III. The days leading up to October 13 453

IV. October 13, 1917: the story of an incredible day 454

V. An informative press review 459

VI. Photographic evidence 474

VII. What happened to the children? 478

VIII. Examination of six possible hypotheses 478

IX. The meaning of Fátima today 484

22. Is Everything Permitted? 487

23. Philosophical Proofs Strike Back 495

I. The Supreme Intelligence 504

II. The One Necessary Being 507

III. The Creator of Time 512

24. Materialist Arguments against the Existence of God 517

DETAILED TABLE OF CONTENTS

I. Skeptical arguments..517

    1. "There is no proof for the existence of God. Otherwise, we would know it already"...517

    2. "The burden of proof is on the person who makes the claim" 518

    3. "The existence of God is impossible to prove one way or the other, so it is better not even to talk about it" 518

    4. "We don't need God to explain the origin or functioning of the Universe, so we can leave him out of the equation"............... 518

    5. "The insignificance of mankind and the Earth proves that man has no special destiny. Therefore, he cannot originate from a god" 519

    6. "Polls show that today few scientists are believers, in any case far fewer than in the general population; this proves that science leads to the belief that God does not exist, and this is logical because God doesn't exist"..................................520

II. Philosophical arguments 520

    7. "Can God create a rock so heavy he can't lift it?"............................520

    8. "Can God commit suicide?".. 521

    9. "A perfect God could not have created an imperfect world. Since the world is obviously imperfect, God does not exist"............. 521

    10. "If God is omniscient and knows the future, man is not free; if, on the other hand, God does not know everything, he is not all-powerful and therefore does not exist" 522

    11. "God is incompatible with chance. If God is perfect and omnipotent, then chance should not exist. But chance does exist, so God does not" 522

    12. "If God is perfect, then he is self-sufficient, and there was no need for him to create the world. But the world exists, so God does not exist" 523

    13. "The Universe is self-sufficient. It does not need a cause" 524

III. Moral arguments 525

    14. "A benevolent God would not tolerate the evil and suffering we see in the world, much less create it. But suffering and evil exist in the world, so God does not exist".. 525

15. "Believers do not provide an example of moral excellence. Their bad and even scandalous lives prove that their religion is just a facade with nothing behind it" 526

IV. Arguments drawn from religion 526

16. "The existence of multiple, contradictory religions on Earth is incompatible with the existence of a single, benevolent God" 526

17. "The fact that there are many gods proves that none of them exist" 527

18. "Religions have always been the cause of violence and wars" 527

19. "The existence of hell proves that God is not good. But since God must be good by definition, neither he nor hell exists" 528

20. "The Bible is just a collection of primitive legends riddled with errors" 530

Conclusion 530

CONCLUSION 533

25. Materialism: An Irrational Belief 535

The pieces of evidence are clear 535

They are intelligible 536

The theoretical implications of this evidence can be tested against the observable Universe 536

The evidence is abundant 536

The evidence comes from many fields of knowledge 536

Many who defended this evidence were persecuted 536

What is the takeaway? 537

For God made man to seek him 538

APPENDIX 539

Appendix 1: Timeline of the Universe 539

Appendix 2: Physical Measurements Great and Small 547

Appendix 3: Benchmarks of Orders of Magnitude in Biology 549

Glossary 551

Index of Personal Names 559

Acknowledgments 565

Detailed Table of Contents 569

Image Credits 581

Image Credits

© SOFAM, Bruxelles [2025] / Œuvre de Jean Léons Huens, Sir Isaac Newton: p. 43;
© AIP Emilio Segrè Visual Archives, Physics Today Collection: p. 178 (3);
© AKG:
  AKG / Alan Dyer / Stocktrek Images: p. 30;
  AKG / Science Photo Library: p. 89 (1);
  AKG-Images / Imagno: p. 88;
© Archives Charmet / Bridgeman Images: p. 475 (4);
© Bettmann Archive / Bettmann distributed by Getty: p. 87;
© Caroline Hardouin: p. 184;
© Christie's Images/ Bridgeman Images: p. 245;
© All rights reserved: p. 10, 12 (1), 12 (2), 13 (2), 14 (1), 14 (2), 15 (1), 15 (2), 16 (2), 40, 129 (1), 129 (2), 129 (5), 129 (7), 129 (8), 129 (10), 129 (11), 178 (2), 202, 217 (1), 320 (1), 475 (5);
© Everett Collection/ Bridgeman Images: p. 320 (2);
© Gabriel Vandervort / AncientResource.com: p. 376;
© Getty images:
  Getty images/ Christopher Lane: p. 11 (1);
  Getty images/ Mondadori Portfolio: p. 476 (8);
  Getty images/ Rick Friedman: p. 217 (3);
© iStock / Getty Images Plus:
  iStock / Getty Images Plus: p. 389;
  iStock / Getty Images Plus/ titoOnz: Cover page, p. 9;
© Julien Faure/ Leextra / OPALE.PHOTO: p. 196;
© Look and Learn / Bridgeman Images: p. 415;
© Los Angeles Times: p. 440 (1), 440 (2);
© Seeley G. Mudd Manuscript Library at Princeton University: p. 190;
© Shutterstock :
  Shutterstock: p. 234;
  Shutterstock / Amanda Carden: p. 375;
  Shutterstock / Christoph Burgstedt: p. 241;
  Shutterstock / Juergen Faelchle: p. 215;
  Shutterstock / Leigh Prather: p. 243;
  Shutterstock / Vitalina Rybakova: p. 63;
© Ted Thai / The LIFE Picture Collection/ Shutterstock: p. 94;
© The Miami Herald: p. 440 (3);
© Wikipedia Commons:
  Public Domain: p. 25 (1), 25 (2), 25 (3), 25 (4), 25 (5), 25 (6), 25 (7), 25 (8), 25 (9), 65, 66, 67, 68, 84 (Library of Congress, United States), 86, 89 (2) (The Caltech Archives), 129 (3), 129 (4), 129 (6), 129 (9), 129 (12), 129 (13), 129 (14) (Library of Congress, United States), 129 (15), 129 (16), 160, 164 (1), 164 (2) (RMY auction), 177, 178 (1) (University of Colorado), 96, 194, 209, 317, 391 (2) (Ministry of Cultural Heritage, Activities and Tourism-Artistic Bulletin), 411, 464, 474 (1), 474 (2), 475 (3), 476 (6), 476 (7).
  Creative Commons: p. 11 (2) (CC BY-SA 2.0), 13 (1) (CC BY-SA 4.0 / Deryck Chan), 17 (CC BY-SA4.0 / Léo René Jean Perez), 18 (1) (CC BY-SA 4.0 / François-Régis Salefran), 18 (2) (CC BY-SA 4.0 / Thaler Tamas), 98 (CC BY-SA 4.0 / National Science Foundation), 145 (CC BY-SA 4.0 / Jochen Burghardt), 217 (2) (CC BY-SA 3.0 / Betsy Devine), 391 (CC BY-SA 3.0 / Velvet).